The physics of vibration

Volume 1

The physics of vibration

Volume 1, containing Part 1, The simple classical vibrator

A. B. PIPPARD, FRS

Cavendish Professor of Physics, University of Cambridge

CAMBRIDGE UNIVERSITY PRESS

Cambridge

London   New York   Melbourne

Published by the Syndics of the Cambridge University Press
The Pitt Building, Trumpington Street, Cambridge CB2 1RP
Bentley House, 200 Euston Road, London NW1 2DB
32 East 57th Street, New York, NY 10022, USA
296 Beaconsfield Parade, Middle Park, Melbourne 3206, Australia

First published 1978

Printed in Great Britain by
J. W. Arrowsmith Ltd.

Library of Congress Cataloguing in Publication Data
Pippard, A. B.
The physics of vibration.
Includes bibliographical references and index.
Contents: v. 1. The simple classical vibrator.
1. Vibration:. I. Title.
QC136.P56   531′.32   77-85685
ISBN 0 521 21899 3

Contents

Note: Equations, diagrams and references are numbered serially in each chapter, and the chapter number is omitted if it is something in the same chapter that is referred to. Thus (4.23) in chapter 7 means equation 23 of chapter 4; (23) in chapter 7 means equation 23 in that chapter.

Cross-references are given as page-numbers in the margin.

Preface

The plan and purpose of this book are outlined in chapter 1, and what is said there need not be repeated here. A preface does, however, offer the opportunity of acknowledging the help I have received during its preparation, and indeed over the years before it was even conceived. I cannot begin to guess how far my understanding and opinions have benefited from the good fortune that has allowed me to spend so much of my working life in the Cavendish Laboratory, surrounded by physics and physicists of the highest quality, and by students who at their best were at least the equal of their teachers. For the resulting gifts of learning, casually offered and accepted for the most part, and now untraceable to their source, I extend belated thanks. More specifically, in the preparation of the material I have relied heavily on Douglas Stewart for the construction of apparatus, Christopher Nex for programming the computer to draw many of the diagrams (and incidentally for two notes on his bassoon in fig. 5.8), Gilbert Yates for help in the theory and practice of electronics, and Shirley Fieldhouse for typing and retyping of the manuscript. My thanks to each of them for much willing and expert help, and to the following who by discussion and criticism, by assistance in carrying out experiments, or by providing material for diagrams have given support and comfort on the way: Dr J. E. Baldwin, Mr W. E. Bircham, Prof. J. Clarke, Dr M. H. Gilbert, Prof. O. S. Heavens, Dr R. E. Hills, Dr I. P. Jones, Mr P. Jones, Mr P. Martel, Prof. D. H. Martin, Mr P. D. Maskell, Dr. P. J. Mole, Mr E. Puplett, Dr G. P. Wallis and Mr D. K. Waterson.

I am not the first scientific author to have struggled with the limitations of alphabets and the restrictions imposed by convention in the choice of symbols, nor am I the first to have to confess in the end that perfect consistency has eluded me. The failure would have been even more obvious without the intervention of the Cambridge University Press whose attention to detail, scrupulous yet kindly, I acknowledge with gratitude.

A. B. PIPPARD

Cambridge 1978

1 General introduction – author to reader

The writing of this book has occupied several years, and now that it is finished it is time to ask what sort of a book it is. For all its length, it turns out to be only the first volume out of two, any hope of covering the topic in a single volume having vanished as the project developed. It must be obvious, therefore, that it is not a textbook in the sense of an adjunct to a conventional course of lectures (there are not even any questions at the ends of the chapters). On the other hand, it is certainly not an advanced treatise, for many of the more difficult topics are treated at a much lower level than is to be found in the specialist works devoted to them. Moreover, I can claim no professional skill in most of them, and this is both a confession and an advertisement. For by writing about them in a way that illuminates for me the essential physical thought underlying what is often a very complicated calculation, I hope to have provided a treatment that will enlighten others in the same unlearned state. The field is very wide, ranging from applied mathematics (non-linear vibration and stability theory) to electrical engineering (oscillators), and taking in masers, nuclear magnetic resonance, neutron scattering and many other matters on the way. Nobody could hope to make himself a master of all these, and few advanced treatises dealing with one topic think fit to mention the analogies with others. Yet it is both useful and enjoyable to recognize that unities lie behind so many diverse phenomena, and I have written for the reader who shares my pleasure in these things. At the same time, I hope no one will be so high-minded as to take offence if occasionally a page or two are devoted to some rather trivial example; there is room in physics for magpies as well as for more systematic collectors, and I am not ashamed to take delight in things for their own sake as well as for their relationship to other things. Besides, you can always skip the boring bits. The cross-references in the margin are there, among other reasons, to make skipping easy – you will be directed back to the passages you have missed if it turns out later that they are relevant to another discussion.

The purpose of the book, then, is to bridge the gap between the normal undergraduate curriculum and detailed treatments of special topics at the research level, and I have aimed at a level of treatment that will be accessible to a student who has mastered the elementary course text and is hungry for more. The emphasis throughout is on the physical processes, and very little is expected in the way of mathematical skill. Complex variable theory at the level of contour integration is only used on those rare occasions when it significantly shortens an argument. For the most part, elementary algebra and calculus provide all that is needed, but in saying this it is only fair to draw attention to the need to think graphically. A differential equation is a specification, such as even a computer can be made to understand, of how to draw a curve, and I take for granted the reader's willingness to think in these terms. Formal solutions are fine if they exist, but exact mathematics, except (possibly even) in the hands of an expert, is a very limited tool. The scientist who scorns the qualitative condemns himself to the quantitative, and his imagination will remain earthbound. The manipulation of ideas with the help of graphs is part of the technique of imaginative physical thinking, without which exact mathematical analysis can easily degenerate into sterile manipulation.† It is with this in mind that I offer the book to the applied mathematician as well as to the physicist and the engineer – in so far as we are all concerned with the real world we need to make a connection in our minds between it and the abstract processes of analysis.

Consistent with this approach is an emphasis on experiment, and not only by reference to accounts in learned journals; too often these resemble the offerings of a supermarket, the final result hygienically wrapped to discourage uncomfortable thoughts of the original raw material. In this volume, however, we shall meet many problems that can be investigated with simple equipment, and I have described a number of these and presented the untreated results of my own measurements with the aim of encouraging you to carry them out for yourself. There is hardly a single one from which I did not learn something that finds an echo in my account of the phenomenon it illustrates; sometimes, indeed, my first thoughts were proved so mistaken by the experiment that the first draft had to be scrapped. This is not to say that a better physicist or a better mathematician would not have got it right first time, without the aid of experiment; but most of us find it easier to get the answer if we already know what answer we want, and if a quick experiment starts us off on the right lines the time spent on it is time well spent.

It may be helpful to say something about the background knowledge of physics that I have assumed. It is not constant throughout the book, but rather I have tried to keep in mind, as I wrote each chapter, the sort of reader who might wish to understand what I wanted to say. Thus chapter 2 is a revision course for someone who is reasonably familiar with simple harmonic motion, with the aim of deepening understanding and at the same

---

† Graphical methods must not be confused with the formal processes of Euclidean or cartesian geometry. In a few places I choose to use geometry to demonstrate a particular exact result; others might prefer to do it by algebra, but the choice is merely a matter of taste, not of principle. I find it helpful to represent the manipulations of complex algebra on an Argand diagram, but have no fault to find with those who get on perfectly well without this aid. But anyone who tackles differential equations without sketching solutions handicaps himself severely.

time indicating some of the problems that are not described by linear differential equations; but it assumes no broad knowledge of other branches of physics. By the time chapter 6 is reached I take for granted a grounding in wave-motion and the general character of such special forms as acoustic and electromagnetic waves. And chapter 8 is written in the expectation that only someone with a background of elementary solid-state and quantum physics will wish to come to grips with the rather difficult physical concepts involved in the calculation of the relaxation time in proton resonance. The justification for this approach is that, although I have written in the style of a continuous narrative, the argument is not a closed integrated logical progress. There is indeed an underlying unity, illustrated by many varied examples, but the sense of unity need not be destroyed by selecting a few and neglecting the rest. My earlier advice to skip what seems trivial or dull applies equally to the too-difficult.

As for the subject matter itself, 'vibration' is taken to encompass almost anything that fluctuates in time more or less periodically. I have not included wave-propagation as an independent branch of the subject, except in so far as the vibrations of strings and solid bodies, and the oscillations of electric fields in cavities, can be considered as standing waves; and a few examples bring in the scattering of waves by resonant obstacles. No attempt, however, is made to treat such matters as wave-groups or diffraction except as incidentals to the main study. The only defence I can offer is that one must draw the line somewhere, and to explore wave-propagation in the same spirit which I have brought to vibration would require another volume at least. The size of the present venture is already regrettable, if only because it has led to a separation of classical and quantal treatments, something that happens all too often in the teaching of physics. Almost inevitably it fosters the feeling that quantum mechanics is intrinsically better than classical mechanics, and that classical mechanics is something real physicists ought to grow out of. This is a disputable proposition. Of course, if they give different answers when applied to a given problem one must expect the quantal treatment to be more nearly correct. There are, however, many problems in vibration where they give the same answer, or where the quantal treatment shows the behaviour to be the same as that of an equivalent classical system. It is then usually easier to think about the classical equivalent, especially if it is complicated. For classical physics is the mode of description appropriate to direct observation, and the imaginative faculties develop through the contemplation of the world around. It is easier to guess the answer to a difficult classical problem than to a difficult quantal problem – indeed, the more sophisticated the intellectual framework the harder it is to extend one's understanding beyond those problems that can be solved exactly. When it is a matter of formulating fundamental laws, quantum mechanics must be accepted as having superseded classical mechanics; but, when specific problems are to be solved, quantum mechanics can prove unconscionably intractable. The physicist or engineer concerned with specific problems must therefore develop his classical skills, analytical and imaginative, to the highest degree, but he must also develop a reliable appreciation of how far he dare thrust his classical argument into quantal territory.

Part 1 of this work, which occupies the rest of this volume, is essentially classical, with only an occasional sortie into problems that are normally regarded as quantal but which yield satisfactorily to classical analysis. In the second volume I propose to take up the quantal theme as Part 2, with examples of strictly quantal phenomena and a careful exposition of the safe limits of the classical approach to the same problems. This will leave the field clear for a treatment, in Part 3, of the vibrations of complex systems, including normal mode analysis, by whatever techniques seem appropriate. Finally, I intend to present accounts of a varied collection of maintained oscillations, especially those with a complicated spatial structure such as aerodynamic vibrations and Gunn oscillators.

But this is in the future, and already I have shown an unbecoming personal touch in revealing my aims and aspirations. It is time to disappear from the scene. But though, following custom, *I* adopt the cloak of invisibility and simultaneously cease to acknowledge the existence of *You*, my reader, there will still be found, as *We*, the assumption of collaboration between writer and reader without which a book might as well remain unwritten.

Part 1

The simple classical vibrator

2 The free vibrator

Introduction: the special role of harmonic vibrators

This chapter is mainly concerned with simple vibrators executing periodic, or nearly periodic, motion undisturbed by external influences. In this context the term *simple* refers to systems which can be characterized by a single variable – the angular displacement of a pendulum, the height of a bouncing ball, the charge on a capacitor connected in series with an inductor. The behaviour of such systems may be displayed as a graph of the value of this variable (which we shall call the *displacement*) against time, as in fig. 1 which shows the behaviour of the three examples cited. The first

Fig. 1. Three vibrators, (*a*) simple pendulum, (*b*) bouncing ball, (*c*) series resonant circuit, with graphs of the variation of displacement with time.

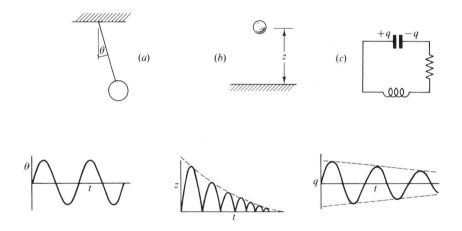

and third are very similar, the oscillations being simple harmonic, i.e. sinusoidal in form, decaying slowly for a massive pendulum swinging in air, but rather more rapidly for a typical LC circuit at room temperature on account of the resistance of the inductor. The decay of the oscillations in the circuit follows an exponential law, so that the charge on one capacitor plate may be expressed as

$$q = q_0 \cos{(\omega' t + \phi)} \, \mathrm{e}^{-t/\tau_\mathrm{a}}, \tag{1}$$

q_0, ω', ϕ and τ_a being constants. If the *time-constant for amplitude decay*, τ_a, is so large that the decay may be neglected, $\mathrm{e}^{-t/\tau_\mathrm{a}}$ being put equal to

7

unity, the oscillation of q is strictly periodic with period $T_0 = 2\pi/\omega'$, since adding T_0 to t in (1) adds 2π to the argument of the cosine and leaves q unchanged; ω' is the *angular frequency* and is the reciprocal of the time required to change the argument of the cosine by unity. The argument, $(\omega' t + \phi)$, is called the *phase* of the oscillation. The moment from which time is measured may usually be chosen at will, and ϕ, the *phase-constant*, is introduced to indicate what point in the cycle of oscillation is reached at the instant $t = 0$. The *amplitude* of the oscillation is denoted by q_0; during a complete cycle the displacement ranges between $\pm q_0$. If the damping is not negligible, but is exponential as in (1) (this is not a universal form of damping, as later examples will show), the oscillations are not strictly periodic, for it is no longer true that $q(t + T_0) = q(t)$. They now obey the modified periodicity relation, $q(t + T_0) = \alpha q(t)$, where α is a constant equal to e^{-T_0/τ_a} and represents the factor by which the amplitude is changed in one period. Between one maximum deflection and the next, on the opposite side of the mean position, a time $\frac{1}{2}T_0$ elapses, and the ratio of the maxima is $\alpha^{-\frac{1}{2}}$, or e^Δ where $\Delta = \frac{1}{2}T_0/\tau_a$; Δ is called the *logarithmic decrement*.

34

At the moment of maximum displacement, when $q = \pm q_0$, no current flows and the energy wholly resides in the capacitor, $E = q_0^2/2C$. If resistive losses are negligible during one cycle the total energy does not change, but, as the capacitor discharges and current flows in the inductor, energy is transferred from the electric field of one into the magnetic field of the other, and then back again in the second half of the cycle. With resistive losses the energy is gradually dissipated as heat; since the amplitude, according to (1), decays as $q_0 e^{-t/\tau_0}$, the energy decays as $q_0^2 e^{-2t/\tau_a}$. If we write this in the form

$$E = E_0 e^{-t/\tau_e},$$

thus defining τ_e, the *time-constant for energy decay*, it is obvious that $\tau_e = \frac{1}{2}\tau_a$. The time needed for the energy to decay by a factor e is τ_e and in this time the system executes T_0/τ_e cycles of vibration. Otherwise expressed, the phase of the oscillation runs through $\omega' \tau_e$ radians while the energy decays by a factor e, and this number is designated by Q and called the *quality factor* or *selectivity*. Systems of high Q vibrate for many cycles before coming to rest.

One more symbol may be introduced at this point, though its utility will not be obvious yet: we define ω'' as $1/\tau_a$. Then, since $\tau_e = 1/2\omega''$,

$$Q = \tfrac{1}{2}\omega'/\omega''. \tag{2}$$

The LCR circuit is the archetypal vibrator, serving as a model for the enormous variety of systems whose behaviour may be more or less adequately represented by a single variable oscillating according to (1). Exponentially damped simple harmonic oscillations play so fundamental a role in so many branches of physics that there is a danger of overlooking, or treating as relatively unimportant, systems which exhibit different forms of oscillation, for example a bouncing ball. To be sure, one could hardly contend that the bouncing ball is important in itself, but one does not have to look far to find examples of non-sinusoidal vibrations that deserve to be treated seriously – the vibrations of the reed of a woodwind instrument and the beating of the heart are examples. The reason for the tendency to neglect such systems of which, as we proceed, we shall meet many other instances, is that, unlike harmonic vibrators, they cannot be accommodated in a single formal treatment; almost every anharmonic system demands its own analysis. One therefore gets a much less tangible return for a given amount of mathematics. Moreover the analysis of anharmonic vibrations is almost always more troublesome.

To appreciate the reason for the peculiar position occupied by harmonic vibrators, let us compare the equation governing the LCR circuit with that for the bouncing ball, as elementary a vibratory system as one may hope to find in the whole menagerie of phenomena that refuse to conform to a common calculus. The ball offers a rare case of easy analysis. So long as it is in flight, its height z obeys the differential equation

$$\ddot{z} = -g. \tag{3}$$

Each flight, between one bounce and the next, is described by the general solution of (3):

$$z = v_0(t - t_0) - \tfrac{1}{2}g(t - t_0)^2, \tag{4}$$

in which t_0 is the instant at which it leaves the ground ($z = 0$) with initial velocity v_0. Each loop of the curve in fig. 1(b) is a parabola with a maximum height, z_m, of $\tfrac{1}{2}v_0^2/g$, and the time of flight between bounces is $2v_0/g$. If Newton's law of impact is obeyed, each bounce reduces v_0 by a factor ε; thus successive values of z_m form a geometrical progression $1 : \varepsilon^2 : \varepsilon^4$ etc., and the successive 'periods' are in the ratios $1 : \varepsilon : \varepsilon^2$ etc. It is easily seen that with a very bouncy ball having ε nearly unity, the envelope of the loops is parabolic. The successive values of z_m would lie on an exponential decay curve if the period remained constant, but the shortening of the period with diminishing amplitude leads to a well-defined terminal point at the bottom of the parabola, a form of behaviour markedly different from exponential decay which in principle goes on for ever.

The apparently elementary formulation of the problem in (3) and the statement of its solution (4) is deceptive, since a wealth of complicated physics is hidden in the rule connecting the value of v_0 in each flight to that which follows, as summed up in the law of impact. One can appreciate this if one attempts to replace (3) by a differential equation which shall govern the whole history as a continuous narrative, and not as a succession of separate episodes. The reader is earnestly advised not to waste any time on so fruitless an exercise; it is enough to recognize that very strange mathematical functions would be required, while by contrast the LCR circuit is governed by a straightforward differential equation possessing, unlike the bouncing ball, the property of *linearity*. This point will be explained presently, but first we must derive the equation for the circuit, starting with the specially easy case where the resistance, R, is small enough to be ignored.

If the capacitor plates hold charges $\pm q$ the potential difference across them is q/C, and this is only supportable by the inductor if the current, i, through it is changing according to the law of induction that $L \, di/dt$ is the magnitude of the induced e.m.f. Since $i = \dot{q}$ the equation for q follows immediately:

$$L\ddot{q} + q/C = 0. \tag{5}$$

Substituting $q = q_0 \cos(\omega' t + \phi)$ as a trial solution (this is (1) without the decay), one sees that (5) is satisfied provided that

$$\omega' = (LC)^{-\frac{1}{2}}. \tag{6}$$

The value of q_0 is irrelevant, and this is where the property of linearity reveals itself with its important consequences. In a linear equation, of which (5) is an example, each term is of such a form that if it is evaluated for trial functions $q(t)$ and $aq(t)$, a being a constant, the results differ only in being scaled by the same multiplier a. Consequently if the equation is satisfied by $q(t)$ it is also satisfied by $aq(t)$. The solutions shown in fig. 1 reflect the difference between a linear and a non-linear system. In the decay of the sinusoidal oscillation of the LCR circuit (which remains a linear system when R is introduced) the magnitude of the oscillations diminishes but their frequency is unchanged – the solution at a later stage is the same as earlier except for a scaling down of the amplitude. On the other hand, the frequency of the bouncing ball increases as its amplitude drops – the time variation is affected along with the magnitude. This is enough to show that if we were to derive a complete equation describing its motion, it would be non-linear. It is worth noting that the converse cannot be assumed without further thought; we shall meet cases later in the **15** chapter, and elsewhere, of systems whose period is independent of amplitude but which are nevertheless non-linear.

Equation (5) is a particular case of a linear equation, containing two different operators: the operation of multiplication by a constant, such as converts q to q/C, and that of double differentiation, such as converts q to $\ddot{q}$. There are many other linear operators, of which examples will be given **44** later. Those in (5) belong to a special class of particular concern to the present argument, with two cardinal properties – *superposition* and *translational invariance*. Equations of motion possessing these properties generate harmonically oscillatory solutions, so that many different physical systems, governed by a wide variety of equations, have certain common features in their behaviour. It is this that gives such general application to the theory of harmonic vibrations, in contrast to systems governed by non-linear equations, where the solutions bear little family resemblance to each other.

Superposition. If $\mathscr{L}(q)$ is a linear operator and $q_1(t)$ and $q_2(t)$ are sufficiently regular functions that $\mathscr{L}(q_1)$ and $\mathscr{L}(q_2)$ are meaningful,

$$\mathscr{L}(aq_1 + bq_2) = a\mathscr{L}(q_1) + b\mathscr{L}(q_2), \qquad (7)$$

a and b being constants. For example the operator $(Ld^2/dt^2 + 1/C)$ in (5) possesses the property of superposition expressed by (7). If $\mathscr{L}$ is used to designate this operator, (5) may be written in the form $\mathscr{L}(q) = 0$. It is obvious that if $q_1(t)$ and $q_2(t)$ are solutions, so also is $aq_1(t) + bq_2(t)$. Because of this all possible solutions can be written in terms of a limited number of special solutions, two in the case of second-order equations like (5). All that is needed is that q_1 and q_2 shall not be multiples of the same function, and then any solution may be generated by giving appropriate values to a and b. Nothing like this holds for a non-linear equation. A second order non-linear equation has as its general solution a functional form containing two arbitrary constants, but different choices of these constants normally generate a family of curves of greater variety than anything that can be resolved into the superposition of two basic forms.

Translational invariance. Any solution $q(t)$ can be shifted bodily along the time axis, becoming $q(t + t_0)$, and still remain a solution whatever the value of t_0. This property is shared by the bouncing ball and the LCR circuit, and by any differential equation in which the coefficients do not involve the independent variable (t in this case). It is an obvious physical necessity in a circuit whose components do not change with time, or in a mechanical system where the forces between constituent parts depend on where they are and what they are doing, but not on the exact instant in time at which the configuration occurs. On the other hand, a circuit whose parameters are being altered (e.g. parametric amplifiers), or a simple pendulum whose string is steadily shortened, may be a linear system yet not possessed of translational invariance in time. The solutions are additive but their character changes with time; the period of the simple pendulum, for example, does not stay constant. Each such system demands special treatment, but there is such a wealth of interest in invariant systems that we must defer this problem till chapter 10.

285   Granted the invariance, it is still necessary to indicate why linear systems are to be singled out, and the reason is not far to seek. In equilibrium the potential energy, V, of a body is at a minimum with respect to small displacements, and the usual form of V will be a quadratic function of displacement. This is not to say that other variations are excluded – a ball resting on a horizontal table has V independent of position, the bouncing ball has V a linear function of height, z, so long as z is positive, but something much stronger for negative z. The norm for V, however, is a smooth function that can be expanded as a power series about its minimum; the linear terms must vanish but only rare circumstances cause the quadratic terms to be absent. So long as the motion to be studied is confined to the range of displacement within which the quadratic expansion is adequate, the total energy, E, of the body may be expressed as a quadratic function of position and velocity. Thus, for motion in one dimension

$$E = T + V, \tag{8}$$

where $T = \frac{1}{2}m\dot{x}^2$, the kinetic energy, and $V = \frac{1}{2}\mu x^2$, the potential energy. For an isolated body E is invariant, so that on differentiating (8) with respect to x we find, since $\mathrm{d}\dot{x}/\mathrm{d}x = \ddot{x}/\dot{x}$,

$$m\ddot{x} + \mu x = 0, \tag{9}$$

a linear equation of the same form as (5). All that this argument says is that if V is a quadratic function of displacement, the force $-\partial V/\partial x$ is a linear function. It is convenient, however, to express this fact in terms such as (8), the quadratic dependence of both kinetic and potential energies on velocity (more generally, momentum) and position coordinates, since a very general development of the theory of linear vibrators, involving many coordinates, may start from here. One can see, for example, how electromagnetic oscillations fit into the common framework; in the LC circuit the electric field $\mathscr{E}$ is proportional to q, the magnetic field $\mathscr{B}$ to $\dot{q}$, and the total energy which has contributions from $\mathscr{E}^2$ and from $\mathscr{B}^2$ takes the form of (8), with electrical energy playing the part of V, magnetic energy the part of T. And this argument does not depend in any marked degree on the possibility of assigning different energy terms to specific circuit components. Any field configuration in free space has energy quadratic in $\mathscr{E}$ and $\mathscr{B}$, and it is therefore readily understandable that electromagnetic oscillations should commonly conform to the harmonic pattern. So long as the fields are not so powerful as to distort the material structures that determine their configuration, or to cause dielectric materials etc. to respond non-linearly, the quadratic dependence is very accurately followed; indeed, among the most perfect examples of harmonic motion electromagnetic oscillations occupy a high place. So important is electromagnetism in the structure of the physical world that this alone would justify paying the greatest attention to harmonic oscillations, even if they were not equally fundamental to small-amplitude mechanical and acoustical vibrations, as well as providing an essential basis for describing transitions of quantized systems.

Anharmonic simple vibrators

There will be time enough to develop a uniform treatment of linear systems, starting with the next chapter, and we may briefly luxuriate in the richer world of non-linear vibrators. Consider a particle moving along a straight line, without dissipation, under the influence of a position-dependent force, so that its equation of motion takes the form:

$$m\ddot{x} = F(x). \tag{10}$$

Obviously (5) is a special case of this equation. There is no essential difficulty about step-by-step integration of (10). If the time scale is divided into equal short steps of length θ, and the value of x at the nth stage denoted by x_n, then $(x_{n+1} - 2x_n + x_{n-1})/\theta^2$ is the first approximation to the value of $\ddot{x}$ at the nth stage. We therefore write as an approximation

$$m(x_{n+1} - 2x_n + x_{n-1})/\theta^2 = F(x_n),$$

i.e.
$$x_{n+1} = 2x_n - x_{n-1} + \theta^2 F(x_n)/m.$$

This equation allows x_{n+1} to be computed from the two preceding values, x_n and x_{n-1}; once the process has been started by assuming two neighbouring values, x_0 and x_1, say, it can be continued indefinitely. By choosing a small enough interval, θ, any desired accuracy may be achieved at the cost of computing time.

If the form of $F(x)$ is such as to constrain the particle within certain limits, it is likely to return ultimately to x_0, and the cycle will then repeat itself. In making this assertion we assume that it returns to x_0 with the same velocity, $\dot{x}$, as it had originally; this, of course, is necessary to conserve energy. Since V is determined by x alone, so too must T and $|\dot{x}|$ be unique functions of x once the total energy has been fixed. The cycle, therefore, repeats itself exactly. In addition each half-cycle, from one limiting point to the other, is a mirror image of the preceding half-cycle, with the velocity at each point reversed in sign. This is an elementary consequence of the time-reversal symmetry of (10) – if t is replaced by $-t$ the equation is unaltered, for only t^2 appears in it. A record of the solution, $x(t)$, can be read equally well forwards or backwards, and is indeed, because of the symmetry of the half-cycles, the same curve either way. Fig. 2(a) is a possible oscillation obeying an equation of the form (10) – fig. 2(b) is not, and when we find such behaviour in a periodic system we must look for breakdown of time-reversal symmetry in the system, usually due to **209** instabilities or other irreversible phenomena (but see also chapter 8).

Fig. 2. Both vibrations are periodic, but only (a) shows time-reversal symmetry.

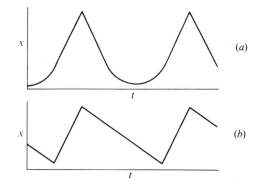

The energy equation (8) often provides a better starting point for integrating the equation of motion, being in fact the first integral of (10). Instead of step-by-step integration the formal solution may be written as an indefinite integral which is then evaluated, by numerical quadrature if necessary. For, since $T = \frac{1}{2}m\dot{x}^2$, (8) may be written in the form

$$\dot{x} = [2(E - V)/m]^{\frac{1}{2}},$$

from which it follows that

$$t = \int \frac{dx}{[2(E - V)/m]^{\frac{1}{2}}}. \tag{11}$$

Two features of the solution are obvious from (11) – once the total energy E is fixed the form of the solution is also fixed, and because an arbitrary integration constant may be added the solution is invariant with respect to shifts in time.

A particle vibrating in a potential well is confined to a region in which $E - V$ is everywhere positive or zero, the limits of its motion being the points where $E = V$. If the integration is started from one of these limits, $t = 0$ at the outset and the value of the integral when the other limit is reached is half the period. Because the integrand is infinite at both limits, a little care is needed, but, this aside, (11) provides a clear procedure for determining the period, as may be illustrated by performing the integral for one of the few cases that is analytically straightforward, the linear vibrator described by (9). Here the force $F(x)$ is $-\mu x$, the minus sign indicating an attractive force towards a point chosen as the origin of x. The potential energy is $\frac{1}{2}\mu x^2$, and

$$t = \left(\frac{m}{\mu}\right)^{\frac{1}{2}} \int \frac{dx}{(x_0^2 - x^2)^{\frac{1}{2}}} = \left(\frac{m}{\mu}\right)^{\frac{1}{2}} \cos^{-1}(x/x_0) + t_0,$$

where x_0 is $2E/\mu$ and t_0 is constant. With ω' written for $(\mu/m)^{\frac{1}{2}}$ the general solution takes the form,

$$x = x_0 \cos[\omega'(t - t_0)].$$

If t_0 is chosen to be either 0 or π/ω', so as to give x one of its limiting values, $\pm x_0$, when $t = 0$, it reaches the other when $\omega' t = \pi$; the periodic time is $2\pi/\omega'$, as we already know from (5) and (6).

If the potential is symmetrical and varies as the nth power of displacement, so that $V \propto |x|^n$, the variation of frequency with energy can be derived from (11). Writing $V = a|x|^n$ and $E = ax_0^n$, we have that

$$t = (\tfrac{1}{2}m/a)^{\frac{1}{2}} \int dx/(x_0^n - |x|^n)^{\frac{1}{2}}.$$

In particular, if the integral is evaluated between 0 and x_0, t is one-quarter of the period, so that

$$T_0 = (8m/a)^{\frac{1}{2}} \int_0^{x_0} dx/(x_0^n - x^n)^{\frac{1}{2}}$$

$$= (8m/a)^{\frac{1}{2}} x_0^{1 - \frac{1}{2}n} \int_0^1 d\xi/(1 - \xi^n)^{\frac{1}{2}}, \quad \text{with } \xi \equiv x/x_0.$$

The integral is a function of n only, not x_0, so that for any given value of n we have that

$$T_0 \propto x_0^{1-\frac{1}{2}n} \propto E^{1/n-\frac{1}{2}}.$$

The only power law that makes T_0 independent of E is the parabolic potential, $n = 2$. The bouncing ball has $n = 1$ and $T_0 \propto E^{\frac{1}{2}}$. At the other extreme, as $n \to \infty$, T_0 tends to become proportional to $E^{-\frac{1}{2}}$; this is the case of a ball moving on a flat surface and bouncing back and forth between vertical walls – the frequency is proportional to the speed.

The determination of the period may be represented graphically, as in fig. 3. If a horizontal line at the energy E is drawn across the graph of $V(x)$, cutting it at the limiting points of the vibration, the vertical distance between the lines is $E - V$, and $(E - V)^{-\frac{1}{2}}$ is readily constructed from this.

Fig. 3. Graphical construction for the period.

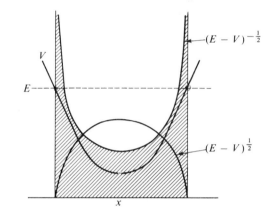

The area under the curve, shown shaded, then represents the period (apart from constants). With a parabolic potential an increase of E increases the width of the shaded area, but decreases its height so as to keep the area constant; the vibrator is *isochronous*, with frequency independent of energy. There is a continuous range of forms for $V(x)$ which lead to isochronous vibrations, for any potential curve $V(z)$ may be 'sheared' without altering the frequency. In fig. 4, the curve B is a sheared version of

Fig. 4. B is a sheared version of A; both potentials yield the same variation of frequency with energy.

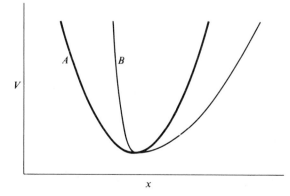

A, both having the same horizontal diameters at the same value of V. The horizontal shift is different for different V, and may be chosen to vary arbitrarily provided the resulting $V(x)$ is single-valued. Both these potential curves yield the same variation of period T_0 with E, for on carrying through the graphical construction of fig. 3 the curve for $(E-V)^{-\frac{1}{2}}$ is itself only sheared, leaving the area under it unchanged. It is clear from this that a specification of how the frequency varies with energy is not enough to fix the form of the potential. In particular, if ω' is independent of E the potential need not be parabolic; any curve derived from a parabola by shearing is a potential in which vibrations are isochronous. The essential property, apart from single-valuedness, is that the diameter must vary as $V^{\frac{1}{2}}$ when V is measured from the lowest point.

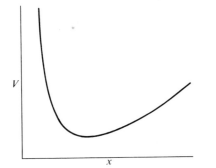

Fig. 5. The isochronous potential ax^2+b/x^2.

As an example consider the potential,

$$V = ax^2 + b/x^2, \tag{12}$$

29

shown in fig. 5. To find the separation of the two positive values of x at which V takes a given value, rearrange (12) as a quadratic equation in x^2:

$$ax^4 - Vx^2 + b = 0,$$

with solutions $\pm x_1$ and $\pm x_2$. The equation shows directly that

$$x_1^2 + x_2^2 = V/a \quad \text{and} \quad x_1^2 x_2^2 = b/a.$$

Hence $(x_1 - x_2)^2 = V/a - 2(b/a)^{\frac{1}{2}} = (V - V_0)/a,$

where $V_0 = 2(ba)^{\frac{1}{2}}$ and is the value of V at the minimum. The diameter, $|x_1 - x_2|$, is proportional to $(V - V_0)^{\frac{1}{2}}$ and the potential (12) is thus one that yields isochronous vibrations.

Of the whole family of isochronous potentials derived from the parabola by shearing, only the parabolic potential itself confers the property of superposition. To illustrate the breakdown of superposition in a special case, consider the half-parabolic potential of fig. 6, with a hard wall at the left, off which the vibrating particle will bounce. The motion of the particle consists of a sequence of positive sinusoidal loops, and two solutions are shown in fig. 7(a) displaced in time but with the same frequency in spite of

Fig. 6. An isochronous potential generating vibrations that are not superposible.

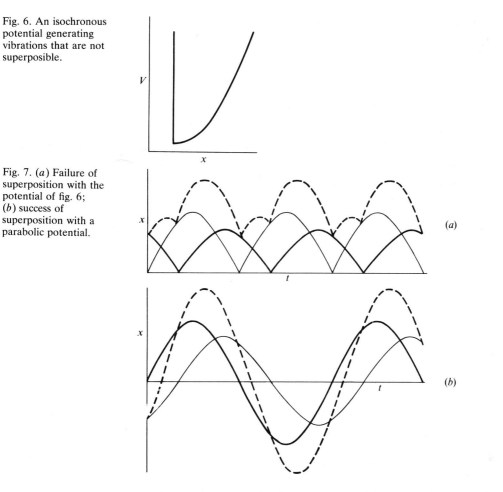

Fig. 7. (*a*) Failure of superposition with the potential of fig. 6; (*b*) success of superposition with a parabolic potential.

(a)

(b)

having different amplitudes. The sum of the two solutions, with its abrupt changes of slope, is obviously no solution in its own right. By contrast, the fully parabolic potential yields solutions which are strictly sinusoidal as well as isochronous, as in fig. 7(*b*). Since any two sinusoids of the same period add to give another sinusoid of the same period, the sum is also a true solution of the equation of motion, and superposition applies.

A convenient way to demonstrate vibrations in a variety of potentials is to roll a steel ball on a curve of the desired form, as in fig. 8; by the use of a groove or of rails the ball may be kept on the curve without excessive friction. So long as the curve is very shallow, and the contribution of the

Fig. 8. Ball rolling on a curve.

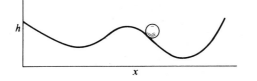

vertical component of velocity to the kinetic energy is negligible, the energy equation may be written in terms of the horizontal coordinate x alone:

$$\tfrac{1}{2}m^*\dot{x}^2 + mgh(x) = E,$$

where $h(x)$ is the height of the curve. In writing m^* instead of m we make allowance for the rotational contribution to the kinetic energy which is also proportional to $\dot{x}^2$. To the extent that the approximation is adequate, the model behaves in accordance with the solution of (8), with V represented by mgh; for example, if $h \propto x^2$, so that the curve is parabolic, the oscillations are isochronous.

But of course this is not strictly true, for the kinetic energy is really $\tfrac{1}{2}m^*\dot{s}^2$, where s is distance measured along the curve, and the curve which yields truly isochronous oscillations is that for which $h \propto s^2$, if s and h are measured from the lowest point. On such a curve the ball rolls so that its displacement s varies sinusoidally with time. The cycloid, not the parabola, is the curve with the desired property, as may be seen by starting from the definition of the cycloid as the curve traced out by a point on the rim of a rolling wheel. In fig. 9 the circle rolls below a horizontal line, but h is measured from the lowest point, reached by P when $\phi = 0$. If the radius of the circle is a,

$$h = a(1 - \cos \phi) = 2a \sin^2 \tfrac{1}{2}\phi. \tag{13}$$

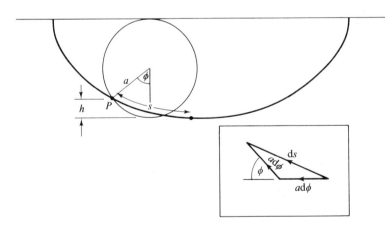

Fig. 9. Geometry of a cycloid.

To find how s depends on ϕ, consider what happens to the point P when the circle rolls through $d\phi$. If its centre remained fixed, the movement of P would be a vector of length $a\, d\phi$ inclined at ϕ to the horizontal. But the circle does not slip and its centre must move through $a\, d\phi$; the first vector therefore has added to it a horizontal vector of the same length, as shown in the inset to fig. 9. The resultant is ds and it is clear that

$$ds = 2a \cos (\tfrac{1}{2}\phi)\, d\phi,$$

so that, on integrating,

$$s = 4a \sin\left(\tfrac{1}{2}\phi\right) = (8ah)^{\frac{1}{2}} \quad \text{from (13).}$$

The relation between s and h is what is required for isochronous oscillation. This property of the cycloid was discovered by Huygens[1] who also showed how a simple pendulum could be constrained by cycloidal cheeks so that its bob described a similar cycloid (fig. 10). The result is amusing and elegant, rather than useful, for an unconstrained pendulum with a small amplitude of oscillation is very nearly isochronous, and it is easier to keep the amplitude constant than to devise practicable tricks to render variations of amplitude harmless.

Fig. 10. Huygens' cycloidal pendulum. With the string constrained by the cheeks, A and B, the bob executes the cyloidal path shown as a broken line.

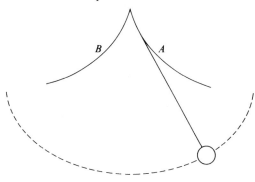

To expand on this last point let us calculate how frequency depends on amplitude for a simple pendulum whose bob moves in a circle. The displacement is described by the angle θ between the string and the vertical, and the energy equation takes the form

$$\tfrac{1}{2}ml^2\dot{\theta}^2 + mgl(1 - \cos\theta) = E,$$

which may be rearranged in the form

$$\dot{y}^2 = \omega_0^2(\sin^2 y_0 - \sin^2 y),$$

in which $y = \tfrac{1}{2}\theta$, $\omega_0^2 = g/l$ and E is written as $2(g/l)\sin^2 y_0$; y_0 measures the amplitude of swing, the pendulum being at rest, with $\dot{y} = 0$, when $y = \pm y_0$. From this the period follows, by the same argument as yielded (11), and with the symmetry of V used to divide the cycle into four equal parts:

$$T_0 = \frac{4}{\omega_0} \int_0^{y_0} \frac{dy}{(\sin^2 y_0 - \sin^2 y)^{\frac{1}{2}}}. \tag{14}$$

The integrand is infinite at the upper limit, but this awkwardness can be removed by a change of variables. Put $\sin\phi = \sin y / \sin y_0$, so that

$$\sin y_0 \cos\phi = (\sin^2 y_0 - \sin^2 y)^{\frac{1}{2}};$$

also $\qquad \sin y_0 \cos\phi \, d\phi = \cos y \, dy,$

2

so that $dy/(\sin^2 y_0 - \sin^2 y)^{\frac{1}{2}} = d\phi/\cos y$. As y ranges from 0 to y_0, ϕ ranges from 0 to $\frac{1}{2}\pi$, and (14) takes the form

$$T_0 = \frac{4}{\omega_0} \int_0^{\frac{1}{2}\pi} \frac{d\phi}{(1 - \sin^2 y_0 \sin^2 \phi)^{\frac{1}{2}}},$$

with an integrand that is finite and expandable in series:

$$T_0 = \frac{4}{\omega_0} \int_0^{\frac{1}{2}\pi} (1 + \tfrac{1}{2}\sin^2 y_0 \sin^2 \phi + \cdots)\, d\phi$$

$$= \frac{2\pi}{\omega_0}(1 + \tfrac{1}{4}\sin^2 y_0 + \cdots) = \frac{2\pi}{\omega_0}(1 + \tfrac{1}{4}y_0^2 + \cdots). \qquad (15)$$

The second term is the first correction to the small-amplitude period $2\pi/\omega_0$, i.e. $2\pi(l/g)^{\frac{1}{2}}$. For a typical 'grandfather' or long-case clock (the author's) with a pendulum 110 cm long swinging through 9 cm overall, $y_0 = \frac{1}{4} \times \frac{9}{110} \doteqdot 0.02$, and $\frac{1}{4}y_0^2$ is about 10^{-4}; compared with the same pendulum swinging with a very small amplitude, this clock would lose nearly 9 seconds a day. Of course such an error is readily compensated by shortening the pendulum by about $\frac{1}{4}$ mm, and the real question one should ask is how constant the amplitude must be to keep fluctuations in the period below a certain prescribed limit. This is answered by taking logarithms and differentiating the second form of (15). To an adequate degree of approximation,

$$\delta T_0/T_0 = \tfrac{1}{2}y_0\delta y_0 = \tfrac{1}{2}y_0^2(\delta y_0/y_0).$$

For the particular clock just mentioned, $\delta T_0/T_0 = 2 \times 10^{-4}\,(\delta y_0/y_0)$; timekeeping to better than 1 second a day means that the amplitude must not vary by more than 8% overall. This is not too stringent a requirement and if most domestic pendulum clocks are not nearly as regular as this it is probably for quite other reasons. Astronomical pendulum clocks have been made to keep time to something better than 1 second a year.[2] Apart from anything else in the design (and these instruments are marvels of craftsmanship) the amplitude of the swing must not only be small but very well stabilized for such precision to be attained.

To go from this nearly harmonic system to the opposite extreme, consider the potential curve with two minima in fig. 11. So long as $E < E_1$, the vibrations are confined to one or other of the minima, being nearly harmonic when the amplitude is very small but being increasingly affected by the maximum at E_1 as E approaches this value. At higher energies the oscillations become symmetrical, with fast sections of the cycle as the body passes over the minima and slow sections both at the ends and in the middle. Just at the critical point, defined by $E = E_1$, the frequency drops to zero, since it takes an infinite time for the body to cross the peak when its velocity momentarily vanishes. To see how the frequency varies in this energy range consider E to be marginally greater than E_1, so that the

Fig. 11. (*a*) Potential curve with two minima; (*b*) schematic variation of frequency with energy in this potential.

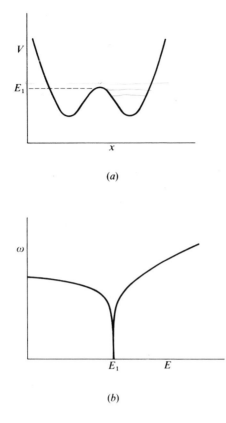

(*a*)

(*b*)

integral (11) takes the form

$$t \sim \int_0^x \frac{\mathrm{d}x}{(x^2+\varepsilon)^{\frac{1}{2}}},$$

with x^2 representing the form of the peak when x is small, and ε the excess energy $E-E_1$. The time taken to travel from the centre to a certain small displacement x_1 is therefore proportional to $\sinh^{-1}(x_1/\varepsilon^{\frac{1}{2}})$, which tends to $\frac{1}{2}\ln(4x_1^2/\varepsilon)$ as ε tends to zero. The time spent near the peak thus ultimately controls the period, and the frequency tends to zero as $(-\ln\varepsilon)^{-1}$. This is an extremely precipitate fall; the same behaviour is found as E tends to E_1 from below.

Using energy conservation to calculate vibration frequency

We return now to harmonic vibrators which are at rest, with their energy wholly potential, when their displacement is greatest, and are moving at their greatest speed, with their energy wholly kinetic, when their displacement is zero. If T and V are given by (8), and the displacement is represented by $a\cos\omega't$, the velocity is $-a\omega'\sin\omega't$. With a maximum displacement of a and a maximum velocity of $a\omega'$, the value of V at maximum displacement is $\frac{1}{2}\mu a^2$, and the value of T at zero displacement is $\frac{1}{2}m(a\omega')^2$. Since these must be equal it follows immediately that $\omega'^2 = \mu/m$, as already found by direct solution of the equation of motion. Note that the amplitude, a, drops out, as expected since the frequency is amplitude-independent.

This method of determining ω' adds nothing to our understanding of what is happening, and although it may be convenient to apply it to simple vibrators it cannot be said to possess any great advantage over direct methods. With more complicated systems, however, it has considerable merit, especially in making approximate estimates, and we shall return to its application in discussing Rayleigh's principle, though not in this volume. Meanwhile let us illustrate its use by calculating the resonant frequency of a Helmholtz resonator, formerly a valuable device for analysing sounds.[3] A more-or-less spherical metal shell of radius R (fig. 12) has one rather large hole A and a small outlet B shaped to fit the ear, which detects weak

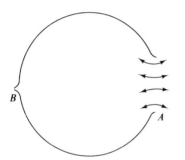

Fig. 12. Section of Helmholtz resonator.

vibrations of pressure within the shell. When exposed to a sound of the right frequency the air within oscillates, flowing to and fro through A as indicated by the arrows. The resonant frequency is essentially determined by the volume of the shell and the area of the hole.† At the maximum inward displacement of the air everything within the shell is uniformly compressed, and the energy is wholly potential; similarly at the maximum outward displacement, when the air within is expanded below atmospheric pressure, which also increases its potential energy. Midway between these two phases the air is at atmospheric pressure, but that part lying close to A is flowing at its greatest speed and carries the energy of vibration in kinetic form. Just how much the kinetic energy is when the mean speed across the aperture is $\dot{z}$ demands some knowledge of the flow pattern, a matter we return to in a moment. For the present, however, it is enough to recognize that the flow speed drops off rapidly away from the aperture, so that the kinetic energy is the same as would be possessed by the mass of gas contained in a fairly short cylinder whose radius r is defined by the hole A,

† That humble musical instrument, the ocarina, is essentially a Helmholtz resonator with a variety of holes to allow different notes to be sounded, and a mouthpiece like that of the recorder to excite it into vibration. The following theory shows why the note rises in pitch as more holes are opened.

and whose length l is as yet undetermined. If this cylinder be imagined vibrating harmonically with small amplitude, a ($\ll l$), its maximum speed is $a\omega'$ and its kinetic energy takes the form:

$$T = \tfrac{1}{2}\pi r^2 l\rho a^2\omega'^2, \tag{16}$$

where ρ is the density of the gas.

To calculate the potential energy at the moment of maximum inward displacement, we note that the volume of the gas in the shell, originally $\tfrac{4}{3}\pi R^3$, has been reduced by $\pi r^2 a$ and has therefore suffered a fractional change $\tfrac{3}{4}r^2 a/R^3$. If the adiabatic bulk modulus of the gas is K (its actual value γP need not be used), the excess pressure within the shell at this instant is $\tfrac{3}{4}Kr^2 a/R^3$, and the work that would be needed to create this strain by slowly forcing the gas in is one-half the product of the pressure and the volume of gas moved. Hence, when the energy is wholly potential,

$$V = \tfrac{1}{2}\times(\tfrac{3}{4}Kr^2 a/R^3)\times\pi r^2 a. \tag{17}$$

Equating (16) and (17) yields the frequency:

$$\omega'^2 = \tfrac{3}{4}Kr^2/\rho lR^3 = \tfrac{3}{4}c_s^2 r^2/lR^3,$$

where c_s is the velocity of sound, $(K/\rho)^{\frac{1}{2}}$. Now the wavelength, λ, in open air of a sound of frequency ω' is $2\pi c_s/\omega'$. The resonance condition for the Helmholtz resonator can therefore be expressed in purely geometrical terms by the formula

$$\lambda = \frac{4\pi(lR^3)^{\frac{1}{2}}}{\sqrt{3}r}. \tag{18}$$

One of a 100-year-old set of resonators in the Cavendish Laboratory museum was confirmed with a signal generator and loudspeaker as resonating at its stated pitch (g below middle C, with a wavelength of 1.77 m); it had $R = 9$ cm and $r = 1.9$ cm, and (18) is satisfied if $l = 3$ cm, a very reasonable-looking value for a hole 3.8 cm in diameter. A good theoretical estimate is easy if R is large enough for the flow through the hole to follow the pattern of flow through a hole in a plane sheet, for this is the same as the electric field radiating from a charged metal disc of radius r, a standard problem in electrostatics. By following through the analogy between field lines and flow lines one can see that $\tfrac{1}{2}l$ is the spacing of plates in an idealized capacitor (see fig. 13) that would have the same capacitance, $8\varepsilon_0 r$, as the isolated disc. In this way l is found to be $\tfrac{1}{2}\pi r$, which is very close to 3 cm in the present example, so that theory and experiment correspond satisfactorily.

It may be noted that determination of vibration frequency by the energy argument is not in principle confined to harmonic vibrations, since in any force field the vibrating object carries its energy alternately in potential and kinetic form. But in practice the relationship between maximum deflection, maximum speed and frequency ($v_{max} = \omega'a$ for harmonic vibrations) is not known without a full solution of the equation of motion, and the technique

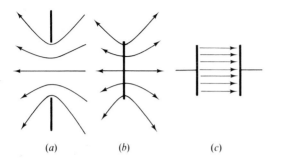

Fig. 13. (*a*) Streamline flow of air through circular hole in a flate plate; (*b*) electric field lines around a charged plate the same size as the hole in (*a*); (*c*) an idealized parallel plate capacitor (no fringing field) having the same capacitance as the capacitance to earth of (*b*).

(*a*)   (*b*)   (*c*)

is therefore virtually useless except in this one case. To illustrate the danger of applying the method naively, consider a simple pendulum swinging up to a horizontal position, so that at the moment of rest its potential energy is *mgl*. If the bob were to move strictly harmonically along its arc, one would write the amplitude as $\frac{1}{2}\pi l$ and the kinetic energy as it passes its lowest point as $\frac{1}{8}m\omega'^2\pi^2 l^2$. Equating these two energies, and remembering that the small-amplitude frequency, ω_0 is $(g/l)^{\frac{1}{2}}$, one estimates the period to be increased by a factor $\frac{1}{2}\pi/\sqrt{2}$, i.e. by 11%, though the true answer is 18%.

Systems with more than one mode

The systems considered so far have been simple in the special sense of having only one significant variable, and simple in the colloquial sense also. There are many others which at first sight are of a complexity that would make one expect their description also to be complex, yet which on investigation turn out to be reducible to simple systems. Take, for example, the case of two identical pendulums hung from a slack line, as in fig. 14.

Fig. 14. Two pendulums coupled by hanging from a slack string.

When one is displaced in a direction normal to the paper and set swinging, after a few swings the other joins in, being excited by the movement of the line supporting it, and the first pendulum begins to lose amplitude; there comes a time when the first is at rest and only the second swinging, and then the process repeats in reverse until all the energy is back where it started – and so on until the vibrations are damped out. Clearly this is not described by a single variable executing sinusoidal vibrations of constant amplitude. But now displace both pendulums by the same amount in the same direction, and they will be found to vibrate together without any exchange of energy; and the same thing happens when they are displaced equally in opposite directions, though the frequency is slightly higher now that they are in antiphase than when they were in phase. The system is capable of executing simple harmonic vibrations in two different modes, and the peculiar exchange of energy first described is a consequence of exciting both simultaneously.

365   This is a phenomenon to which we shall devote much attention later, and it is enough to note, to clarify the issue in a preliminary way, that if x_1 and x_2 are the displacements of the two pendulums, $(x_1 + x_2)$ and $(x_1 - x_2)$ both behave like the coordinate of a simple vibrator and can execute only harmonic oscillations:

$$(x_1 + x_2) = A_1 \cos (\omega_1 t + \phi_1), \tag{19}$$

$$(x_1 - x_2) = A_2 \cos (\omega_2 t + \phi_2). \tag{20}$$

These can coexist independently, and any motion of the system can be represented by a superposition of the two modes with arbitrary choice of the constants $A_{1,2}$ and $\phi_{1,2}$ (always subject to the amplitude being small enough for the linear approximation to hold). Thus the two *normal modes* which undergo harmonic motion are the symmetric mode, in which $x_2 = x_1$ and A_2 but not A_1 vanishes, and the antisymmetric mode, in which $x_2 = -x_1$ and A_1 but not A_2 vanishes. If both modes are excited equally, as when $A_1 = A_2 = 1$, and if $\phi_1 = \phi_2 = 0$, adding and subtracting (19) and (20) show that

$$x_1 = \tfrac{1}{2}(\cos \omega_1 t + \cos \omega_2 t) = \cos \bar{\omega} t \cos \delta t$$

and
$$x_2 = \tfrac{1}{2}(\cos \omega_1 t - \cos \omega_2 t) = \sin \bar{\omega} t \sin \delta t,$$

where
$$\bar{\omega} = \tfrac{1}{2}(\omega_1 + \omega_2) \quad \text{and} \quad \delta = \tfrac{1}{2}(\omega_2 - \omega_1).$$

If the two modes have nearly the same frequency, $\delta \ll \bar{\omega}$, so that x_1 and x_2 both oscillate with mean frequency $\bar{\omega}$ but with amplitudes that vary slowly, that of x_1 as $\cos \delta t$, that of x_2 as $\sin \delta t$, the peaks of amplitude alternating between one pendulum and the other.

The two pendulums represent a physical system whose configuration is determined by two coordinates, x_1 and x_2, and which has two normal modes of vibration. A stretched string needs a continuous function, effectively an infinite number of discrete variables, to describe its shape at any instant, and is capable of executing harmonic vibrations at a series of different frequencies. In this particular case the normal mode frequencies are all multiples of the lowest, *fundamental*, frequency as can be readily demonstrated on a piano. Depress the key of C, two octaves below c' (middle C), gently so as to raise the damper without sounding the note. Then strike the octave above, c, and release it – the low string will be heard sounding an octave above its own natural pitch. Similarly with the g next above, when the low C-string will be excited to vibrate in its third harmonic mode and sound the note g; and so on through c′ and the e′, g′ and c″ above it. Intermediate notes excite no response from the low string. In its fundamental mode the string vibrates as a simple loop with a single antinode at its centre (fig. 15), and the other modes excited in the way described have 2, 3, 4, 5, 6 and 8 loops (the seventh harmonic sounds about halfway between a′ and b′ and is not represented in the diatonic scale; it

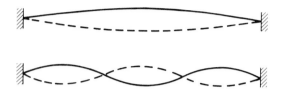

Fig. 15. Fundamental and third harmonic vibrations of a stretched string. The full and broken lines show the extremes of displacement half a cycle apart.

cannot be excited by any string on the piano, but if you have a steady tenor or higher voice you may be able to excite it by singing loudly at the C-string). The string has a virtually unlimited series of modes, each of which by itself confers strictly harmonic vibrations on each element of the string. When several are simultaneously excited, as by the piano hammer, the superposition of all the modes with their different frequencies and standing-wave patterns causes each element of the string to execute a different non-sinusoidal motion. Even more complicated are the vibrations of drums and bells, but the same resolution into normal modes is possible, and each normal mode has the character of a simple vibration and can be represented as the harmonic motion of a single variable. All these systems obey linear equations of motion, on which the resolution into normal modes depends. Non-linear systems with many variables are quite different and do not conform to any simple generalized behaviour, but each has to be treated separately. We shall discuss a few examples of this extremely difficult field at appropriate places later.

225, 385

A system with two independent normal modes that deserves some consideration is the simple pendulum swinging not in a plane but in any direction and arranged, for example, as in fig. 16, so that the frequency depends on the direction of its swing. The normal modes can be seen by inspection – if it swings in and out of the paper it is effectively of length l_1, if

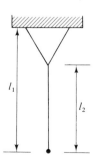

Fig. 16. A simple pendulum supported on a loop so as to vibrate with different frequency in two normal planes.

in the plane of the paper its effective length is l_2. With these directions as x- and y-axes, motion in each of the two normal modes follows one of the equations:

$$x = A_1 \cos(\omega_1 t + \phi_1),$$
$$y = A_2 \cos(\omega_2 t + \phi_2),$$

(21)

where $\omega_1^2 = g/l_1$ and $\omega_2^2 = g/l_2$. The combination of these two motions gives rise to the well-known Lissajous figures, of which examples are shown in fig. 17. If the ratio ω_1/ω_2 is a rational fraction of the form N/M (N and M having no common factor) the figure is closed after N cycles in the x-direction and M in the y-direction; the whole pattern is contained within a rectangle of sides $2A_1$ and $2A_2$. If ω_1/ω_2 is irrational the figure never closes and ultimately every point within the rectangle is traversed. These figures are most easily demonstrated with a cathode-ray tube, by applying separate alternating voltages of variable frequency to the X and Y plates.

The same behaviour is shown by a body confined to motion in two dimensions, and acted upon by springs so that its potential energy is a positive definite quadratic function of its displacement from equilibrium; any arrangement of springs leads to this, for small displacements at least. If the axes are chosen arbitrarily the potential energy takes the form $ax^2 + by^2 + cxy$, but by choosing the orientation of the axes appropriately the cross-term, cxy, can be eliminated to leave V in the form

373

$$V = a'x^2 + b'y^2.$$

The force acting on the displaced body is a two-dimensional vector $F(x, y)$ having components

$$F = (-\partial V/\partial x, -\partial V/\partial y) = (-2a'x, -2b'y).$$

With this coordinate system the variables are separated in the equations of motion:

$$m\ddot{x} = F_x = -2a'x$$

and

$$m\ddot{y} = F_y = -2b'y,$$

so that the normal mode frequencies are given by $\omega_1^2 = 2a'/m$, $\omega_2^2 = 2b'/m$. The Lissajous figures result from the simultaneous excitation of both.

In the special case of an isotropic two-dimensional vibrator (e.g. the conical pendulum), $a = b$ and $\omega_1 = \omega_2$. Then F is in the direction of the displacement and the system can oscillate in a straight line in any direction:

$$r = A \cos(\omega t + \phi),$$

where A is any constant vector in the plane. But any combination of x and y oscillations with arbitrary amplitude and phase difference is equally possible, giving rise to motion in an elliptical path. Let the phase difference

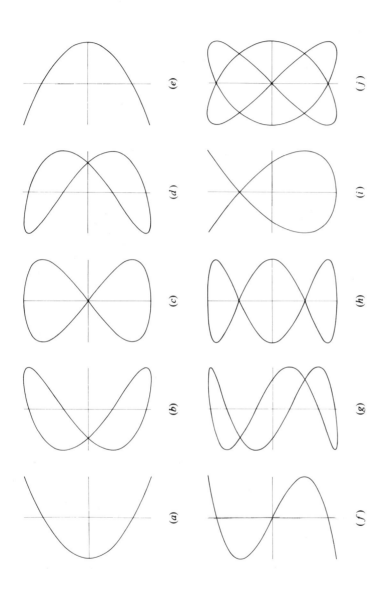

Fig. 17. Lissajous figures. Top row (a)–(e), $\omega_1 = 2\omega_2$ and $\phi = 0, \frac{1}{4}\pi, \frac{1}{2}\pi, \frac{3}{4}\pi, \pi$. Bottom row (f)–(h), $\omega_1 = 3\omega_2$ and $\phi = 0, \frac{1}{4}\pi, \frac{1}{2}\pi$; (i) and (j) $\omega_1 = 1.5\omega_2$ and $\phi = 0, \frac{1}{4}\pi$.

be 2ϕ and choose the origin of time so that the motion can be written in the form

$$x = A_1 \cos(\omega t - \phi) = A_1(\cos \omega t \cos \phi + \sin \omega t \sin \phi),$$

$$y = A_2 \cos(\omega t + \phi) = A_2(\cos \omega t \cos \phi - \sin \omega t \sin \phi).$$

Then $x/A_1 + y/A_2 = 2 \cos \omega t \cos \phi$ and $x/A_1 - y/A_2 = 2 \sin \omega t \sin \phi$;

hence $(x/A_1 + y/A_2)^2 \sin^2 \phi + (x/A_1 - y/A_2)^2 \cos^2 \phi = \sin^2 2\phi.$

The time-variable has been eliminated to yield the equation of the trajectory, which is quadratic and readily shown to be an ellipse. The reader may work out for himself the relationship between A_1, A_2 and ϕ on the one hand, and the shape and orientation of the ellipse on the other. The formulae are rather complicated and not particular useful.[†]

The conical pendulum has the property that for small displacements all trajectories are closed and have the same period. The oscillatory radial motion is thus isochronous, though it is not harmonic. Let us separate this radial motion from the rotation, examining the bob from a frame of reference that rotates in step with it. The angular momentum $L = mr^2\Omega$ when the radial displacement is r and the instantaneous angular velocity Ω, and since L is a constant of the motion, $\Omega = L/mr^2 \propto 1/r^2$. To the observer in the frame which also rotates at the variable angular velocity Ω, the bob is subject to a centrifugal force $mr\Omega^2$ as well as the restoring force $-\mu r$ due to the string, and therefore appears to be moving in a straight line under the influence of the force field

$$F(r) = mr\Omega^2 - \mu r = L^2/mr^3 - \mu r,$$

which is derivable from a potential

$$V(r) = \tfrac{1}{2}\mu r^2 + L^2/2mr^2.$$

This has the same form as (12) which was earlier cited as an example of an isochronous potential. We are now in a position to write down the form of the vibration in this particular potential, for it is just the radial motion of a conical pendulum moving in an ellipse. By choosing the x- and y-axes to coincide with the axes of the ellipse we arrange that the two oscillations in

---

[†] The analysis of circular or elliptical motion into two harmonic oscillations along orthogonal axes is only useful in a linear system. A simple example shows how the argument may fail for a non-linear vibrator; a conical pendulum swinging in a circle with the string making an angle θ to the vertical has a frequency $(l\sec\theta/g)^{\frac{1}{2}}$, becoming greater as θ increases, while oscillations in a plane have lower frequencies at larger amplitudes. The former obviously is not a simple composition of two of the latter oscillations. In the same way if we were to set a body moving on a smooth bowl formed as a figure of revolution from a cycloid, we might naively expect the isochronous property to give rise to simple closed orbits whatever the shape or amplitude; but we should be wrong.

(21) are in phase-quadrature – the x-displacement is greatest as the bob crosses the x-axis $(y = 0)$, and similarly for the y-displacement:

$$x = C_1 \cos \omega t, \qquad y = C_2 \sin \omega t.$$

Hence
$$r^2 = C_1^2 \cos^2 \omega t + C_2^2 \sin^2 \omega t,$$

which can be written in the form

$$r = C_1(1 + B \sin^2 \omega t)^{\frac{1}{2}},$$

where C_1 can take any positive value and $B(=C_2^2/C_1^2 - 1)$ any value between -1 and $+\infty$. The same result may of course be reached directly by integrating (11).

Dissipation

The simple vibrators discussed so far have been idealized to the extent of neglecting the inevitable losses that in practice cause the amplitude of a passive vibrating system to decay. The use of the term 'passive' distinguishes one class of vibrating systems from another, 'active' vibrators which are supplied by an energy source. In a passive system the energy of vibration may be converted into heat but not vice versa, and therefore decay is the only mode of change;[†] by contrast the active system may build up in amplitude until the dissipation matches the power input or the character of the system is changed by some catastrophe. The dependence of dissipation rate on amplitude of motion in a passive vibrator may take different forms, of which the simplest from a mathematical point of view is that exemplified by an ohmic resistor in an oscillating circuit or viscous damping of a mechanical vibrator. In both cases the force responsible for damping is proportional to the first derivative of the displacement. For the circuit of fig. 1(c) the displacement is measured by q, $\dot{q}$ is the current through R and $R\dot{q}$ is the potential difference across R; if q and $\dot{q}$ are positive, the electric field between the capacitor plates and the field in the resistor have the same sign, so that (5) must be supplemented with another positive term:

$$L\ddot{q} + R\dot{q} + q/C = 0, \tag{22}$$

which may be written in what will henceforth be taken as the standard form:

$$\ddot{\xi} + k\dot{\xi} + \omega_0^2 \xi = 0. \tag{23}$$

In this particular case the generalized displacement, ξ, stands for q, $k = R/L$ and $\omega_0^2 = (LC)^{-1}$. Similarly the pendulum bob in fig. 1(a), when

---

† For the present we ignore the interchange between random thermal motions and vibration of the system, which is of great importance in small systems and in delicate instruments, but rarely noticeable otherwise. This question will be discussed in chapter 4 and subsequently.

89, 156, 361

displaced through a positive angle θ and moving outwards with a positive velocity $l\dot{\theta}$, experiences a restoring force proportional to θ and a force from any viscous fluid surrounding it proportional to $\dot{\theta}$ and in the same sense as the restoring force. Its equation of motion therefore has the form (23) also.

The popularity of (23) as an expression of damped vibratory motion arises partly from the great importance of electrical circuits for which it really is a very good representation, but partly also from the linearity of the equation which makes it mathematically tractable. There is something of a tendency among physicists to try to reduce everything to linearity, and this may be defended if the linear approximation is not too fanciful and if the alternative (of tackling the real problem) involves analysis of a different order of difficulty. Since we shall be discussing linear or linearized systems so much in this book let us recognize at the outset, by means of a few examples, that the reality may not always conform to what we might wish, rather more so with damping forces than with the restoring force in small-amplitude vibrations. Thus a mechanical vibrator immersed in a viscous fluid, and moving so slowly that the flow round it is laminar, will indeed suffer a retarding force proportional to its velocity. But if it is moving fast enough to create a turbulent wake the force will be more nearly proportional to v^2; while if it is rubbing against some fixed surface the frictional force may be independent of the speed and merely directed oppositely to the motion. These are elementary examples of non-linear damping effects, and more complicated cases are by no means unusual; for example the frictional force may be substantially higher when there is no slipping and fall as soon as slipping begins.

41, 336

Each case demands its own special treatment if a complete analytical solution is to be found, but a very adequate approximation may be derived from energy considerations if the rate of damping is not too great. Before giving this we shall find complete solutions for two models which will serve for comparison with the approximate result. The first, which is important in its own right, and therefore deserves rather full discussion, is the linear equation (23) whose solution is the exponentially damped harmonic oscillation expressed by (1). Substituting $\xi = \xi_0 \cos(\omega't + \phi) e^{-t/\tau_a}$ in (23), one finds that the terms in $\cos(\omega't + \phi)$ and $\sin(\omega't + \phi)$ both vanish and the equation is satisfied when

either $\qquad \left. \begin{array}{l} \omega'^2 = \omega_0^2 - \tfrac{1}{4}k^2, \\ \tau_a = 2/k, \end{array} \right\}$ if $k < 2\omega_0$, $\qquad$ (24)

or $\qquad \left. \begin{array}{l} \omega' = 0, \\ \tau_a = 2/[k \pm (k^2 - 4\omega_0^2)^{\frac{1}{2}}], \end{array} \right\}$ if $k > 2\omega_0$. $\qquad$ (25)

The non-oscillatory solution (25) arises when the damping is too high (e.g. $R^2 > 4L/C$) to permit vibrations, and the system is said to be overdamped. The complete solution with two adjustable constants then takes the form

$$\xi = \xi_1 e^{-t/\tau_1} + \xi_2 e^{-t/\tau_2}, \qquad (26)$$

where τ_1 and τ_2 are the two values appearing in (25). A typical curve of the time-variation of ξ in an overdamped system is shown in fig. 18, where $\xi(0)$ has been made equal to zero by choosing $\xi_2 = -\xi_1$.

In the limit of very heavy overdamping, τ_1 approaches $1/k$ and τ_2 approaches k/ω_0^2. For the LCR circuit $\tau_1 = L/R$, the time-constant for L and R in series, while $\tau_2 = RC$, the much longer time-constant for C and R in series. When an impulse is applied to the system at rest, $\xi_2 = -\xi_1$ in (26); the amplitudes of the two terms match initially, and the initial rise is therefore dominated by ξ_1, with the much larger slope. After one or two times τ_1, however, ξ_2 takes over and the fall back to zero is very slow, being governed by τ_2. It is easy to see the physical processes at work here. The initial impulsive e.m.f. which establishes a current has finished before any significant charge has appeared on the capacitor. Immediately afterwards, therefore, the capacitor plays no part and the decay of the current is controlled by L and R, with time-constant L/R. A little later, when the current is small, the capacitor, which by now has a charge, will begin to take over, and by the time the current has fallen to zero it will be dominant. For at this moment the charge on the capacitor reaches its peak value, and thereafter discharge occurs; and when R is large the discharge rate, governed by the time-constant RC, is slow so that inductive effects are negligible.

The underdamped system, obeying (24), is oscillatory with a Q-value that is determined by use of (2). Since $\omega'' = 1/\tau_a = \frac{1}{2}k$,

$$Q = \omega'/k = (\omega_0^2/k^2 - \tfrac{1}{4})^{\frac{1}{2}}. \tag{27}$$

Thus the LCR circuit, for which $\omega_0^2 = (LC)^{-1}$ and $k = R/L$, has a Q-value given by $(L/R^2C - \tfrac{1}{4})^{\frac{1}{2}}$ or approximately (when $Q \gg 1$) ω_0L/R or $(\omega_0RC)^{-1}$.

A particular case arises at the point of critical damping, $k = 2\omega_0$, at which $Q = 0$ and (24) gives place to (25). If an underdamped system is initially at rest and is struck sharply at $t = 0$ so as to initiate vibration, the subsequent motion is described by the equation

$$\xi = \xi_0 \sin \omega' t \cdot e^{-t/\tau_a}, \tag{28}$$

since this choice of phase ($\phi = -\pi/2$) ensures that ξ vanishes at the beginning. Now as k is increased towards $2\omega_0$, τ_a decreases to $1/\omega_0$ and ω' to zero, so that an ever smaller fraction of a cycle of $\sin \omega' t$ is accomplished before the exponential term kills the vibration. In the limit, as the critical condition is reached, it is permissible to replace $\sin \omega' t$ by $\omega' t$, so that

$$\xi = \dot{\xi}_0 t\, e^{-\omega_0 t},$$

the constant $\dot{\xi}_0$ being the initial value of $\dot{\xi}$ due to the impulse. The form of this function is very similar to fig. 18.

With an instrument like a moving coil galvanometer, whose damping is largely provided by induced currents flowing in an external resistor, and is therefore adjustable, it is customary to arrange that it is close to critical damping, for it then responds most rapidly. If the damping is less it

Fig. 18. Displacement of
an overdamped vibrator
struck while at rest, at
$t = 0$; the two time-
constants here differ by a
factor of two.

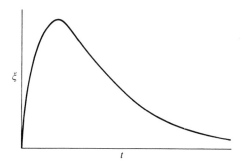

oscillates several times before reaching a steady reading; if it is more the longer of the two time-constants in (25) causes it to be sluggish. The galvanometer incidentally is a very useful device for submitting the theory of linear vibrators to experimental check, for the decrement is easily measured, and the damping is adjustable in a controlled way by connecting different resistors across the coil. If the coil is deflected through an angle θ, the e.m.f. induced by its motion is proportional to $\dot{\theta}$ (the constant of proportionality, depending on the strength of the magnetic field and the dimensions of the coil, does not concern us for this purpose), and the current in the circuit of total resistance R is therefore proportional to $\dot{\theta}/R$, as is the consequent torque on the coil. The equation of angular motion therefore takes the form

$$I\ddot{\theta} + (\lambda_0 + A/R)\dot{\theta} + \mu\theta = 0, \tag{29}$$

where I is the moment of inertia of the coil and μ the torsion constant of the suspension; λ_0 takes care of the air-damping present at all times, which leads to dissipation even on open circuit $(R = \infty)$. This equation has the same form as (23) and we make the following identifications:

$$k = (\lambda_0 + A/R)/I,$$

$$\omega_0^2 = \mu/I.$$

8 The logarithmic decrement $\Delta = \pi/\omega'\tau_a = \frac{1}{2}\pi k/(\omega_0^2 - \frac{1}{4}k^2)^{\frac{1}{2}}$ from (24);

hence $(\lambda_0 + A/R)/(\mu I)^{\frac{1}{2}} = 2\Delta/(\Delta^2 + \pi^2)^{\frac{1}{2}}$.

This is a convenient formula to compare with experiment, since a graph of $\Delta/(\Delta^2 + \pi^2)^{\frac{1}{2}}$ against $1/R$ should be linear over the whole range of damping up to the critical condition. The result of a test is shown in fig. 19 which substantiates the theory very satisfactorily. It is worth noting that the measurements have been taken to very high damping conditions, with the ratio of successive swings on opposite sides as high as 600. But although this ratio cannot be measured very accurately, errors become much less significant when the logarithm of a large number is taken to form Δ, and still less significant when $\Delta/(\Delta^2 + \pi^2)^{\frac{1}{2}}$ is formed, since this quantity tends to a constant, unity, as the decrement rises without limit while k approaches its critical value $2\omega_0$.

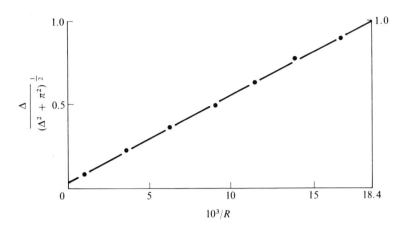

Fig. 19. Experimental verification of variation of damping with circuit resistance in a moving-coil galvanometer. The instrument has a resistance of $10\,\Omega$, and the graph shows that critical damping ($\Delta = \infty$) occurs when the total circuit resistance is $54\,\Omega$ (i.e. $44\,\Omega$ connected across the terminals). To obtain these points it was necessary to keep the deflection of the light spot (at 1 m) less than 10 cm; with larger deflections the air damping increased very rapidly.

The second dissipative model, this time non-linear, can be dealt with quickly. If the frictional force is independent of speed, the vibrating body is subject to a constant retarding force so long as it continues moving in the same direction; as soon as it reverses the force also reverses. With constant frictional force F, and with restoring force $-\mu x$, proportional to displacement, as the only other force acting, the equation of motion for a body of mass m takes the form

$$m\ddot{x} + \mu x \pm F = 0, \qquad (30)$$

the sign in front of F being the same as the sign of the velocity $\dot{x}$. By moving the origin of x to $\mp F/\mu$ (i.e. writing $x' = x \pm F/\mu$), (30) is cast into the standard form for undamped sinusoidal oscillations, analogous to (5):

$$m\ddot{x}' + \mu x' = 0.$$

The complete solution now follows immediately – each half-cycle between positive and negative extremes is of sinusoidal form and of the same frequency, but alternate half-cycles are centred on different values of x, as shown in fig. 20; the vibration stops abruptly at a turning point where the

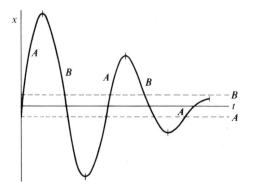

Fig. 20. Frictional damping. The segments labelled A are sinusoids vibrating about the horizontal line A, and similarly for B.

restoring force is no longer as large as *F*. Until that moment is reached, the turning points are successively $2F/\mu$ nearer the origin, and in one complete cycle the amplitude decreases by $4F/\mu$. In contrast to the first model (damping force $\propto$ velocity) which gives exponential decay, the amplitude now decays in a linear fashion.

The difference between the two models is readily demonstrated with a rigid pendulum, about 1 m long. If the damping is provided by fixing a stiff wire onto the pendulum, and letting it dip into a can of treacle, the exponential decay obtained (fig. 21) shows the force to be proportional to velocity. But if, instead, a long wire be stretched horizontally so that the

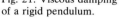
Fig. 21. Viscous damping of a rigid pendulum.

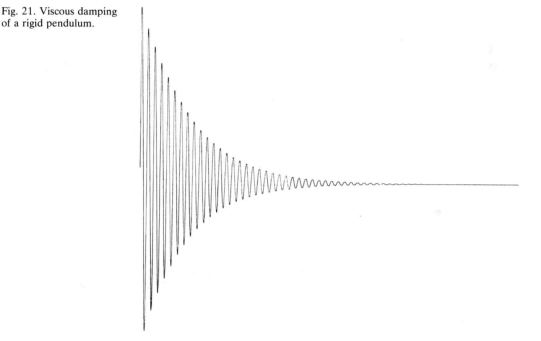

pendulum rod rubs lightly against it, the resulting frictional damping gives the almost linear decay of fig. 22. The behaviour recorded in fig. 23 was obtained by letting a rather stiff wire attached to the pendulum rub against the fixed horizontal wire. When the amplitude is large slipping occurs, with rapid damping due to friction, but at a smaller amplitude the two wires stick together, and there is enough flexibility in the wire attached to the pendulum to permit vibration to continue with little damping, though at a frequency that is 8% higher because of the extra restoring force of the wire. Exactly the same effect has been observed with superconducting discs

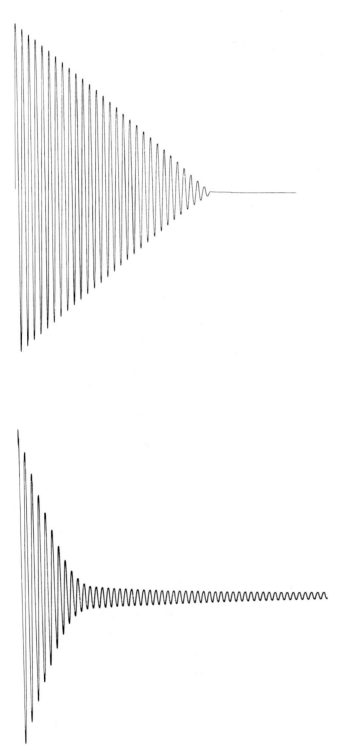

Fig. 22. Frictional damping of a rigid pendulum.

Fig. 23. Pendulum with frictional damping only at large amplitude.

oscillating in a magnetic field; so long as the magnetic flux can move relative to the metal a frictional force is observed, but when the disc oscillates with only small amplitude the flux is pinned and provides an elastic restoring force. Damping curves like fig. 24 are obtained and may be analysed to give a measure of the pinning and frictional forces involved when magnetic fields penetrate and move within a superconductor.

Fig. 24. Oscillation of a superconducting disc in a non-uniform field (P. C. Wraight[4]) exhibiting the same behaviour as fig. 23.

The behaviour represented by figs. 21 and 22 may now be compared with the predictions of an approximate analysis based on the rate of dissipation of energy. If the damping force is weak (though varying with velocity) it will not greatly perturb the sinusoidal form of the vibrations, and we may take $x = x_0 \sin \omega' t$ still to represent them, while bearing in mind that x_0 now decreases slowly. If x_0 is taken as constant over one cycle the velocity, and hence the damping force, $F(x)$, are determined everywhere. The work done in a cycle is proportional to $\oint F(x)\,dx$, where x is taken through a complete cycle between $\pm x_0$; alternatively we may calculate how F varies with time and write the work done as $\oint F(t)\dot{x}(t)\,dt$. It is this that determines the decrease in the energy stored in the vibrator, and hence the change in x_0 during the cycle. Clearly if x_0 changes substantially in a cycle the argument can be only approximate.

Let us apply the argument to the two models already treated. First, when damping is viscous the mechanical version of (23) takes the form

$$m\ddot{x} + \lambda \dot{x} + \mu x = 0. \tag{31}$$

If the oscillation be assumed to have the form $x_0 \cos \omega_0 t$, where $\omega_0^2 = \mu/m$, the velocity $\dot{x}$ is $-\omega_0 x_0 \sin \omega_0 t$, and the damping force F is $\lambda \omega_0 x_0 \sin \omega_0 t$. The rate of dissipation of energy is $-F\dot{x}$, i.e. $\lambda \omega_0^2 x_0^2 \sin^2 \omega_0 t$, and the energy dissipated in one cycle is therefore given by

$$\Delta E = -\int_0^{2\pi/\omega_0} \lambda \omega_0^2 x_0^2 \sin^2 \omega_0 t \, dt = -\pi \lambda \omega_0 x_0^2.$$

This may be compared with the stored energy, $E = \frac{1}{2}\mu x_0^2$, to give the decay of energy per cycle in the form

$$\Delta E/E = -2\pi\lambda\omega_0/\mu.$$

Since $E \propto x_0^2$, we may take the logarithmic decrement (per half-cycle), $-\frac{1}{2}\Delta x_0/x_0$, to be $-\frac{1}{4}\Delta E/E$, i.e. $\frac{1}{2}\pi\lambda\omega_0/\mu$. Alternatively, the rate of decay of amplitude with time is obtained by dividing this by half the period, π/ω_0, to yield

$$\frac{1}{x_0}\frac{dx_0}{dt} = -\frac{1}{2}\lambda\omega_0^2/\mu = -\frac{1}{2}\lambda/m.$$

Hence
$$x_0 \propto e^{-t/\tau_a} \quad \text{where } \tau_a = 2m/\lambda. \tag{32}$$

Now the parameter k introduced in (23) is represented in (31) by λ/m, so that the value of τ_a given by (32) is $2/k$, the same as that obtained in (24) by direct solution of the equation. The exact coincidence is a fluke, for in the energy calculation variations of frequency with damping, as expressed in (24), were neglected; and certainly there is no hint in the energy calculation of the transition to the overdamped state when λ reaches the critical value $2(m\mu)^{\frac{1}{2}}$.

In the second model, with frictional damping, the energy loss is most readily computed by integration over x rather than t, for during one cycle the particle moves through $4x_0$ against a constant retarding force F, and therefore loses energy $\Delta E = -4Fx_0$. But $E = \frac{1}{2}\mu x_0^2$, so that $\Delta E = \mu x_0 \Delta x_0$. Hence $\Delta x_0 = -4F/\mu$, once again in coincidence with the exact treatment. These two examples give us confidence in using the energy argument even when the decrement is not particularly slow, and indeed though not exact it is a most useful first attack on what may otherwise be a tedious analytical problem.[5] It may be especially valuable for hysteretic systems, in which the force is not determined at each instant by the position and velocity, but depends on the whole previous history. Whatever the underlying physical process, the dissipation per cycle is given by the area enclosed by a graph of force versus displacement as the latter runs through a complete cycle.

It may be observed that when F is independent of v, the energy loss per cycle $\Delta E \propto x_0$, while when $F \propto v$, $\Delta E \propto x_0^2$. It is easily seen by writing down the integral for the energy loss that if $F \propto v^n$, n being positive, $\Delta E \propto x_0^{n+1}$. The form of decay when $n > 1$ follows immediately, for we write

$$\Delta x_0/x_0 \propto \Delta E/E \propto x_0^{n-1};$$

hence
$$dx_0/dt \doteq \Delta x_0/T_0 \propto x_0^n,$$

and
$$x_0 \propto (t-t_0)^{1-n}, \quad \text{where } t_0 \text{ is a constant.} \tag{33}$$

For example, a vane moving in a light mobile fluid experiences a force proportional to v^2 when moving fast enough to set up a turbulent wake, and a vibrator so damped will decay as $1/(t-t_0)$; if the system is started when $t = 0$ and in the first period, of duration T_0, drops to an amplitude α

times the initial amplitude, t_0 must be given the value $-\alpha T_0/(1-\alpha)$. One would not expect (33) to hold as the amplitude gets small, since turbulent flow will give place to laminar, with a force proportional to v; the subsequent decay will then be exponential.

Maintained vibrations

We have already drawn a distinction between passive and active vibrators, the latter being supplied with energy from an external source. This may be insufficient to overcome completely the internal dissipation, in which case the system need not differ in any marked degree from a passive vibrator, though its decay time may be longer than could easily be achieved without external aid. If the provision of energy is steadily increased, τ_a will also increase until, at a critical rate of supply that exactly balances the internal dissipation, the vibrations will maintain themselves at constant amplitude. Beyond this critical condition the amplitude grows, with τ_a in (1) negative so that in a linear system the increase is exponential. This growth cannot, of course, continue indefinitely; something must break or change, or the energy source prove insufficient. Non-linearity is therefore an ultimately inevitable feature of a spontaneously oscillating system. Yet until non-linearity prevails there need be no mathematical discontinuity to distinguish active from passive systems. That is to say, the standard form (23) has the same type of solution whether k is positive or negative – changing the signs of k and t together leaves the equation unaltered, and mathematically time-reversal is no great matter. In real life it is of the greatest consequence, however, and we are bound to regard as altogether different in kind systems whose equations of motion can nevertheless all be treated in a common framework. A great variety of oscillators will be discussed in more or less detail in later chapters and in the next volume. Let us note here some of the types with which we shall be concerned, cataloguing them under different headings in order to put some rough order into what could become a bewildering list. The first two headings comprise systems which can be brought into formal correspondence with simple passive vibrators.

330   (*a*) *Negative resistance devices.* If the current through a circuit element, such as a semiconducting tunnel diode, does not increase steadily with the potential difference across it, but has a region of negative slope as in fig. 25, it is possible to bias the device at N and incorporate it in a resonant circuit where it will counteract the lossy effects of the circuit resistance. Related to this type of device are those in which other mechanisms of loss are reversed

345   in sign; in masers and lasers, for instance, artificial dielectrics are created by rearranging the populations of different energy levels of molecules, so as to turn the normal dielectric loss into dielectric gain.

314   (*b*) *Feedback oscillators.* A signal derived from an oscillating system is amplified and fed back into the system with such a phase as to encourage the oscillation to grow (*positive feedback*). Fig. 26 illustrates such an arrangement. Ignoring all but the essentials, we may think of a current i in

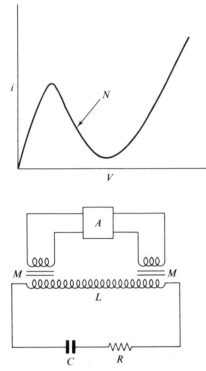

Fig. 25. Current–voltage characteristic of a tunnel diode, showing negative slope region, N.

Fig. 26. Use of amplifier (A) and feedback to maintain oscillation in a resonant circuit.

the resonant circuit inducing an e.m.f. $M\,\mathrm{d}i/\mathrm{d}t$ in the input circuit of the amplifier. If this generates an output current $\mu M\,\mathrm{d}i/\mathrm{d}t$, the e.m.f. induced back into the resonant circuit is $\mu M^2\,\mathrm{d}^2 i/\mathrm{d}t^2$, which must be added to the e.m.f. $L\,\mathrm{d}i/\mathrm{d}t$ due to the inductance itself. Corresponding to (22) we now have

$$\mu M^2 \ddot{q} + L\ddot{q} + R\dot{q} + q/C = 0. \tag{34}$$

Without going into the solution of a cubic equation we may see how this can lead to spontaneous oscillation very much in the same way as introducing a negative resistance. For if q be supposed to oscillate sinusoidally as $q_0 \sin \omega' t$,

$$R\dot{q} = R\omega' q_0 \cos \omega' t$$

and

$$\mu M\ddot{q} = -\mu M^2 \omega'^3 q_0 \cos \omega' t.$$

Both terms have the same phase (i.e. cosine rather than sine), and if the connections to the amplifier are made so that μ is positive the feedback term counteracts the resistive term.

A related system, though one that is essentially non-linear, is the escapement mechanism of pendulum clocks and balance-wheel watches. The drive here is automatically synchronized with the oscillatory system, as in the feedback oscillator, but the driving force is not proportional to the oscillatory amplitude; instead, a certain minimum amplitude is needed to

323

actuate the escapement, and above this amplitude the energy supplied is roughly independent of amplitude.

It is at about this point in the catalogue that we discover classes of maintained vibrators which cannot usefully be thought of as resembling passive vibrators running backwards in time.

359   (*c*) *Relaxation oscillators*. This is a rather ill-defined term which is used to cover apparently very different systems. Perhaps the nearest we can get to characterizing the type is by observing that there are intrinsically unstable systems which are prevented from reaching a stable configuration by a sudden change in the behaviour of one element of the system, having the effect of returning them to their original unstable state; each subsequent attempt to reach equilibrium suffers the same fate and the system therefore oscillates. It might be contended that a bouncing ball is a relaxation oscillator in this sense, but a more convincing example is the 'linear time-base' circuit of fig. 27. The capacitor charges through the resistor until the breakdown voltage of the gas tube is reached, whereupon the capacitor discharges through the ionized gas; once the discharge is over, the ionized molecules recombine and the charging starts again. The only stable state for the circuit containing the capacitor and the cell is that in which the capacitor is fully charged to the cell voltage, V_0, and this state is barred by the breakdown of the tube.

Fig. 27. The 'linear time-base' circuit using breakdown in a gas-filled tube.

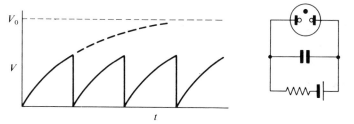

The on–off switch of a thermostat, as in a domestic electric oven or a tank for tropical fish, can be considered as constituting a relaxation oscillator. There are two different equilibrium conditions, too cold (when the switch is off) and too hot (when it is on), and both are precluded through the action of the switch. These are examples of what we shall call *astable* systems, in contrast to the *unstable* systems of, say, class (*b*) above.† A feedback oscillator circuit can in principle be set in a condition of unstable equilibrium (e.g. no current in the resonant circuit of fig. 26 or in the input circuit of the amplifier, and such steady current in the output circuit of the amplifier as results from zero input current); but any chance fluctuation sets it oscillating with ever-increasing amplitude. With the astable system, on the other hand, no such condition of even transient rest can be found.

336   The stick–slip phenomenon may also be put in the class of relaxation oscillators, though it may not represent a strictly astable situation. This is a

† This is not quite the normal use of the word 'astable', but vibrators that have been called astable have enough in common with these systems for the application not to be misleading.

subtlety that need not be pursued here – it is the discontinuous character of the phenomenon itself that matters. This arises when the frictional force on a sliding body, though more-or-less independent of speed once sliding has started, is less than the force required to get it moving from rest. If such a body is dragged along a surface by a spring it tends to stick until the spring has stretched enough to dislodge it, when it jumps forward, the tension of the spring being now greater than the sliding friction; but as the spring tension is reduced by this accelerated movement the body slows down and eventually sticks again. The process is then repeated. It is this effect that causes a door-hinge and a blackboard chalk to squeak, a stainless steel spoon to judder along the bottom of an aluminium saucepan, and, more beneficially, a violin bow to set the strings in vibration. The photograph of fig. 28 shows a magnified image (in a scanning electron microscope) of the pits left by a steel scriber drawn swiftly and lightly across an aluminium sheet. Here the sticking is associated with plastic deformation of the aluminium creating a ridge which holds the scriber until the force on it is enough to make it jump and begin to gouge out the next pit.

Fig. 28. Scanning electron micrograph of the track left by a steel scriber moving from left to right on an aluminium plate. The pits are $\frac{1}{4}$ mm apart, and the bright edges show piled-up aluminium scraped from the troughs.

(d) *Aerodynamic oscillations.* We use this term to cover more than aerodynamic effects, but these are best known. Wind blowing past a tall chimney sheds a stream of vortices alternately from each side, as in fig. 29 which shows the phenomenon on the laboratory scale with oil rather than air. So long as the chimney is immovable the frequency of vortex formation

Fig. 29. Vortex pattern behind a cylindrical obstacle (F. Homann[6]).

is not very well defined, but the vortices exert a lateral force and if the chimney can move they may set it swaying at its own natural frequency. In its turn the movement of the chimney tends to control the formation of vortices which, on becoming synchronized with the sway, are able to make it grow still further, possibly in the end causing collapse. A similar effect set the Tacoma bridge into such oscillations as to break it, and on a much gentler scale the singing of telegraph wires and the production of sound in a flute, recorder and flue pipe of an organ are all of the same basic nature.

3 Applications of complex variables to linear systems

Complex exponential solutions to equations of motion

Vibrators having the properties of linearity and translational invariance may obey a considerable variety of equations. We have already met in chapter 2 a few examples of linear differential equations with constant coefficients; (2.22) and (2.34) are of second and third order respectively, and there is no difficulty in devising circuits with equations of higher order than this. In other systems there may be effects not so easily described by differential coefficients; by the use of delay lines it is possible to introduce forces determined by what the system was doing at some earlier time, yet still linear in these quantities. What we expect of a linear system whose parameters do not change with time is that its equation of motion will be expressible in the form **30, 40**

$$\int_{-\infty}^{\infty} \xi(t')K(t'-t)\,dt' = 0. \tag{1}$$

which includes all the cases so far cited, if we allow the function K to take appropriate forms, including the Dirac δ-function and its derivatives. The δ-function itself, $\delta(t'-t)$, with its sharp peak of unit area at the moment when $t'-t$ vanishes, generates the value $\xi(t)$ from the integral. Putting $K(t'-t) = \dot{\delta}(t'-t)$, the first derivative of the δ-function, generates $\dot{\xi}(t)$; and so on for higher derivatives. Similarly if K is given the form $\delta(t'-t+t_0)$, the integral generates the value of ξ at $t-t_0$, and in this way a delay can be introduced. In general $K(t'-t)$ can take a suitable form to construct from $\xi(t)$ a summation (or, if K is a smooth function, an integral) over values of ξ and its derivatives both at the instant t and at any other instants. The linear property is clearly exhibited and so is the time-independence of the physical parameters, for nothing in (1) is related to any absolute time but only to what is occurring at the moment t and at other moments specified in relation to t.

The form of (1) is such that exponential solutions are to be looked for; on substituting e^{pt} for $\xi(t)$, we have immediately

$$e^{pt} \int_{-\infty}^{\infty} e^{px}K(x)\,dx = 0, \quad \text{where } x = t'-t, \tag{2}$$

44

and any value of p that causes the integral to vanish allows (1) to be satisfied at all times. The peculiar property that singles out the exponential function is this – the exponential curve is similar to itself everywhere, in the sense that shifting it bodily through t_0 has the same effect as multiplying it at each point by a constant, e^{pt_0}. As a special case of this property of similarity, the nth derivative bears a constant relationship at every point to the value of the function itself: $d^n\xi/dt^n = p^n\xi$. If the derivatives, delayed values etc. combine at any instant, t, to satisfy (1), they are bound to do so at all times since every term is scaled by the same amount in moving to another value of t.

It is one thing to recognize that self-similar exponential solutions to (1) are likely, but one must also consider other possibilities. It may be that no value of p exists which causes the integral in (2) to vanish; or that there are other solutions which are not self-similar. It may be shown, however, that all solutions of (1) may be cast in exponential form, provided p is permitted complex values.[1] For example, the sinusoidal solutions we have met in simple cases, of the form $\xi = \cos(\omega t + \phi)$, may be constructed by adding exponential solutions in which $p = \pm i\omega$:

$$\cos(\omega t + \phi) = A\,e^{i\omega t} + B\,e^{-i\omega t},$$

if
$$A = \tfrac{1}{2}e^{i\phi} \quad \text{and} \quad B = \tfrac{1}{2}e^{-i\phi}. \tag{3}$$

Since the displacement, ξ, is a real quantity any expression of it in terms of complex values of p must also include the complex conjugate; for example, in (3) $B\,e^{-i\omega t}$ is the complex conjugate of $A\,e^{i\omega t}$. The values of p, if complex, always appear in pairs, p and p^* together. For if e^{pt} is a solution of (1), e^{p^*t} is a solution of the complex conjugate equation; but ξ and K are real functions describing measurable physical quantities, so that the equation and its complex conjugate are identical. Eq. (2.23) will serve to illustrate the point. If the trial solution, $\xi = A\,e^{pt}$, is substituted the equation is reduced to an algebraic equation,

$$p^2 + kp + \omega_0^2 = 0,$$

yielding two possible values for p,

$$p = \tfrac{1}{2}[-k \pm (k^2 - 4\omega_0^2)^{\frac{1}{2}}]. \tag{4}$$

Let us keep ω_0 constant while allowing k to vary between $\pm\infty$. So long as $k^2 < 4\omega_0^2$ the two solutions are complex conjugates, but for larger values of k both are real. The solution of course matches (2.24) and (2.25) which were derived by the use of real functions alone, a somewhat more laborious process; it is one of the merits of complex exponentials in describing linear oscillations that they reduce differential equations to algebraic equations and minimize the effort of finding solutions.

The complex p-plane

As k changes, the solution of (2.23) represented by (4) describes a trajectory on the complex plane (fig. 1) whose essential features can be deduced directly from (2.23) before even proceeding to the solution. The product of the two values of p is ω_0^2, and so long as they are complex conjugates their product is pp^*, the square of the radius vector from the origin. Hence the complex solutions lie on a circle of radius ω_0, which is the natural frequency of the loss-free circuit. The overdamped solutions lie on the real axis, one inside and one outside the circle to keep the product equal to ω_0^2.

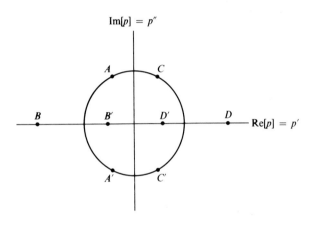

Fig. 1. The complex p-plane. Solutions of (2.23) lie in pairs, indicated by the same letters, on the circle of radius ω_0, or on the real axis; the left-hand side represents decaying modes, the right-hand side growing modes. A, A' are underdamped, B, B' overdamped, C, C' growing oscillatory modes, D, D' exponentially growing modes.

For a given value of k any linear combination of the two solutions is also a solution, and the most general solution takes the form

$$\xi = A\, e^{p_1 t} + B\, e^{p_2 t}. \tag{5}$$

Since ξ must be real, A and B must also be real, though otherwise arbitrary, in the overdamped case; in the underdamped case, where $p_2 = p_1^*$, $B = A^*$ but A may be complex. In both cases there are two arbitrary real constants in the solution, either A and B or the real and imaginary parts of A. These constants must be determined in any actual situation by the boundary conditions; if, for instance, at time $t = 0$ the displacement is ξ_0 and the velocity $\dot{\xi} = v_0$, we have from (5) that

$$A + B = \xi_0 \quad \text{and} \quad p_1 A + p_2 B = v_0,$$

which are enough to determine A and B. Physically this means that a simple oscillator set moving at a given speed from a given position is thenceforward fully determined in its motion. More complicated systems, described by higher order differential equations, will normally reduce, on substituting the trial solution e^{pt}, to algebraic equations of the same order, with as many characteristic values of p as the order of the highest differential coefficient; the number of arbitrary constants will be the same and so also will be the number of boundary conditions required to specify the behaviour completely. All complex values of p must occur in pairs, p and p^*, and obviously if the order is odd at least one solution must have a real value of p.

To return to the quadratic equation and its representation in fig. 1, it will be observed that the left-hand half of the circle, which arises when k is positive, describes decaying oscillations. Thus, two typical conjugate values of p may be written as $p' \pm ip''$, p' and p'' being real, and if the complex amplitude is written as $A\,e^{i\phi}$, A being real, the general solution has the form

$$\xi = A\,e^{i\phi}\,e^{(p'+ip'')t} + A\,e^{-i\phi}\,e^{(p'-ip'')t}$$

$$= 2A\,e^{p't}\cos{(p''t + \phi)}. \tag{6}$$

Negative values of p' clearly denote a decaying oscillation; positive values, on the right-hand side of the diagram, arise when k is negative and the oscillations build up exponentially. There is no mathematical reason to deny continuity to k between positive and negative values, and we have noted in chapter 2 systems in which external energy sources allow negative k to be realized; the whole of the diagram may be regarded therefore as accessible in principle. Its symmetry about the imaginary axis and its continuity lead to its crossing this axis normal to it. On the axis p is purely imaginary and describes a steady oscillation at frequency ω_0; as the lossy term k is added the decay represented by p' appears as a first-order effect in k, but the change in frequency is of second order. It is only when there is considerable decay during one cycle that the frequency suffers any significant change.

This is a rather general result, as may be seen by considering a differential equation of arbitrarily high order:

$$a + b\ddot{\xi} + c\ddddot{\xi} + \cdots = \alpha\dot{\xi} + \beta\dddot{\xi} + \cdots,$$

in which we have segregated even and odd terms. On substituting $\xi = e^{pt}$ the algebraic equation results:

$$a + bp^2 + cp^4 + \cdots = \alpha p + \beta p^3 + \cdots.$$

Pure imaginary values of p are only possible if either the left or right of the equation vanishes; a differential equation describing a non-dissipative system contains only even or only odd differential coefficients. Let us suppose that the system under consideration has even coefficients in its loss-free state, all the coefficients on the right vanishing, so that the introduction of weak losses is represented by small values of α, β etc. Then if a typical pair of loss-free solutions is $\pm p_0$, p_0 being imaginary,

$$a + bp_0^2 + cp_0^4 + \cdots = 0.$$

The first-order effect of the losses may be obtained by equating the first-order change in the left-hand side to the value of the right-hand side when $p = p_0$:

$$(2bp_0 + 4cp_0^3 + \cdots) \delta p \approx \alpha p_0 + \beta p_0^3 + \cdots.$$

Since both sides now contain only odd orders, both brackets are pure imaginary and δp is therefore real. The trajectory of p must leave the imaginary axis in a normal direction.

In talking of a trajectory one implicitly assumes that the losses are controlled by a single physical parameter, i.e. that α, β etc. all vary in a well-defined way as the lossy mechanism (e.g. η, the viscosity of the oil in a number of dashpots) is systematically changed. The real and imaginary parts of p are then functions of η, and the trajectory is defined parametric-ally by $p'(\eta)$ and $p''(\eta)$. If, however, the coefficients α, β etc. can be varied independently the trajectory of p on the complex plane spreads out to cover whole areas (domains), any one point in which may be realized by appropriate adjustment of the lossy mechanisms. What the analysis just presented implies is that these domains narrow down to sharp points as they approach the imaginary axis so that one can only reach the axis, while remaining within a domain, in a normal direction.

This point may be illustrated by means of the simple pendulum shown in fig. 2, which is constrained to vibrate in one plane and has damping vanes attached to the bob and at a point B on the string. The mass of the vane at B can be made negligible, so that the equations of motion of B, whose displacement is x_1, and C, whose displacement is x_2, take the form

$$mg[x_1/l_1 - (x_2 - x_1)/l_2] + \lambda_1 \dot{x}_1 = 0$$

and
$$m\ddot{x}_2 + \lambda_2 \dot{x}_2 + mg(x_2 - x_1)/l_2 = 0.$$

Here λ_1 and λ_2 define the viscous drag of the dashpots at B and C. When

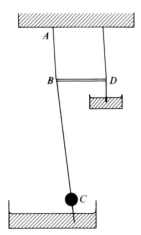

Fig. 2. The simple pendulum ABC has a light rod carrying a damping vane BD attached at B, and another damping vane attached to the bob. $AB = l_1$ and $BC = l_2$.

x_1 is eliminated from these equations the motion of the bob is found to be governed by the third-order equation:

$$\dddot{x}_2 + (\omega_3 + \omega_4)\ddot{x}_2 + (\omega_2^2 + \omega_3\omega_4)\dot{x}_2 + \omega_1^2\omega_3 x_2 = 0,$$

in which

$$\omega_1^2 = g/(l_1 + l_2), \quad \omega_2^2 = g/l_2, \quad \omega_3 = mg(l_1 + l_2)/\lambda_1 l_1 l_2$$

and

$$\omega_4 = \lambda_2/m.$$

Correspondingly there are three values of p, obeying the equation

$$p^3 + (\omega_3 + \omega_4)p^2 + (\omega_2^2 + \omega_3\omega_4)p + \omega_1^2\omega_3 = 0. \tag{7}$$

There is always one real solution describing a rapidly decaying transient, which we ignore.

It is convenient to discuss the oscillatory solutions of this equation in two stages, first setting $\lambda_2 = 0$ so that only the damping vane at B is operative and p follows a trajectory in the complex plane:

$$p^3 + \omega_3 p^2 + \omega_2^2 p + \omega_1^2\omega_3 = 0. \tag{8}$$

The damping represented by ω_3 appears in two terms but they are not, of course, independently variable. The trajectory of p, like that for a simple harmonic vibrator, is still defined by a single parameter, though its form is markedly altered by ω_3 appearing twice in a third-order equation; there are now two situations, rather than one, where lossless vibration can occur. One is when ω_3 is very large (virtually no damping at B) and the pendulum swings as a simple pendulum from A ($p = \pm i\omega_1$); the other when $\omega_3 = 0$ (very heavy damping) and only the second and fourth terms in (8) matter; $p = \pm i\omega_2$, the pendulum swinging from B, again without decrement. The trajectory between these points may be found by computing solutions of (8) for a range of values of ω_3, or by a graphical construction based on rearranging (8) in the form

$$\frac{p^2 + \omega_1^2}{p^2 + \omega_2^2} = -\frac{p}{\omega_3}. \tag{9}$$

At every point on the trajectory, i.e. for every value of ω_3, the two sides must have the same argument (the value of θ when a complex number is represented as $a\,e^{i\theta}$, a being real). On the complex plane of fig. 3, with origin at O, M represents $(-\omega_1^2)$ and N represents $(-\omega_2^2)$. Choose a point X on the perpendicular bisector of MN and draw a circle C_1 centred on X to pass through M and N; draw C_2 centred on the origin and with the same radius, to cut the positive real axis in Q. Construct QR equal to MN and extend OR to cut C_1 in S_1 and S_2. Then S_1 and S_2 are points on the trajectory of p^2 between M and N;† other points are found by moving the

† *Proof.* By construction, the triangles NMX and ROQ are congruent. Therefore $\widehat{QOR} = \widehat{NXM} = 2\,\widehat{NSM}$, where S is any point on C_1. The argument of S_1 is $\widehat{QOR}$, so that if S_1 represents p^2, the argument of p is $\frac{1}{2}\widehat{QOR}$, i.e. $\widehat{NS_1M}$. But if S_1 represents p^2, $MS_1 = p^2 + \omega_1^2$ and $NS_1 = p^2 + \omega_2^2$. The argument of $(p^2 + \omega_1^2)/(p^2 + \omega_2^2)$ is therefore $\widehat{NS_1M}$, the same as the argument of p; consequently it is possible to find a real value of ω_3 such that (9) is satisfied by this value of p.

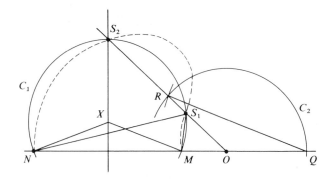

Fig. 3. Graphical construction of the solution of (8).

centre X to different positions, and the whole trajectory of p^2 above the real axis is shown as a broken line. Its mirror image below the axis completes the diagram. Every value of p^2 generates two values of p, and the single loop for p^2 becomes a double loop for p, having symmetry about both real and imaginary axes, as in fig. 4.

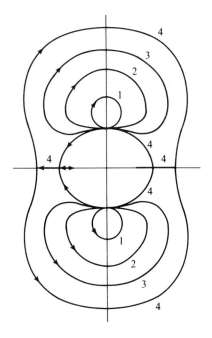

Fig. 4. Solutions of (8) for four different positions of B in fig. 2: (1) vane close to the top; (2) and (3) vane progressively lower; (4) vane low enough to allow critical damping – the trajectory consists of two loops joined by part of the real axis. The arrows show the direction of motion along the trajectory as the damping is increased.

A number of trajectories, corresponding to different positions of the damping vane on the pendulum, are shown in fig. 4. Note how they are all normally incident on the imaginary axis. The total length, $l_1 + l_2$, and therefore ω_1 have been kept constant while l_2 and ω_2 are varied. In contrast to a pendulum whose bob alone is damped and whose trajectory, like that in fig. 1, exhibits one critical point where it meets the negative real axis, this system exhibits either none (if $l_1 < 8l_2$) or two. The difference clearly lies in the ability of the pendulum to swing without damping when B is locked by a very viscous dashpot, unless B actually coincides with the bob.

When both damping mechanisms are present, with λ_1 and λ_2 independently variable, p can occupy a domain whose bounds are fixed by the following considerations. Rearrange (7) in two alternative ways:

$$\omega_3(p^2 + \omega_4 p + \omega_1^2) + p(p^2 + \omega_4 p + \omega_2^2) = 0, \tag{7'}$$

$$(p^3 + \omega_3 p^2 + \omega_2^2 p + \omega_1^2 \omega_3) + \omega_4 p(p + \omega_3) = 0. \tag{7''}$$

The vanishing of each bracketed expression defines a trajectory, those in (7') being parametrized by ω_4 and those in (7'') by ω_3. The trajectories are drawn in fig. 5. When $\omega_3 = 0$ and ω_4 is variable, p follows the trajectory C_2; while when ω_3 is made very large it is the bracket defining C_1 that dominates (7') and C_1 is therefore the trajectory of p. It may now be noted that since C_1 and C_2 do not intersect there is no intermediate value of ω_3 which can allow any point on either C_1 or C_2 to satisfy (7'). As ω_3 is varied from 0 to ∞, ω_4 being constant, the trajectory of p moves from C_2 to C_1 without passing through either. By the same argument, as ω_4 is varied from 0 to ∞, ω_3 being constant, p moves from C_3 to C_4 without intersecting either. The domain available to p is thus the shaded region bounded by these curves. It may be noted in particular that as λ_1 and λ_2 are raised from zero in any way, the trajectory of p must leave the axis normally, as implied by the general argument given previously.[†]

Fig. 5. Domain in which the solutions of (8) fall when ω_3 and ω_4 are positive; the complex conjugate solutions, mirrored below the real axis, are omitted. The circular arcs C_1 and C_2, of radius ω_1 and ω_2, are described parametrically by the equations:

$$p^2 + \omega_4 p + \omega_{1,2}^2 = 0$$

$$(0 < \omega_4 < \infty).$$

C_3 is described by the equation, $p^3 + \omega_3 p^2 + \omega_2^2 p + \omega_1^2 \omega_3 = 0$ $(0 < \omega_3 < \infty)$, and is the same as one of the trajectories in fig. 3. C_4 is described by the equation $p + \omega_3 = 0$ $(0 < \omega_3 < \infty)$.

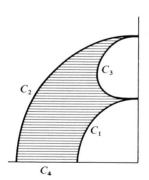

† If λ_1 and λ_2 can be negative, other domains become accessible to p. Thus it is possible to choose λ_1 and λ_2 with opposite signs so that the pendulum vibrates at constant amplitude, p being pure imaginary but not $i\omega_1$ or $i\omega_2$. Further analysis of this system is left to the interested reader.

Complex frequency; conventions

If one is more interested in lightly damped vibrators than in heavily or overdamped systems, it is convenient to take as the standard form of solution $e^{i\omega t}$ or $e^{-i\omega t}$ rather than e^{pt}; in other words to introduce a new variable, the complex angular frequency, defined by the relation $\pm i\omega \equiv p$. Real values of ω describe undamped vibrations, complex values decaying or growing vibrations. There is no guide but convention to the choice of sign in the definition, since p and p^* always occur together in the solution of equations in real variables. If $p_1 = p' + ip''$ and $p_2 = p_1^* = p' - ip''$, the alternative forms of ω are as follows:

(i) if $i\omega \equiv p$, $\qquad \omega_1 = p'' - ip'$, $\qquad \omega_2 = -p'' - ip' = -\omega_1^*$.

(ii) if $-i\omega \equiv p$, $\qquad \omega_1 = -p'' + ip'$, $\qquad \omega_2 = p'' + ip' = -\omega_1^*$.

For a decaying vibration p' is negative, so that the imaginary part of ω is positive in (i), negative in (ii). The real part of ω takes either sign in both (i) and (ii); when numerical values of ω are quoted it is conventional to state the real part as a positive number.

Corresponding to these definitions there are differences in the representations of trajectories on the complex plane. Thus if convention (i) is adopted, a trajectory on the p-plane is converted to one on the ω-plane by multiplying p by $-i$, i.e. rotating the diagram through $\pi/2$ in a clockwise sense; but if convention (ii) is adopted the diagram must be rotated anti-clockwise. In addition the sign of the real part of ω may be altered, i.e. the diagram may be reflected in the imaginary axis. The three representations are illustrated in fig. 6.

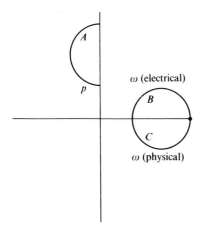

Fig. 6. Relation between p and ω; the p-trajectory is represented on the ω-plane by either B or C, according to the convention adopted.

The labelling of convention (i) as electrical and (ii) as physical springs from customary usage in electrical engineering on the one hand, and quantum mechanics on the other. In addition the electrical engineer, to avoid confusion with the symbol i as representing current, designates $\sqrt{(-1)}$ as j. The engineer, inserting a trial oscillatory solution into, for example,

(2.23), uses $e^{j\omega t}$, while the physicist uses $e^{-i\omega t}$. We shall be concerned with both fields and shall keep to the appropriate custom as far as possible, while adhering to the physical convention as the norm. The reader may like to think of j as being the same as $-i$, both being square roots of -1. It must be remembered, however, that when j is used decaying oscillations are represented by points in the upper half of the complex ω-plane, while when i is used they fall in the lower half.

The complex frequency ω may be broken up into its real and imaginary parts by writing it as either $\omega' - i\omega''$ or $\omega' + j\omega''$, the signs being chosen so that positive values of ω'' represent decaying oscillations. It has already been noted in (2.2) that, since $\tau_a = 1/\omega''$ and $\tau_e = \frac{1}{2}/\omega''$, the quality factor Q is $\frac{1}{2}\omega'/\omega''$. Slowly decaying vibrations, with high Q, are described by complex ω with very small arguments ($\tan^{-1}(\omega''/\omega') \ll 1$); but if $\omega''/\omega' = 1$, with the vector representing ω lying at $\pi/4$ to the real axis, the amplitude decays by a factor $e^{2\pi}$, i.e. 535, during one cycle. A considerable sector of the complex plane therefore corresponds to such heavy damping that the vibratory character of the system would hardly be discernible.

The vibration diagram

The typical solution (5) for an underdamped vibration, in which the two terms are complex conjugates, may be rewritten in terms of frequency as

either
$$\xi = A\,e^{-i\omega t} + A^*\,e^{i\omega^* t},$$

or
$$\xi = A\,e^{j\omega t} + A^*\,e^{-j\omega^* t}.$$

Because the second term in each case is the complex conjugate of the first, it is normal to take it for granted. Thus when it is stated that the vibration of a physical system is described by $A\,e^{-i\omega t}$, it is to be assumed that only the real part is meaningful:

$$\xi = A\,e^{-i\omega t}$$

signifies
$$\xi = \operatorname{Re}[A\,e^{-i\omega t}] = \tfrac{1}{2}[A\,e^{-i\omega t} + A^*\,e^{i\omega^* t}].$$

The amplitude and phase of the vibration are both contained in the complex amplitude, A, but it must be remembered that the physical and electrical conventions involve different interpretations. If $A = a\,e^{i\phi}$, a being real, according to the physical convention $\xi = \operatorname{Re}[a\,e^{-i(\omega t - \phi)}]$; this is a vibration having maximum positive displacement when $\omega t - \phi = 0$, i.e. at a time ϕ/ω, so that vibrations of greater ϕ *lag* behind those of lesser ϕ. Conversely, according to the electrical convention an oscillation $A\,e^{i\phi}$ implies that $\xi = \operatorname{Re}[a\,e^{i(\omega t + \phi)}]$, and ϕ is a phase *lead*.

The possibility of describing a vibration by means of a single complex exponential leads to a helpful diagrammatic representation, based on the complex plane but with a quite different interpretation. In the previous discussion a point on the complex plane represented the complex frequency – now we use it to represent the phase and amplitude of the vibration itself. If ξ has the form $A\,e^{-i\omega t}$, or $a\,e^{-i(\omega t-\phi)}$, ω being real (constant amplitude of vibration),† the corresponding point on the plane is found by drawing a vector of length $|A|$, i.e. a, from the origin at an angle $(\omega t-\phi)$ from the positive horizontal axis in a clockwise direction. As time proceeds, the vector rotates clockwise at an angular velocity ω, and its projection on the real axis describes harmonic motion of amplitude $|A|$. Different vibrators having the same frequency, ω, but different phase angles ϕ are drawn as vectors making different angles with the horizontal, but all sweeping around together at the same speed. In describing the behaviour of different parts of a complicated system (such as an electrical circuit) which have different amplitudes and phases but a common frequency, it is normally only the relative phases that matter, the angles between the different vectors rather than their absolute orientation. All that need be shown in the diagram, then, to represent any one vibration, is the complex amplitude A, which is drawn as a fixed vector of length a making an anticlockwise angle ϕ with the horizontal axis; it is to be understood that $e^{-i\omega t}$ has been suppressed. This is the *vibration diagram* (or *phasor diagram* – there is no generally accepted terminology). It is illustrated in fig. 7, which also shows in elementary form one of the

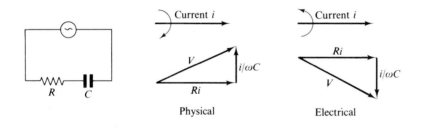

Physical   Electrical

Fig. 7. Vibration diagram for the voltages in an electrical circuit, according to the physical and electrical conventions. The rotation direction is such that the current i, which has the same phase as Ri, the voltage across the resistor, leads the generator voltage V.

principal uses of the diagram in forming the resultant of a number of superimposed vibrations, by the simple process of vector addition. If the suppressed $e^{-i\omega t}$ is included just for the moment, each vector rotates at angular velocity ω, but while all components to be added have the same frequency, the vectors all rotate together, keeping the resultant constant and itself rotating at the common angular velocity to generate, as its real part, a vibration that is still harmonic. When the components differ in frequency the configuration changes as they turn relative to one another, and the resultant displacement oscillates in a non-sinusoidal manner. We shall discuss a variety of instances in the next chapter.

† If ω is complex, $\omega'-i\omega''$, $e^{-i\omega t}$ describes an equiangular spiral, the curve whose polar equation has the form $r = a\,e^{\mu\theta}$. The equiangular spiral is a self-similar curve in that bodily rota- tion though ϕ is equivalent to a uniform scale change, $r \to r\,e^{\mu\phi}$. The vibration diagram described in this section, though in principle applicable when $\omega'' \neq 0$, is only really useful when $\omega'' = 0$.

Circuit analysis

The use of vibration diagrams for analysing electrical circuits, as in fig. 7, has largely been superseded by algebraic analysis, a technique too well known to need more than passing comment here. To write the impedance of a capacitor as $1/j\omega C$, or of an inductor as $j\omega L$, is to express the same relationships between e.m.f. and current as do the vibration diagrams. Both depend on the differential equations describing the behaviour of the circuit elements:

$$\text{for a capacitor,} \quad C\dot{V} = i,$$

$$\text{for an inductor,} \quad L\dot{i} = V.$$

If V and i are oscillatory and represented by $V\,e^{j\omega t}$ and $i\,e^{j\omega t}$, it is only necessary to replace each time-differentiation by the multiplier $j\omega$ to obtain the phase and amplitude relationships for the circuit elements:

$$j\omega CV = i, \quad \text{or} \quad Z \equiv V/i = 1/j\omega C$$

and $$j\omega Li = V, \quad \text{or} \quad Z \equiv V/i = j\omega L.$$

This substitution is just as valid when ω is complex as when it is real, but it should be noted that while V and i remain in phase for a resistor, the quadrature relation for a capacitor or inductor is broken; the tangent to an equiangular spiral (see footnote on p. 54) is not normal to the radius vector.

    To see how circuit analysis may replace the manipulation of differential equations, consider the series resonant circuit of fig. 2.1(c), reproduced in fig. 8(a). If the circuit is broken, at XX for instance, and the impedance Z between these points calculated, a complex frequency, ω, may be found at which $Z = 0$; at such a frequency current can flow in the circuit without establishing a potential difference across XX, which may therefore be

Fig. 8. Resonant circuits: (a) series resonant, when fed at XX, (b) parallel resonant, (c) generalized form of (b).

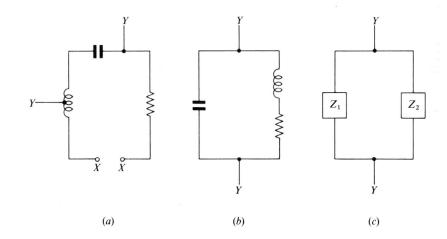

(a)   (b)   (c)

connected together. The circuit is thus proved capable of free oscillation at any frequency which causes its impedance to vanish. In this case

$$j\omega L + R + 1/j\omega C = 0,$$

or
$$L\omega^2 - jR\omega + 1/C = 0,$$

which is the same as is obtained from (2.22) when d/dt is replaced by $j\omega$. It **30** is worth noting that since the total impedance round the circuit vanishes at resonance, the impedances, and therefore the admittances, along the two paths between any two points, YY say, must be equal and opposite. When an e.m.f. is applied to YY the resultant admittance, which determines the current, vanishes at resonance. The resonance condition may therefore be expressed in terms of a parallel resonance circuit, as in fig. 8(b) or, more generally, 8(c). If YY are not connected to an external source the only oscillations of e.m.f. that can occur must be such that the net current leaving or entering at Y is zero, which is achieved when $Z_1 = -Z_2$.

The method may be applied with advantage to the less trivial problem presented by the pendulum in fig. 2, whose properties are modelled by the circuit of fig. 9. When R_1, which is analogous to λ_1, is small the circuit

Fig. 9. Equivalent circuit for the pendulum of fig. 2.

behaves like L, C_1 and R_2 in series; when it is large, like L, C_1, C_2 and R_2 in series. As R_1 is varied from 0 to ∞, the resonance frequency goes from a lower to a higher value, and the damping passes through a maximum. We now break the circuit at XX and calculate its impedance, which we equate to zero to determine the resonances:

$$1/j\omega C_1 + j\omega L + R_2 + (1/R_1 + j\omega C_2)^{-1} = 0,$$

or
$$-j\omega^3 - \omega^2(\omega_3 + \omega_4) + j\omega(\omega_2^2 + \omega_3\omega_4) + \omega_1^2\omega_3 = 0,$$

where
$$\omega_1^2 = 1/LC_1, \quad \omega_2^2 = (C_1 + C_2)/LC_1C_2,$$

$$\omega_3 = 1/C_2R_1, \quad \omega_4 = R_2/L.$$

This is the same as (7), but with p replaced by $j\omega$.

Work and power

So long as addition is the only operation carried out on complex exponentials no serious problems arise, since the real part of a sum of complex numbers is the same as the sum of each real part separately. With other operations, however, it is essential to remember that only the real part has meaning, and in general the safe course is to take the real part at the outset, before any other operation is performed. There are certain manipulations, however, which can be accomplished on the complex numbers by knowing the appropriate rules, and of these multiplication is the most important. Suppose an oscillatory force $F_0 \cos(\omega t + \phi_1)$ acts on a body whose velocity component parallel to the force is $v_0 \cos(\omega t + \phi_2)$ – both having the same frequency but with a phase difference. The rate, P, at which work is done by the force is the product of force and velocity, correctly given by

$$P = F_0 v_0 \cos(\omega t + \phi_1)\cos(\omega t + \phi_2)$$
$$= \tfrac{1}{2}F_0 v_0 [\cos(2\omega t + \phi_1 + \phi_2) + \cos(\phi_1 - \phi_2)]. \tag{10}$$

The first term in the square brackets oscillates at frequency 2ω about a mean value zero, and represents energy being alternately transmitted to and taken from the body; the second term is a constant power transfer which may have either sign, depending on the phase difference, $\phi_1 - \phi_2$. For the mean power transfer only the second term matters,

$$\bar{P} = \tfrac{1}{2}F_0 v_0 \cos(\phi_1 - \phi_2). \tag{11}$$

In terms of the complex representation, F is expressed as $(F_0 e^{-i\phi_1}) e^{-i\omega t}$ and v as $(v_0 e^{-i\phi_2}) e^{-i\omega t}$, the amplitudes being placed in brackets. The real force is $\tfrac{1}{2}(F + F^*)$ and the real velocity $\tfrac{1}{2}(v + v^*)$, so that

$$P = \tfrac{1}{4}(F + F^*)(v + v^*)$$
$$= \tfrac{1}{2}\operatorname{Re}[Fv + Fv^*] \quad \text{or} \quad \tfrac{1}{2}\operatorname{Re}[Fv + F^*v]. \tag{12}$$

Each term in the brackets is the same as the corresponding term in (10) so that, in particular, the mean power transfer is to be found by forming

$$\bar{P} = \tfrac{1}{2}\operatorname{Re}[Fv^*] \quad \text{or} \quad \tfrac{1}{2}\operatorname{Re}[F^*v]. \tag{13}$$

Waves

In many oscillatory systems the vibrations can best be described in terms of standing waves formed by the reflection and superposition of travelling waves. The complex representation derived for simple vibrators is readily extended to wave-motion. If the medium is linear in its properties, uniform so that any one part is similar to all others, and unchanging, so that translational invariance applies to time and space coordinates alike, all waveforms can be derived from exponential functions, though now functions of both time and space coordinates. To a certain degree this fact is obscured by the common habit in elementary texts of discussing almost exclusively the so-called Wave Equation, as if there were not an infinite number of differential and integral equations with wavelike solutions. This particular equation, with the one-dimensional form

$$\frac{\partial^2 \xi}{\partial x^2} - \frac{1}{v_0^2}\frac{\partial^2 \xi}{\partial t^2} = 0, \tag{14}$$

describes the variation with x and t of a physical quantity ξ, which may be the transverse displacement of a stretched string, the longitudinal displacement of elements of air in a trombone, or the electric field between two parallel wires, to list but a few. It happens to have a general solution of particularly simple form

$$\xi = f_1(x - v_0 t) + f_2(x + v_0 t), \tag{15}$$

in which f_1 and f_2 are arbitrary functions, requiring only such properties of continuity and differentiability as will make (14) meaningful when they are substituted. If only f_1 is present the equation describes a displacement imposed on the system which runs along at velocity v_0 in the direction of increasing x, without changing its shape; this is because the simultaneous increase of t by t_0 and of x by $v_0 t_0$ leave the argument unchanged; if at time t the displacement at x was ξ, the same displacement will reach $x + v_0 t_0$ at time $t + t_0$. The second function, f_2, in (15) represents a disturbance travelling in the opposite direction at the same speed, and the two are independent, being able to pass one another as though each alone were present.

This is a non-dispersive wave, and indeed a very important example since, among others, the electromagnetic wave in free space is of this type. But waves in real materials tend to be dispersive, the shape of a disturbance does not propagate unchanged, and the simple wave equation is inadequate. A much more general linear equation may be constructed by analogy with (1). If the behaviour at any point is determined by what is going on actually at that point or in the neighbourhood, at that of some related instant of time, the linear superposition of all influences may be conflated into an integral equation of motion:

$$\int_{-\infty}^{\infty} \int_{-\infty}^{\infty} \xi(x', t') K(x' - x, t' - t) \, dx' \, dt' = 0,$$

of which a special type may be a partial differential equation involving derivatives other than those appearing in (14), and possibly including mixed derivatives like $\partial^{n+m}\xi/\partial x^n \, \partial t^m$. Such equations, like (14), have solutions in exponential form and the most general solution can be expressed as a sum of exponentials. A typical harmonic component may be written (adopting the physical convention)

$$\xi = A \, e^{i(kx - \omega t)}, \tag{16}$$

in which the *wave-number* k and the frequency ω may in general be complex, and A is a constant which may also be complex. When this is substituted in a partial differential equation the same transformation to an algebraic equation occurs as with the equation of an oscillator. For

$$\partial \xi / \partial x = ikA \, e^{i(kx - \omega t)} = ik\xi,$$

$$\partial \xi / \partial t = -i\omega A \, e^{i(kx - \omega t)} = -i\omega \xi;$$

in general
$$\partial^{n+m}\xi/\partial x^n \, \partial t^m = (ik)^n (-i\omega)^m \xi. \tag{17}$$

Thus (14) when treated in this way becomes, quite simply,

$$-k^2\xi + \omega^2\xi/v_0^2 = 0,$$

i.e. $$k = \pm\omega/v_0.$$

(18)

For each choice of frequency (achieved with a stretched string, for example, if ω is to be real, by waggling any one point on the string with regular sinusoidal motion), a wave propagates with velocity $v = \omega/k$, since this is the relation that must exist between changes of x and t if the argument $(kx - \omega t)$ is to stay constant. In this case $v = v_0$ and is constant – every sinusoidal wave, whatever its angular frequency ω or wavelength $2\pi/k$, travels at the same speed, a special case of the non-dispersive property that all disturbances travel unchanged.

In general, however, substitution of (16) in an equation containing several differential coefficients leads to a more complicated algebraic relationship between k and ω; an equation of the nth degree generates a polynomial of the same degree in ω and k whose vanishing gives the dispersion curve – the dependence of k on ω. A simple example is provided by waves of bending on an elastic rod, the lateral displacement ξ being governed by the differential equation (there is no occasion to derive it here):[2]

$$\partial^4\xi/\partial x^4 + a^4\,\partial^2\xi/\partial t^2 = 0, \text{ where } a \text{ is a constant.} \quad (19)$$

On substituting (16) the dispersion equation is seen to take the form:

$$k^4 = a^4\omega^2,$$

i.e. $$k = \pm a\omega^{\frac{1}{2}} \text{ or } \pm ia\omega^{\frac{1}{2}}. \quad (20)$$

The two real solutions, $\pm a\omega^{\frac{1}{2}}$, show that sinusoidal waves can propagate in either direction with the same speed at a given frequency, but that higher frequency waves go faster ($v = \omega/k \propto \omega^{\frac{1}{2}}$) the wavelength varying in proportion to $\omega^{-\frac{1}{2}}$ rather than $1/\omega$ as in non-dispersive waves. In addition there are non-propagating disturbances, for the imaginary forms of k produce solutions of the form $e^{-i\omega t}\,e^{\pm a\sqrt{\omega}x}$, in which the amplitude of the oscillations grows or decays exponentially along the rod while every part vibrates in phase. In this system an arbitrary disturbance soon changes its shape, and only a sinusoidal wave continues with the same form.

As with simple oscillations, the complex wavelike solutions need to be combined if they are to describe real disturbances, but there are now more possibilities. Since ξ in (16) is the solution of an equation with real coefficients, ξ^* is also a solution; thus if (ω, k) defines a solution, so also does $(-\omega^*, -k^*)$. Moreover, if waves can propagate in the same form in both directions, $(\omega, -k)$ and $(-\omega^*, k^*)$ are also solutions. The equivalence of both directions (which does not necessarily hold for waves travelling in a rotating body or in the presence of a magnetic field) exhibits itself in the differential equation by the presence only of even differential coefficients, $\partial^2/\partial x^2$, $\partial^4/\partial x^4$ etc., which are unchanged when x is replaced by $-x$; in consequence only even powers of k appear in the dispersion equations. If the medium is loss-free ω and k can be real, and real disturbances can be described in terms of the four solutions $(\pm\omega, \pm k)$. Thus we can construct:

travelling waves

$$e^{i(kx\pm\omega t)} + e^{-i(kx\pm\omega t)} = 2 \cos (kx \pm \omega t), \tag{21}$$

standing waves

$$e^{i(kx+\omega t)} + e^{i(kx-\omega t)} + e^{-i(kx+\omega t)} + e^{-i(kx-\omega t)} = 4 \cos kx \cos \omega t, \tag{22}$$

or, more generally, by an arbitrary combination of travelling waves in both directions, but of different amplitudes, partial standing waves.

If the medium is lossy, ω and k cannot both be real; in different situations either may be taken to be real and the other complex – more exceptionally both may be complex but there is no need to discuss examples of this last. It should be appreciated that the dispersion relation defines either ω or k in terms of the other, and it is therefore open to choose which, if any, shall be real. If a vibrator of unvarying amplitude generates a wave on a long string, it is obvious that one must take ω as real so that the time variation in $e^{i(kx-\omega t)}$ is constant in amplitude. Consequently k is complex, $k'+ik''$ say, and the variation along the string shows exponential attenuation if k'' is positive, e^{ikx} becoming $e^{ik'x} e^{-k''x}$. A wave travelling in the opposite direction with ω unchanged has both parts of k reversed in sign, so that the amplitude along the string is still attenuated in the direction of propagation.

This prescription will not work, however, for the free vibration of a standing wave. The half-wave pattern of the lowest mode of a string, as shown in fig. (2.15) can be constructed according to (22) if there are no losses, and the boundary condition that $\xi = 0$ at both ends is satisfied if $kx = \pm\pi/2$ at the ends. Physically, one can imagine a travelling wave running to and fro, and being reflected with phase reversal at the ends, so that the incident and reflected waves always add to give zero displacement there. Provided the double length is an exact multiple of a wavelength there is no discontinuity anywhere and the wave equation is satisfied. But as soon as there is attenuation the travelling wave after traversing the string once in each direction has lost amplitude and can no longer join up without discontinuity. The boundary conditions can in fact only be satisfied by taking k to be real, and ω complex. This is not surprising, for we expect such a standing wave, unsustained by any source, to decay. If we write ω as $\omega'-i\omega''$, and take k as real, (22) may be rewritten as the sum of four complex components:

26

$$e^{i(kx-\omega t)} + e^{i(-kx-\omega t)} + e^{i(-kx+\omega^* t)} + e^{i(kx+\omega^* t)}$$

$$= 2 \cos kx(e^{-i\omega t} + e^{i\omega^* t})$$

$$= 4 e^{-\omega'' t} \cos kx \cos \omega' t.$$

The vector diagram is also useful here in visualizing different combinations, and here again there is no need to keep the conjugate complex functions in the diagram. By convention waves moving in opposite directions are ascribed the same ω and opposite signs of k. Travelling waves of amplitude A are thus expressed as $A\,e^{i(\pm kx-\omega t)}$ and represented by a single vector $A\,e^{ikx}$ or $A\,e^{-ikx}$, according to the direction of motion; this vector shows the amplitude and phase (at $t=0$) of the oscillation at the point x. In a lossless medium the oscillations at different points are represented by vectors of the same length but pointing in different directions, giving a graphic description of the effect of a travelling wave in causing every point to vibrate equally, but with a progressive phase shift. A partial standing wave is expressed as $A\,e^{i(kx-\omega t)}+B\,e^{i(-kx-\omega t)}$, where A and B may be complex, and represented as the sum of two vectors $A\,e^{ikx}+B\,e^{-ikx}$. As one moves along the line, increasing x, the vector of length A turns clockwise and that of length B anticlockwise, so that at some point the resultant takes its maximum value, $|A|+|B|$, and half a wavelength away its minimum $|A|-|B|$. Every point oscillates sinusoidally, but the amplitude varies with position. With the complete standing wave, when $|A|=|B|$, there are nodes, half a wavelength apart, where no oscillation occurs. The variation of amplitude with position in a partial standing wave can be written down immediately. If x is measured from a point where the amplitude is greatest, with the vectors in line, anywhere else they are inclined to one another at an angle $2kx$, and solution of the triangle gives for the resultant amplitude R:

$$|R|^2 = |A|^2 + |B|^2 + 2|A|\,|B| \cos 2kx. \tag{23}$$

It is the *intensity*, the square of the amplitude, that varies with position as a sinusoid superimposed on a steady mean level. When intensity is plotted the maxima and minima have the same shape, but in a plot of amplitude the minima are sharper than the maxima, becoming cusps when $|A|=|B|$ and the standing wave is complete. This is illustrated in fig. 10.

Fig. 10. Variation over half a wavelength of the amplitude and intensity in a partial standing wave. The amplitude is given by $|e^{ikx}+0.6\,e^{-ikx}|$, the intensity by the square of this, $1.36+1.20\cos 2kx$. Note that the minimum of the amplitude curve is sharper than the maximum.

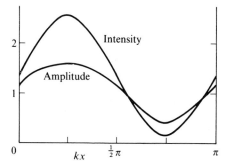

Circular polarization

Finally, let us note yet another application of complex algebra in the description of a circularly polarized wave or vibration. If the end of a stretched string is moved in a circle at constant angular velocity ω, a circularly polarized disturbance runs down it. At any instant the string is displaced into a helical form, and as time proceeds the helix moves bodily down the string at the wave velocity. An observer watching the motion of a single point sees it moving in a circle. Since the disturbance is purely transverse, at any point it may be represented by a two-dimensional vector, just the thing that a complex number is most convenient to describe. To establish the coordinate system, we choose for a *real plane* any plane containing the direction of propagation (e.g. with a horizontal string we might choose the horizontal plane containing it); at each point on the string, x, we draw the *transverse plane* normal to the string, whose intersection with the real plane defines the real axis. Then e^{ikx} represents a unit vector everywhere, but the direction of the vector rotates progressively along the direction of propagation, so that the ends of all the vectors lie on a stationary helix. Looking along the x-axis we see the vector turning anti-clockwise as x increases; this is a left-handed screw. By the same token e^{-ikx} describes a right-handed screw. Wave propagation is described by $e^{i(kx-\omega t)}$, which can be viewed either as the bodily rotation of the helix in a clockwise sense without translation (the effect of multiplying by $e^{-i\omega t}$ throughout) or as bodily translation at velocity ω/k without rotation (since changing x and t in the ratio ω/k leaves the argument unaltered). The fact that both descriptions are equivalent is only a way of saying that the helix is yet another self-similar curve, as it must be described by an exponential function.

This notation is particularly valuable for circularly polarized transverse electromagnetic waves, where the electric and magnetic field vectors can both be expressed as complex numbers. It should be noted that the complex number applied to transverse waves in this way is to be seen as a physically meaningful symbol in its entirety – the displacement it describes is a vector of two components. The original use, on the other hand, was as a convenient representation of a scalar displacement, for which the complex number contains too much information; only the real part therefore was needed. The vector interpretation need not be confined to circular polarization, since a plane-polarized wave may be analysed into two circularly polarized waves of opposite sense moving in the same direction:

$$A\,e^{i(kx-\omega t)} + A\,e^{-i(kx-\omega t)} = 2A\cos(kx-\omega t).$$

This describes a progressive wave with an amplitude A which in general is complex and the same everywhere; it is therefore a plane-polarized wave with the plane of polarization defined by the argument of A.

4 Fourier series and integral

Synthesis of waveforms; Fourier series

The theory of Fourier series and the Fourier integral pervades mathematics and physics, and there is no lack of texts to which the student can refer for as much detail as he needs.[1] None of the applications in this book demands deep knowledge of mathematical technique, but the idea of analysing arbitrary functions into a series of sinusoids is so helpful that it is worth spending some time assimilating the concepts, so that they become a natural mode of thought, qualitative as much as quantitative. We shall concentrate on illustrating the application of Fourier methods to problems in vibration, with no more formal mathematics than is needed to make the arguments plausible. For the most part the discussion will be confined to functions of a single variable, such as the displacement of a vibrating system considered as a function of time. Once this is well assimilated there is no essential difficulty in extending the analysis to functions of many variables.

Periodic functions with a waveform other than sinusoidal may be built up by adding to a fundamental sinusoid others forming a harmonic series, that is, with frequencies that are integral multiples of the fundamental. Fig. 1 shows the progressive addition of terms of the Fourier series

$$y = 0.363 \sin x + \tfrac{1}{2} \sin 2x + 0.404 \sin 3x + \tfrac{1}{4} \sin 4x + \cdots$$

$$= \sum_{1}^{\infty} \frac{1}{n} \left[1 - \frac{\sin\left(\tfrac{1}{2}n\pi\right)}{\tfrac{1}{2}n\pi} \right] \sin nx. \tag{1}$$

As terms are added the waveform is seen to approach the zig-zag represented by the complete series (1). Certain properties of the function are evident from (1); since $\sin nx$ is an odd function, y also must be an odd function, $y(-x) = -y(x)$, and the period is obviously 2π, since 2π is the smallest quantity that can be added to x while leaving $\sin nx$ unchanged, for all integral values of n. The vanishing of y for all x between $\pi/2$ and $3\pi/2$ is not obvious, but the sharp discontinuity at $x = 0$ may be inferred without difficulty. It arises from the first term in (1) and to discuss its origin we need only consider the simpler series:

$$y = \sum_{1}^{\infty} \frac{1}{n} \sin nx. \tag{2}$$

63

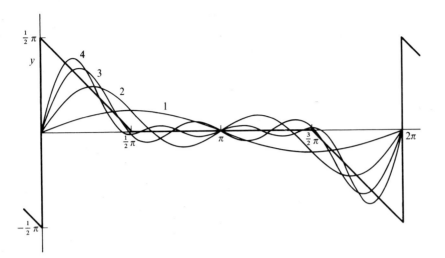

Fig. 1. Progressive stages in the Fourier synthesis of the saw-tooth function shown as a heavy line. The four curves are the first term, the first two, the first three and the first four, according to (1).

When x is very small, $\sin nx \sim nx$, at least for values of n up to something like $1/x$, and each term in (2) contributes an amount x to the value of y. If the series is terminated at a certain N, it is always possible to choose x so small that the approximation holds, and therefore the truncated series gives rise to a gradient equal to N at the origin; as N goes to infinity so does the gradient.

It may be observed in fig. 1 that the curve tends to overshoot its mark after the steep rise from the origin. This is an inevitable concomitant of the representation by a Fourier series of a discontinuous function, and it persists however many terms are employed, though confined to a steadily narrower region near the discontinuity. It can be demonstrated by use of (2). The truncated series,

$$y_N = \sum_1^N \frac{1}{n} \sin nx, \tag{3}$$

has a maximum when $dy_N/dx = 0$, i.e. when $\sum_1^N \cos nx = 0$. Now $\sum_1^N \cos nx$ is the real part of the geometric series $\sum_1^N e^{inx}$, whose sum is $e^{ix}(1 - e^{iNx})/(1 - e^{ix})$. The smallest value of x that causes the real part to vanish is π/N, which value, inserted into (3), shows that y_N rises to a maximum of $\sum_1^N 1/n \sin(\pi n/N)$. As N tends to infinity it becomes permissible to replace the sum by an integral; writing $\pi n/N$ as z, we have that

$$y_{max} \sim \int_0^\pi \frac{\sin z}{z} \, dz = 1.179 \times \pi/2. \tag{4}$$

Beyond the maximum y oscillates with diminishing amplitude about the value $\pi/2$. Since the oscillations are confined to a range of x that tends to zero as $1/N$, and since their amplitude remains finite, they have no practical significance and this mathematical curiosity, the Gibbs phenomenon, will be referred to no more.

The synthesis in fig. 1 is an explicit example of the representation of a function by a Fourier series, the function in this case being

$$
\begin{aligned}
y(x) &= \frac{\pi}{2} - x \quad \text{when } 0 < x < \pi/2, \\
&= 0 \qquad \text{when } \pi/2 < x < \pi, \\
\text{and} \qquad y(-x) &= -y(x).
\end{aligned}
\tag{5}
$$

In general, when the function is periodic, with $y(x+2\pi)=y(x)$, but with no other symmetry properties, both sines and cosines must be used in the series:

$$
y(x) = A_0 + \sum_1^\infty A_n \cos nx + \sum_1^\infty B_n \sin nx,
\tag{6}
$$

A_0 being the mean value of y. Fig. 2(a) shows an example involving both:

$$
y = \sin x + \tfrac{1}{5}\cos 2x + \tfrac{1}{5}\sin 4x.
$$

The function is, of course, periodic but has no point about which it is symmetric or antisymmetric.

If y is a complex variable, the A_n and B_n must also be complex. By introducing new coefficients,

$$
a_n = \tfrac{1}{2}(A_n + iB_n), \qquad a_{-n} = \tfrac{1}{2}(A_n + iB_n),
\tag{7}
$$

(6) may be recast in the tidier and altogether more convenient form

$$
y(x) = \sum_{-\infty}^{\infty} a_n\, e^{inx}.
\tag{8}
$$

This is the Fourier expansion of a function whose period is 2π. Applied to a periodic vibration, $\xi(t)$, of period T_0, with the property that $\xi(t+T_0)=\xi(t)$, (8) takes the form

$$
\xi(t) = \sum_{-\infty}^{\infty} a_n\, e^{in\omega_0 t},
\tag{9}
$$

where ω_0 is the fundamental frequency, $2\pi/T_0$.† If $\xi(t)$ is a real observable, A_n and B_n are real but a_n is in general complex, though (7) shows that $a_{-n} = a_n^*$. Only a_0, the mean value of ξ, and the values of a_n for positive n need then be specified to define $\xi(t)$ completely; a_1 is the amplitude of the *fundamental*, and a_n the amplitude of the nth *harmonic* (i.e. we call the first harmonic the fundamental – this is usual modern practice but there are exceptions in the literature).

† When it is clear from the context, as here, that the vibration is of constant amplitude, so that ω is real, we shall drop the prime from ω'.

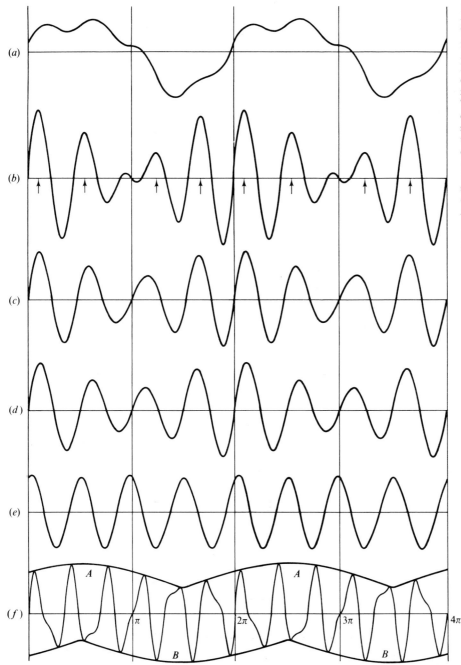

(a)

(b)

(c)

(d)

(e)

(f)

Fig. 2(a) $\sin x + \frac{1}{5}\cos 2x + \frac{1}{5}\sin 4x$. (b) $\sin 4x + \sin 5x$. The arrows show the irregular spacing of peaks, and especially the extra half-period when the envelope goes through zero. (c) $\sin 4x + \frac{2}{5}\sin 5x$. The envelope does not drop to zero, and the spacing of peaks is less uneven, though still obviously irregular. (d) $\sin 4x + \frac{1}{5}(\sin 3x + \sin 5x)$. Like (c) but with spacing evened out. (e) $\cos 4x + \frac{1}{5}(\sin 3x + \sin 5x)$. Frequency modulation at almost constant amplitude. (f) $\sin 6x + \frac{2}{5}\sin 11x$. Cusped envelope.

Once it is recognized that a function $y(x)$ may be represented by the Fourier series (8), it is easy to determine the values of a_n; for

$$\frac{1}{2\pi}\int_0^{2\pi} y(x)\,e^{-imx}\,dx = \frac{1}{2\pi}\sum_{n=-\infty}^{\infty}\int_0^{2\pi} a_n\,e^{i(n-m)x}\,dx = a_m, \qquad (10)$$

the only term in the summation to survive integration over one period being that for which $n = m$. It requires a certain mathematical delicacy to prove the precise conditions for the validity of a Fourier expansion, but there is no need to discuss this here; it is sufficient to remark that periodic physical observables that are everywhere finite and either continuous or, if discontinuous, only at a finite number of points in the cycle, can always be Fourier-expanded, as in (8) or (9). It may be verified by substitution of (5) in (10) that the series (1) is what it purports to be.

Beats and related phenomena

Consider next the superposition of sinusoids that do not form a harmonic series but are rather close together in frequency. Two vibrations of the same amplitude produce the well-known beat pattern illustrated in fig. $2(b)$, where the frequencies have been chosen in the ratio $5:4$. In general, for frequencies ω_1 and ω_2,

$$\cos \omega_1 t + \cos \omega_2 t = 2\cos\frac{\omega_1+\omega_2}{2}t\cos\frac{\omega_1-\omega_2}{2}t,$$

the first cosine representing a rapid oscillation at the mean frequency $\frac{1}{2}(\omega_1+\omega_2)$, the second a slow modulation of the amplitude with maxima occurring whenever $\frac{1}{2}(\omega_1-\omega_2)t = n\pi$, i.e. at a frequency $|\omega_1-\omega_2|$. If the two components have different amplitudes the depth of modulation is less but the beat frequency is the same. Fig. $2(c)$ shows the combination

$$y = \sin 4x + \tfrac{2}{5}\sin 5x.$$

It is helpful to visualize the combination as the imaginary part of a complex diagram, fig. 3, in which one vector of unit length rotates at an angular velocity of 4 units while another of length $\frac{2}{5}$, attached to it, rotates at 5 units. Since it turns, relative to the first vector, at 1 unit, this is the frequency of recurrence of maxima in the amplitude every time the vectors are momentarily in line.

Fig. 3. Vector diagram for fig. 2(c).

When the two frequencies are very close together, the resultant vector changes its magnitude only a little in each revolution, and one may then attach some significance to the definition of the instantaneous frequency as the rate of change of phase angle of the resultant. If the stronger component, $a_1 e^{-i\omega_1 t}$, has the lower frequency, then at the moment when they are in line, to give a maximum resultant, the two oscillations cooperate in giving the end point its maximum angular velocity. The linear velocity of the end point is $a_1\omega_1 + a_2\omega_2$ (a_1 and a_2 being taken as real and positive) and its angular velocity

$$\omega_{max} = \frac{a_1\omega_1 + a_2\omega_2}{a_1 + a_2} = \omega_1 + \frac{a_2(\omega_2 - \omega_1)}{a_1 + a_2}.$$

By the same argument when the amplitude is at its minimum

$$\omega_{min} = \frac{a_1\omega_1 - a_2\omega_2}{a_1 - a_2} = \omega_1 - \frac{a_2(\omega_2 - \omega_1)}{a_1 - a_2}.$$

The variation of instantaneous frequency can be discerned in fig. 2(c) from the uneven spacing of the zeros. The effect is most marked when $a_1 = a_2$, for then the zeros in the beat pattern are points of phase reversal, as can be clearly seen in fig. 2(b); ω_{min} momentarily falls to $-\infty$.

The variation of instantaneous frequency is eliminated when the beat pattern results from the modulation of a dominant component (*carrier*) by symmetrically disposed sidebands, as, for example, in fig. 2(d) which illustrates the function

$$y = \sin 4x + \tfrac{1}{5}(\sin 3x + \sin 5x).$$

The vector diagram, fig. 4, shows how the two sidebands combine to alter the magnitude but not the phase angle of the carrier.

If now the phase of the carrier is advanced by $\pi/2$, to give the function

$$y = \cos 4x + \tfrac{1}{5}(\sin 3x + \sin 5x),$$

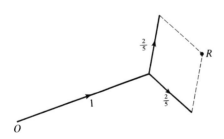

Fig. 4. Vector diagram for fig. 2(d).

the resulting oscillation is that shown in fig. 2(e), which suffers little change in amplitude but is frequency-modulated. The vector diagram, fig. 5, shows how this comes about, the two sidebands combining at right angles to the carrier to change its phase considerably but its amplitude to only a minor degree. It is easy to calculate the range of variation of frequency when the

Fig. 5. Vector diagram for fig. 2(e).

sidebands are weak, but the more general case, of interest in f.m. radio transmission, involves a more thorough analysis which need not be embarked on here.

A final example of the superposition of a small number of sinusoids is worth showing as a curiosity which has been known to cause puzzlement. Two frequencies differing by a factor near 2 combine to give a cusped beat pattern as in fig. 2(f) which is the function

$$y = \sin 6x + \tfrac{2}{5} \sin 11x.$$

If the frequencies were in the ratio $2:1$ and the oscillations were phased so that their positive peaks added, their negative peaks would have opposite sign. This would displace the envelope upwards, as at point A in the diagram. Because of the departure from the ratio $2:1$, there comes a moment, as at B, where the relative phases are reversed and the envelope is displaced downwards. It should be noted that the 'local mean' value of y does not oscillate with the beat frequency – that can only happen if a component at the beat frequency is present in y. At A, where the envelope is displaced upwards, the positive peaks are spiky and the negative peaks blunt to the point of being doubled; this distortion compensates for the envelope shift and maintains the mean level of y at zero.

The Fourier integral

Let us now add together a larger number of sinusoids, equal in amplitude and in spacing of frequencies, and covering altogether a fairly small range of frequency. It is easiest to appreciate how the resultant arises by drawing a vibration diagram. In the simplest case the components are adjusted so that at some instant they are all in phase (fig. 6(a)). At other times there is a progressive phase variation from one to the next (fig. 6(b)). Thus if the nth component is $a \, \mathrm{e}^{-\mathrm{i}(\omega + n\varepsilon)t}$, ε being the frequency difference between

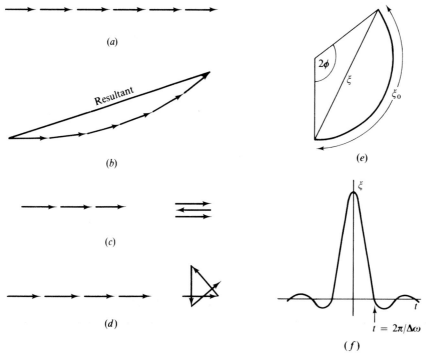

Fig. 6. Addition of several sinusoids of equal amplitude and evenly spaced frequencies. (*a*) All in phase; (*b*) a little later, with a constant phase difference between components; (*c*) subsidiary maximum with 3 components having phases 0, π and 2π; (*d*) subsidiary maximum for 4 components with phases in multiples of 133°; (*e*) construction for a continuous spectrum; (*f*) resultant obtained from (*e*), according to (12).

neighbouring components, the resultant amplitude is given by

$$\xi = \sum_0^{N-1} a\,\mathrm{e}^{-i(\omega+n\varepsilon)t} = a\,\mathrm{e}^{-i\omega t}\sum_0^{N-1}\mathrm{e}^{-in\varepsilon t} = a\,\mathrm{e}^{-i\omega t}\frac{1-\mathrm{e}^{-iN\varepsilon t}}{1-\mathrm{e}^{-i\varepsilon t}}, \qquad (11)$$

the same geometrical series as we have already met in a different context. The presence of $\mathrm{e}^{-i\omega t}$ makes the point that if we always draw the first vector in the same direction, horizontal in this case, the whole diagram must be thought of as spinning at angular velocity ω. The instantaneous resultant is the closing vector, and it is obvious from the diagram that it has the same phase as the central component, and is therefore spinning at a frequency $\omega + \frac{1}{2}N\varepsilon$. The resultant vibrates at this mean frequency, with an amplitude that is modulated according to the development with time of the length of the closing vector. When only two components are present this is a sinusoidal modulation, as shown in the simple beating pattern of fig. 2(*b*). With three components the condition that all are in phase recurs regularly at intervals T such that $\varepsilon T = 2\pi$, but between each pair of principal maxima there is a subsidiary maximum of only one-third the amplitude, as illustrated in fig. 6(*c*). Similarly with four components there are two subsidiary maxima (fig. 6(*d*)). If the number of components is increased to N while their spacing is reduced to keep the total frequency spread, $N\varepsilon$, constant the principal maxima move further apart ($T = 2\pi/\varepsilon \propto N$), but the number of subsidiary maxima increases to $N-2$, and the spacing of subsidiary maxima remains very nearly constant. As N is allowed to rise indefinitely, the pattern around each principal maximum quickly approaches a limiting form which is found by replacing the polygons of fig. 6 by the arc of a circle. If the frequencies of the components in this 'top-hat' spectrum lie evenly in the range $\omega_0 \pm \frac{1}{2}\Delta\omega$ the length of the arc remains constant at ξ_0 while its angular length (2ϕ in fig. 6 (e)) changes as $\Delta\omega \cdot t$; the radius of the arc is $\xi_0/2\phi$ and the closing vector is given by ξ, where

$$\xi/\xi_0 = \sin\phi/\phi = \sin\left(\tfrac{1}{2}\Delta\omega \cdot t\right)/\tfrac{1}{2}\Delta\omega \cdot t. \tag{12}$$

This result, illustrated in fig. 6(f), is well known in Fraunhofer diffraction theory, and indeed the theory of the interference pattern of N equally spaced slits, with $N-2$ subsidiary maxima between the principal maxima, is exactly the same as that developed here, leading in the limit to an expression like (12) for the diffraction pattern of a finite aperture.

An analytical derivation of the result for a continuous distribution of components is easily given. The components lying in an infinitesimal frequency range between ω and $\omega + d\omega$ all have the same phase, ωt, and combine to give a contribution whose amplitude is a fraction $d\omega/\Delta\omega$ of the maximum resultant, ξ_0, that arises when all components are in phase. Their contribution is therefore $\xi_0\, d\omega\, e^{-i\omega t}/\Delta\omega$, and the summation over all components takes the form

$$\xi = \frac{\xi_0}{\Delta\omega}\int_{\omega_0-\frac{1}{2}\Delta\omega}^{\omega_0+\frac{1}{2}\Delta\omega} e^{-i\omega t}\, d\omega = \xi_0\, e^{-i\omega_0 t}\sin\left(\tfrac{1}{2}\Delta\omega \cdot t\right)/\tfrac{1}{2}\Delta\omega \cdot t, \tag{13}$$

the same as (12), except that the spinning of the diagram at the mean frequency ω_0 was omitted there. It is included in (13), which shows the mean oscillation $\xi_0\, e^{-i\omega_0 t}$ modulated by the function $\sin\phi/\phi$, as in fig. 6(f).

The synthesis expressed in (13) may immediately be generalized. Let the sinusoidal components present a continuous range of frequencies, $-\infty < \omega < \infty$, with amplitude and phase described by a complex variable $f(\omega)$, whose meaning is that the components in an infinitesimal range $d\omega$ combine to give a resultant $f(\omega)\, e^{-i\omega t}\, d\omega$. Then the resultant of all components varies with time according to $\xi(t)$, where

$$\xi(t) = \int_{-\infty}^{\infty} f(\omega)\, e^{-i\omega t}\, d\omega. \tag{14}$$

The function $\xi(t)$ is the *Fourier integral transform* of $f(\omega)$. Fourier transforms possess the property of reciprocity – if ξ is the Fourier transform of f, then (with certain small modifications) f is the Fourier transform of ξ. To make the matter precise, let us write down $f_1(-p)$, the Fourier transform of $\xi(t)$ in terms of a third variable, $-p$,

$$f_1(-p) = \int_{-\infty}^{\infty} \xi(t)\, e^{ipt}\, dt = \int_{-\infty}^{\infty}\int_{-\infty}^{\infty} f(\omega)\, e^{i(p-\omega)t}\, d\omega\, dt, \quad \text{from (14).} \tag{15}$$

Integrate first with respect to t, taking the limits as $\pm T$ which will later be allowed to tend to infinity. Then

$$f_1(-p) = 2 \int_{-\infty}^{\infty} \left\{ f(\omega)\, d\omega \lim_{T \to \infty} \left[\frac{\sin (p - \omega) T}{p - \omega} \right] \right\}.$$

Now the quantity in square brackets has the form $\sin \phi / \phi$ as in fig. 6(f), with a peak value T when $\omega = p$, and dropping to zero when $|p - \omega| = \pi / T$. As $T \to \infty$, the area under the curve remains constant at π, but the width decreases to zero. Provided that $f(\omega)$ is a continuous function (like almost all physical variables), its variation over the range $p \pm \pi / T$ becomes negligible, so that we write

$$f_1(-p) = 2\pi f(p),$$

or

$$f(p) = \frac{1}{2\pi} f_1(-p) = \frac{1}{2\pi} \int_{-\infty}^{\infty} \xi(t)\, e^{ipt}\, dt, \text{ from (15).}$$

To express this result in more usual terminology, rename $\xi(t)$ as $g(t)$ and replace p by ω. Then,

if

$$g(t) = \int_{-\infty}^{\infty} f(\omega)\, e^{-i\omega t}\, d\omega,$$

$$\left. \right\} \quad (16)$$

it follows that

$$f(\omega) = \frac{1}{2\pi} \int_{-\infty}^{\infty} g(t)\, e^{i\omega t}\, dt.$$

This, out of several equivalent forms of the Fourier transform theorem, is the one we shall take as standard. Since, mathematically speaking, ω and t are only symbols for variables, they may be interchanged without altering the truth of the proposition,

i.e.

$$f(t) = \frac{1}{2\pi} \int_{-\infty}^{\infty} g(\omega)\, e^{i\omega t}\, d\omega, \quad (17)$$

or

$$f^*(t) = \frac{1}{2\pi} \int_{-\infty}^{\infty} g^*(\omega)\, e^{-i\omega t}\, d\omega, \quad (18)$$

the latter, (18), having the same form, apart from the factor $1/2\pi$, as the first equation of (16).

It follows, therefore, that if $f(\omega)$ is a top-hat spectrum of frequencies generating $g(t)$, a $\sin \phi / \phi$ envelope, then $g(\omega)$, a $\sin \phi / \phi$ spectrum, will generate $f(t)$, a top-hat envelope, i.e. a wave train of constant amplitude but finite length. Let us verify this by explicit calculation, starting with a spectrum centred on ω_0, all components being in phase at time $t = 0$, and with an amplitude distribution according to $\sin [t_0(\omega - \omega_0)]/(\omega - \omega_0)$, in which t_0 is a constant defining the width of the spectral distribution. Then the resulting amplitude follows from (14):

$$g(t) = \int_{-\infty}^{\infty} \frac{\sin [t_0(\omega - \omega_0)]}{\omega - \omega_0}\, e^{-i\omega t}\, d\omega = e^{-i\omega_0 t} \int_{-\infty}^{\infty} \frac{\sin \phi}{\phi}\, e^{-i\phi t/t_0}\, d\phi,$$

on substituting ϕ for $t_0(\omega - \omega_0)$. To evaluate the integral, note that $\sin \phi/\phi$ is an even function, so that only the even part of $e^{-i\phi t/t_0}$, that is, $\cos(\phi t/t_0)$, need be considered. Further

$$\sin \phi \cos(\phi t/t_0) = \tfrac{1}{2}\{\sin[(1+t/t_0)\phi] + \sin[(1-t/t_0)\phi]\}, \qquad (19)$$

so that
$$g(t) = \tfrac{1}{2} e^{-i\omega_0 t} \int_{-\infty}^{\infty} \left(\frac{\sin \alpha_1 \phi}{\phi} + \frac{\sin \alpha_2 \phi}{\phi}\right) d\phi,$$

where $\alpha_1 = 1 + t/t_0$ and $\alpha_2 = 1 - t/t_0$.

Now $\int_{-\infty}^{\infty} (\sin \alpha\phi/\phi)\, d\phi = \pi$ if $\alpha > 0$, $-\pi$ if $\alpha < 0$. For such t that confer opposite signs on α_1 and α_2 the two contributions to $g(t)$ will cancel; $g(t)$ therefore only takes a non-zero value when t lies between $\pm t_0$:

$$g(t) = \pi e^{-i\omega_0 t} \quad \text{if } -t_0 < t < t_0, \qquad (20)$$

a limited train of constant amplitude as predicted by Fourier's theorem.

The length of the train is determined by the frequency range of its components; the central peak of the frequency spectrum is bounded by zeros when $t_0(\omega - \omega_0) = \pm \pi$, and therefore is $2\pi/t_0$ wide, while the train of oscillations is $2t_0$ long. Denoting these measures by $\Delta\omega$ and Δt, we have in this case

$$\Delta\omega\, \Delta t = 4\pi, \qquad (21)$$

independent of the choice of t_0. The same result follows from the original integration of a top-hat spectrum $\Delta\omega$ wide to form a train of oscillations (12) with its central peak $4\pi/\Delta\omega$ long. The qualitative reason for this result is easily understood. If a set of sinusoids with frequencies in a limited range $\Delta\omega$ are added together, and all combine in phase to give a maximum at some instant, a time $1/\Delta\omega$ must elapse before the extreme components have become dephased by one radian, and somewhat longer before they can add to give a resultant substantially smaller than their maximum. Since, as follows immediately from a sketch of the phase-amplitude diagram, the amplitude is symmetrical about the in-phase point, the total length of the train must be something in excess of $2/\Delta\omega$, i.e. $\Delta\omega\, \Delta t > 2$. Obviously, the precise value of $\Delta\omega\, \Delta t$ in any particular instance, where there may not be zeros to provide a convenient definition of $\Delta\omega$ and Δt, must depend on the definitions adopted, but the close link between $\Delta\omega$ and Δt is no less inevitable for possessing a small measure of numerical imprecision.

A particular case of the foregoing example involves taking the mean frequency ω_0 in (20) as zero. A pulse of constant magnitude in the time interval between $\pm t_0$ can be resolved into Fourier components centred on $\omega = 0$, with an amplitude distribution given by $\sin(t_0\omega)/\omega$, having the central zeros at $\pm\pi/t_0$. As the length of the pulse is reduced, so the width of the spectrum is increased, until in the limit we may describe an instantaneous impulse as made up of a uniform spectrum; all frequencies are equally represented and are phased so as to add up at one, and only one, moment of time – at all other times they mutually annihilate one another.

It can be seen immediately from the second equation of (16) that if the *impulse*, defined as $\int_{-\infty}^{\infty} g(t)\,dt$, remains constant at some value P as $g(t)$ is compressed into an infinitesimal time interval at $t=0$, the transform $f(\omega)$ takes the limiting constant value $P/2\pi$.

In these examples the components were phased so as to give the maximum possible resultant at one instant, and this is normally the best synthesis of a given set of components if the aim is to generate the shortest train of oscillations. Consider, for example, a Gaussian spectrum of components, with the amplitude varying as $\exp[-t_0^2(\omega-\omega_0)^2]$, and falling to $1/e$ of the peak at frequencies $\omega_0 \pm 1/t_0$. When these are compounded so as to be in phase at $t=0$, the resultant takes the form

$$g(t) = \int_{-\infty}^{\infty} e^{-t_0^2(\omega-\omega_0)^2}\, e^{-i\omega t}\, d\omega. \tag{22}$$

Now the exponents in the integrand may be combined and rearranged in the form

$$t_0^2(\omega-\omega_0)^2 + i\omega t = [t_0(\omega-\omega_0) + \tfrac{1}{2}it/t_0]^2 + i\omega_0 t + \tfrac{1}{4}t^2/t_0^2,$$

and if z is defined as $t_0(\omega-\omega_0) + \tfrac{1}{2}it/t_0$, (22) takes the form

$$g(t) = e^{-\frac{1}{4}t^2/t_0^2}\, e^{-i\omega_0 t} \int_{-\infty}^{\infty} e^{-z^2}\, dz/t_0$$

$$= [(2\pi)^{\frac{1}{2}}/t_0]\, e^{-\frac{1}{4}t^2/t_0^2}\, e^{-i\omega_0 t}. \tag{23}$$

This is a train of oscillations having the mean frequency ω_0 and an envelope which, like the spectral distributions, is also Gaussian. If we take the width to $1/e$ as defining $\Delta\omega$ and Δt, we have

$$\Delta\omega = 2/t_0 \qquad \Delta t = 4t_0$$

and

$$\Delta\omega\,\Delta t = 8. \tag{24}$$

The numerical value of $\Delta\omega\,\Delta t$ depends, as already remarked, on the form of $f(\omega)$ and on the definitions of $\Delta\omega$ and Δt, which here are rather more generous than is customary. More usually $\Delta\omega$ is taken as the root mean square deviation of the intensity distribution in the frequency spectrum. If the amplitude distribution is $f(\omega)$, the mean frequency $\bar{\omega}$ is defined by the equation

$$\bar{\omega} \equiv \int_{-\infty}^{\infty} \omega ff^*\, d\omega \Big/ \int_{-\infty}^{\infty} ff^*\, d\omega,$$

and the width $\Delta\omega$ by the equation

$$(\Delta\omega)^2 \equiv \int_{-\infty}^{\infty} (\omega-\bar{\omega})^2 ff^*\, d\omega \Big/ \int_{-\infty}^{\infty} ff^*\, d\omega.$$

With the Gaussian distribution, $f(\omega) = \exp[-t_0^2(\omega-\omega_0)^2]$, evaluation of the integrals shows that $\bar{\omega} = \omega_0$ and $\Delta\omega = \tfrac{1}{2}t_0$, only one-quarter of the width as defined in the text.

By the same argument Δt would be t_0 and (24) would take the form

$$\Delta\omega\,\Delta t = \tfrac{1}{2}. \tag{25}$$

In Schrödinger's wave mechanics, a particle with energy E has associated with it a wave of frequency $\omega = E/\hbar$. To describe a particle that was known to pass a certain point during an interval Δt it is necessary to associate with it a wave train no longer than this, which is a combination of sinusoidal waves lying in a frequency range $\Delta\omega \sim 1/\Delta t$. The energy of the particle is correspondingly uncertain, in accordance with Heisenberg's Uncertainty Relation

$$\Delta E\,\Delta t \sim \hbar. \tag{26}$$

The relation (25) between the width of the envelope and the spectral range represents the best possible. If there is no instant when the components are all in phase the envelope becomes wider, as may be illustrated by assigning at $t=0$ a phase which varies as $\beta^2(\omega-\omega_0)^2$. It should be noted that a phase shift proportional to the first power, $(\omega-\omega_0)$, only moves the peak in time without preventing all components from adding in phase at some instant; but the quadratic shift proposed stops this from happening. We now have

$$g'(t) = \int_{-\infty}^{\infty} e^{-t_0^2(\omega-\omega_0)^2}\, e^{-i[\omega t + \beta^2(\omega-\omega_0)^2]}\, d\omega,$$

which is susceptible to the same procedure as was used in evaluating (22), with t_0^2 replaced by $t_0^2+i\beta^2$. Consequently

$$g'(t) = \left(\frac{2\pi}{t_0^2+i\beta^2}\right)^{\frac{1}{2}} e^{-\frac{1}{4}t^2/(t_0^2+i\beta^2)}\, e^{-i\omega_0 t}$$

$$= \left(\frac{2\pi}{t_0^2+i\beta^2}\right)^{\frac{1}{2}} e^{-\frac{1}{4}t_0^2 t^2/(t_0^4+\beta^4)}\, e^{-i[\omega_0 t - \frac{1}{4}\beta^2 t^2/(t_0^4+\beta^4)]}.$$

The envelope is still Gaussian, as in (23), but the amplitude at the centre is reduced, the local frequency is not constant and equal to ω_0 throughout, and the envelope is wider. The new value of Δt is $(t_0^2 + \beta^4/t_0^2)^{\frac{1}{2}}$, from which it is clear that for a given choice of frequency spread as determined by t_0, the narrowest envelope results from phasing all components together, making β vanish.

In this example the envelope is still symmetrical, but in general when the components are not arranged so that they can all combine in phase an unsymmetrical envelope results. An important case is an envelope which is zero up to some instant which we take as the origin of t, then rises sharply and afterwards decays exponentially:

$$\begin{aligned} g(t) &= e^{-\alpha t}\, e^{-i\omega_0 t} \quad &&\text{when } t>0, \\ &= 0 \quad &&\text{when } t<0. \end{aligned} \tag{27}$$

When this form of $g(t)$ is substituted in the second equation of (16), the spectral distribution follows immediately:

$$f(\omega) = \frac{1}{2\pi} \int_0^\infty e^{-(\alpha - i\Delta)t} \, dt = 1/2\pi(\alpha - i\Delta), \qquad (28)$$

where $\Delta = \omega - \omega_0$. The amplitude $|f(\omega)|$ varies as $(\alpha^2 + \Delta^2)^{-\frac{1}{2}}$ and the argument as $\tan^{-1}(\Delta/\alpha)$. The significance of the latter is that, since the components between ω and $\omega + d\omega$ have the form $f(\omega) e^{-i\omega t} \, d\omega$, the argument of $f(\omega)$ is the phase of these components at $t = 0$. These functions are shown in fig. 7; the spectrum represented by $f(\omega)$ in (28) is called *Lorentzian*. We shall meet this particular transform on several occasions – it dominates the mathematical analysis of resonant systems.

107, 129

Fig. 7. The exponentially damped oscillation and its Fourier transform.
(*a*) Illustrating (27), being actually $-\text{Im}[g]$, i.e. $e^{-\alpha t} \sin \omega_0 t$; (*b*) the amplitude of $f(\omega)$ as given by (28); (*c*) the phase of $f(\omega)$.

(*a*)

(*b*)

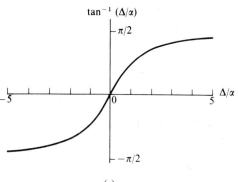

(*c*)

One final example in this section applies the same ideas to wave propagation.† Suppose an impulse is applied at time $t = 0$ to the point $x = 0$ in an infinite one-dimensional medium, as, for example, by striking a piano string with a hammer. If, as in this case, the waves are non-dispersive, the kink created by the impulse runs away unchanged in both directions, but if they are dispersive the disturbance changes shape. One may now ask, in particular, what will be observed at a very great distance where the waves have had plenty of opportunity to alter their phase relationships. The formal solution may be written down immediately:

$$g(x, t) = a \int_{-\infty}^{\infty} e^{i(kx - \omega t)} \, dk, \quad \text{where } \omega = \omega(k). \qquad (29)$$

This is the superposition of harmonic waves of all wave-numbers, the uniform spectrum of amplitudes and the absence of phase variation with frequency ensuring that at time $t = 0$ the displacement $g(x, 0)$ $(= a \int_{-\infty}^{\infty} e^{ikx} \, dx)$ vanishes everywhere except at the origin, where it is infinite; the integrated displacement $\int_{-\infty}^{\infty} g(x, 0) \, dx$ is finite and equal to $2\pi a$. In (29), therefore, we have an expression correctly describing the initial conditions and built up of wave-motions which are correct solutions of the equation of motion of the medium, provided the right dispersion law $\omega(k)$ is used.

To see what happens at a great distance X we may employ the vibration diagram to integrate (29), given that

$$g(X, t) = a \int_{-\infty}^{\infty} e^{i(kX - \omega t)} \, dk,$$

and that this can be represented as the vector addition, by head-to-tail joining, of innumerable vectors of which a typical one is $a \, e^{i(kX - \omega t)} \, dk$;

$$\left. \begin{array}{ll} \text{length of elementary vector,} & ds = a \, dk, \\ \text{phase of elementary vector,} & \phi = kX - \omega t. \end{array} \right\} \qquad (30)$$

Note that since many different frequencies are involved, the rotation $e^{-i\omega t}$ is kept in the expression for the phase. The vibration diagram changes shape and spins incessantly, but we shall construct it at some chosen time t.

In fig. 8 is shown a typical variation of ω with k for a dispersive medium. A point (k_1, ω_1) is found at which the slope of the tangent is X/t; ϕ is constant on this tangent. We start drawing the vibration diagram with vectors representing points around (k_1, ω_1). If the dispersion curve followed the straight line, all vectors would have the same phase and the vibration diagram would be itself straight, with a very large resultant. This

† Since we do not intend to present a systematic account of the theory of wave-motion, no attempt is made to develop the elementary concepts of wave and group velocity. This section should be viewed simply as an illustration of how Fourier analysis may be applied to an interesting phenomenon. Moreover, since we shall need the theory of Cornu's spiral elsewhere, we have here a convenient excuse for presenting it.

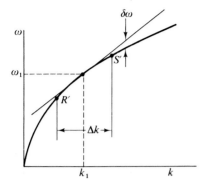

Fig. 8. Dispersion curve $\omega(k)$ for a wave-system.

is what happens with a non-dispersive medium, where $k = \omega/v$, v being constant; if X and t are such that $X/t = v$, the resultant is very large – otherwise it is zero, i.e. the pulse passes the point X at time X/v. Because of dispersion, however, the vibration diagram is rolled up into a form that is readily deduced when X is large. The significance of the intercept $\delta\omega$ shown in fig. 8 is that the vector contributed by this point on the curve has a phase differing by $\delta\omega \cdot t$ from that for a point on the line. When X is large t also is large, and a small intercept may imply a phase difference many times 2π. It is then adequate, as will become clear in a moment, to think only of the contributions from the immediate vicinity of k_1 and to suppose that $\delta\omega \propto (k - k_1)^2$. Hence, by use of (30), the vibration diagram has the properties:

(1) $\phi - \phi_1 \propto (k - k_1)^2$, where ϕ_1 is the phase of the vector due to k_1, or
  $d\phi \propto (k - k_1)\,dk$;

(2) $s = a(k - k_1)$,
  or $ds \propto dk$.

The curvature of the vibration diagram at any point is $d\phi/ds$, and is therefore proportional to s, which is distance measured along the curve from the point representing k_1. The curve with this property is Cornu's spiral, drawn in fig. 9, which is well known in the theory of Fresnel diffraction where it arises out of exactly parallel considerations.[2] Because,

Fig. 9. Cornu's spiral.

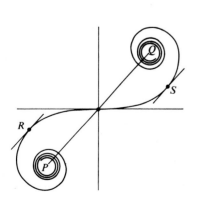

when X is large, the spiral runs through many turns and comes close to its end-points while the quadratic approximation to $\delta\omega$ is still valid, it hardly matters precisely how $\delta\omega$ behaves further out.

Detailed analysis of the spiral can be found elsewhere, all that is needed here being the two facts that the closing vector PQ lies at an angle $\pi/4$ to the central vector, and that its length is the same as the length round the curve between R and S, where the tangents are parallel to the closing vector. From the constant angle between PQ and the central vector it follows that as the diagram spins the resultant spins at the same rate as the central vector, i.e. at ω_1. As time proceeds, an observer at X sees a resultant vibration whose frequency ω_1 itself changes gradually, being always that frequency at which $d\omega/dk = X/t$. The gradient, $d\omega/dk$, is the *group velocity*, u, of the wave; it is clear that the prescription for the gliding frequency experienced by the observer is that a frequency ω arrives at that time after the impulse that it would take a particle moving at velocity $u(\omega)$ to reach the observer. To determine the amplitude, find the points R' and S', on fig. 8, corresponding to R and S in fig. 9, where $t\,\delta\omega = \pi/4$, and let R' and S' be separated by Δk. Then the resultant amplitude is the same as if all contributions in the range Δk were added in phase, everything outside being ignored. From (30) it follows that $|g(X, t)| = a\,\Delta k$.

The propagation of magneto-ionic waves in the upper atmosphere illustrates the phenomenon.[3] Lightning strokes in, say, the Southern Hemisphere initiate a sharp electromagnetic pulse which through the action of the ionized upper layers is guided along the Earth's magnetic field, rising high above the equator before descending in the Northern Hemisphere. The magneto-ionic wave has a dispersion law $k^2 \propto \omega$, the constants of proportionality depending on ion density etc. Consequently the group velocity $u \propto \omega^{\frac{1}{2}}$, and over this long path the high frequencies travel faster and arrive first. A radio antenna connected straight to an audio-amplifier and loudspeaker signals the arrival of a dispersed pulse by emitting a descending whine (*whistler*) which traverses the audible range of frequencies in something like a second.

By contrast, waves in deep water obey a dispersion law $k \propto \omega^2$, so that $u \propto 1/\omega$, or $u \propto \lambda^{\frac{1}{2}}$ where λ is the wavelength, $2\pi/k$.[4] When a local storm far out in the ocean whips up the water suddenly, the watcher on a distant coast may first become aware of it by the arrival of a train of regular long-wavelength rollers, giving way to shorter and usually more irregular waves as the slower components reach shore.

Noise

Let us turn to an extreme situation in which many sinusoids are added with random phase relationships, such as our ears experience in heavy traffic or other situations described as noisy. The description of noise is important in the theory of communications systems, and, more generally, whenever organized signals are weak enough to need amplification to the point that they are confused by the irregular fluctuations inseparable from physical mechanisms.[5] There are many subtle difficulties attending the mathematical description of random noise which rule out a systematic treatment here. What follows is sufficient to give the background to later applications, but does not aspire to anything approaching rigour.

A vibration diagram representing all the separate components at any instant would consist of a large number of vectors of different length pointing at random, and their sum would be found by drawing them head to tail as in fig. 10. As time proceeds, each vector rotates at its characteristic frequency. If they all have the same frequency, the pattern spins unchanged and the resultant is a sinusoid of constant amplitude. If the components fall in a narrow range of frequencies, the resultant will at any instant spin at something close to the mean frequency, and the displacement will vary in an approximately sinusoidal fashion – approximate because of the slow variation of amplitude as the vibration diagram gradually changes in shape. With a wider range of frequencies the changes of amplitude will be more rapid. Given long enough, we may suppose that the relative phases run through all possible combinations, and we might ask what will the mean amplitude be, and how frequently will the resultant be found to have some other specified amplitude.

Fig. 10. Addition of vectors representing oscillations of random amplitude and phase.

It is more convenient, however, especially when the range of frequencies is so wide as to destroy the sinusoidal character of the resultant, to abandon the concept of amplitude of vibration, and consider instead the mean-squared value of the displacement, which will be termed the *mean intensity*, $\bar{I}$. Working in real numbers, a single component of the form $a \cos (\omega t + \phi)$ has an instantaneous intensity $I(t)$ given by $a^2 \cos^2 (\omega t + \phi)$, and since the mean value of $\cos^2 \theta$ is $\frac{1}{2}$, $\bar{I} = \frac{1}{2} a^2$. With a number of superimposed sinusoids we similarly write for the instantaneous intensity

$$I(t) = \left[\sum_i a_i \cos (\omega_i t + \phi_i) \right]^2$$

$$= \sum_i a_i^2 \cos^2 (\omega_i t + \phi_i) + \sum_i \sum_{j \neq i} a_i a_j \cos (\omega_i t + \phi_i) \cos (\omega_j t + \phi_j). \tag{31}$$

If there are a very large number of components, so that the phase angles ϕ_i

are evenly distributed throughout the range 0 to 2π, the mean value of $\cos^2(\omega_i t + \phi_i)$ is $\frac{1}{2}$, while the double summation vanishes; for if we choose a particular value of i and first carry out the summation over j, $a_i \cos(\omega_i t + \phi_i)$ remains constant while $a_j \cos(\omega_j t + \phi_j)$ runs through the gamut, with positive and negative values equally likely. Consequently

$$\bar{I} = \tfrac{1}{2} \sum_i a_i^2, \tag{32}$$

which may also be written as $\frac{1}{2}N\overline{a^2}$, or $\Sigma_i \bar{I}_i$ if we write $\bar{I}_i$ as the mean value of $a_i^2 \cos^2(\omega_i t + \phi_i)$. This is the well-known result that randomly phased vibrations are to be combined by summation of intensities. In particular, N equal vibrations combined in random phase lead to a root mean square displacement $N^{\frac{1}{2}}$ times that due to a single vibration.

If energy is associated with the vibrating quantity, as when the amplitude is the current in an inductor or the stretching of a spring, (32) may be interpreted as a statement that the total energy of the system is the sum of the energies of each component alone. This important result is not confined to randomly phased Fourier components. It depends on the fact that any two components of different frequency, however they are phased at any special instant, will in the course of time run evenly through all possible phase differences relative to one another. The time-average of I, as expressed by (31), therefore has no contribution from the double summation. With a continuous spectrum of frequencies present we may select those lying within a range $d\omega$ and define $I_\omega \, d\omega$ as $\frac{1}{2}\Sigma_i a_i^2$ for those components lying with $d\omega$. I_ω so defined is the *spectral intensity* or *power spectrum*, and represents the energy per unit frequency range.†

It must be emphasized that the result expressed in (32) represents the long-term average, with every opportunity given for the vibration diagram to take up all possible configurations. It does not imply that if we form a

---

† So long as we represent the noise signal as a sum of real oscillatory functions the above definition of I_ω is satisfactory. With the more convenient representation in terms of complex exponentials, such as the Fourier integral (14), a little care is needed, since the components at $\pm\omega$ are not randomly phased, but related by the reality condition. Thus $a \, e^{-i\omega t}$ is accompanied by $a^* e^{i\omega t}$, the two together describing a real oscillation of amplitude $2|a|$ and mean intensity $2aa^*$. It is therefore correct to write $I_\omega \, d\omega$ as $2\Sigma_i a_i a_i^*$, restricting the summation to those components whose frequencies are positive and lie within the range $d\omega$. This will be the normal usage in what follows; if we revert to representation by real functions we shall draw attention to the fact. There is, one must admit, scope for misunderstanding here on account of different conventions adopted by different users of complex notation. So long as only one frequency is present, as in many electrical applications and as was assumed in chapter 3, $A \, e^{-i\omega t}$ describes an oscillation of amplitude $|A|$. In Fourier theory, however, the presence of the conjugate $A^* e^{i\omega t}$ means that the amplitude at frequency ω is $2|A|$. No misunderstanding of physical principle is likely to arise from failure to recognize the convention in any particular case, but quantitative errors involving factors of 2 and 4 are all too easily incurred.

short-term average of I it will be more or less the same at all times. Take the case of a large number of components lying in a narrow frequency range, so that the resultant is quasi-sinusoidal but with varying amplitude. It is easy in this case to appreciate the meaning of the short-term average, which may be taken over a period several cycles long, yet not long enough to allow the pattern to change appreciably. Over an extended period this short-term average changes, but although it fluctuates about a mean value given by (32) it does not remain close to this value – on the contrary, we shall find that the fluctuations are comparable with the mean value itself, and do not become any less when more vibrations are compounded.

There is no need to confine oneself to this special case in determining the range of the fluctuations. When the frequency range is narrow the vibration diagram of fig. 10 changes shape slowly on the time scale of the spinning of each vector, while when it is wide the spinning and the shape changes proceed at comparable rates. In both cases, however, the range of configurations available to the diagram is the same and in the long term the probability of a given resultant displacement is not affected by whether the behaviour is quasi-sinusoidal or totally devoid of periodic character.

The determination of this probability is resolved into the question: given a vibration diagram consisting of N vectors, not necessarily equal in length, joined end-to-end with random phase, what is the chance that the resultant will have a certain length? To answer the question, we may imagine proceeding by drawing the diagram many times, using the same set of vector lengths but a different choice of phase angles, and marking the result by a dot in each case, so that in the end a pattern of dots is obtained, with a density $\rho(r)$ varying with distance from the origin. Then the chance that any one choice of phase angles will yield a resultant between r and $r + dr$ may be written as $P_r \, dr$, which expresses the fraction of the dots lying within the annulus between r and $r + dr$:

$$P_r \, dr = r\rho(r) \, dr \bigg/ \int_0^\infty r\rho(r) \, dr. \qquad (33)$$

To find the form of ρ and P_r, we ask what happens to ρ if one more vector, of length a, is added to the N already there, again with arbitrary phase angle; clearly every dot will be shifted to some point on a circle of radius a, centred on its original position, each one being displaced in a different direction. The radius of the circle is small on the scale of the distribution as a whole, if N is large, since the distribution has a mean radius about $N^{\frac{1}{2}}$ times the length of the average vector. In any region where ρ is constant, virtually nothing happens on adding the new vector, since for every dot that crosses a line drawn in the plane there is another that will cross in the opposite direction. Where, however, ρ varies with position in a manner described by the vector grad ρ, and a line is drawn normal to grad ρ, there will be an excess of dots crossing the line from the denser to the rarer region; we can therefore introduce a flux vector, **V**, describing the number crossing a line of unit length, and write

$$\mathbf{V} = -D \text{ grad } \rho. \tag{34}$$

The magnitude of the constant D need not be considered, since in the end its meaning becomes clear. This equation is the same as expresses Fick's law for diffusion, and in fact the movement of the dots as N increases is exactly analogous to Brownian motion or the diffusion of solute molecules resulting from random thermal agitation.

Variations of **V** from point to point are responsible for accumulation or depletion of the density of dots, according to the equation of continuity,

$$\text{div } \mathbf{V} = -\partial\rho/\partial N. \tag{35}$$

Combining (34) and (35) results in the diffusion equation,

$$D \text{ div grad } \rho \equiv D\nabla^2\rho = \partial\rho/\partial N, \tag{36}$$

whose solution $\rho(r, N)$ shows how the distribution of a given number of dots depends on N. Since when $N = 0$ all the dots necessarily lie at the origin, we seek a solution having this initial form, and the answer to this problem is well known from the theory of diffusion:

$$\rho = \frac{C}{N} e^{-r^2/4DN}, \tag{37}$$

in which C is a constant depending on the number of dots in the pattern, i.e. the number of trials with different choices of the phase angles. A typical set is shown in fig. 11. As N is increased the radial scale increases as $N^{\frac{1}{2}}$ (since r^2/N is the only combination containing r) and the density scale decreases as $1/N$ to compensate for the area occupied by the dots increasing as N.

Fig. 11. Each point represents the resultant of adding the same large number of vectors with a different random choice of phases.

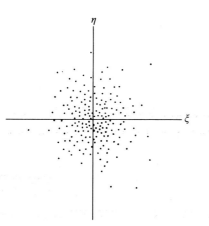

Since every point in fig. 11 represents the closing vector in a complex vibration diagram, one must project the whole distribution onto the real axis to give it physical significance. Thus the probability, $P_\xi \, d\xi$, of observing a displacement between ξ and $\xi + d\xi$ is proportional to the number of dots in a strip of width $d\xi$ parallel to the η-axis. The integration of (37) with respect to η is easy, since $r^2 = \xi^2 + \eta^2$ and the variables separate; there is no need even to carry through the integration over the η-coordinate since it yields a constant; for a given value of N

$$P_\xi \propto e^{-\xi^2/4DN}. \tag{38}$$

From this the mean intensity follows:

$$\bar{I} = \overline{\xi^2} = \int_{-\infty}^{\infty} \xi^2 P_\xi \, d\xi \Big/ \int_{-\infty}^{\infty} P_\xi \, d\xi = 2DN, \tag{39}$$

as may be seen by partial integration of the denominator; $\int_{-\infty}^{\infty} e^{-z^2} \, dz = [z \, e^{-z^2}]_{-\infty}^{\infty} + 2 \int_{-\infty}^{\infty} z^2 \, e^{-z^2} \, dz$, and the quantity in square brackets vanishes. Now we know from (32) that $\bar{I} = \tfrac{1}{2} N a^2$, so that we must assign to D the value $\tfrac{1}{4} a^2$. In the light of this (38) can be rewritten in the form

$$P_\xi = e^{-\tfrac{1}{2}\xi^2/\bar{I}} / (2\pi \bar{I})^{\tfrac{1}{2}}, \tag{40}$$

the normalizing denominator being assigned such a value as ensures that $\int_{-\infty}^{\infty} P_\xi \, d\xi = 1$. The (Gaussian) form of P_ξ is shown in fig. 12.†

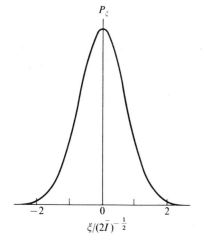

P_ξ

−2   0   2

$\xi/(2\bar{I})^{-\tfrac{1}{2}}$

Fig. 12. The Gaussian function (40).

† In the derivation of (40) it may have seemed that each vector must have the same length a, but this is unnecessary. If there is a distribution of lengths the only change is that $D = \tfrac{1}{4} \overline{a^2}$, but since the form of (40) is independent of such details this is irrelevant, and (40) describes the distribution of the resultant when any set of vectors is summed with random phase angle.

As a measure of the extent of variation to which the resultant is subject we may use the standard deviation, σ, of I from its mean:

$$\sigma \equiv \overline{(I - \bar{I})^2} = \int_{-\infty}^{\infty} (\xi^2 - \bar{I})^2 P_\xi \, d\xi = 2\bar{I}^2.$$

No matter how many oscillations are compounded, the variations in intensity are comparable to the mean.

It is easy to demonstrate the general character of random noise by displaying the output of a high-gain amplifier on an oscilloscope, for noise is always present in electrical circuits and can be amplified to any desired level. If the amplifier has a very wide band the noise output is entirely devoid of sinusoidal character and the characteristic time in which the amplitude can change appreciably is determined by the highest frequency that the amplifier can handle. The typical record of such noise ('grass') in fig. 13(a) was taken with a low-pass filter limiting the frequency range to virtually everthing below 10 kHz. This may be compared with fig. 13(b), taken with a narrow-band filter, so that only frequencies in the range 13.6–13.9 kHz were accepted. The quasi-sinusoidal character of the latter is evident, as are the wide variations of amplitude which we have seen are inescapable when random vibrations are compounded. It appears from the record that 1 ms is a fair measure of the time taken for the amplitude to change significantly. In this time the extreme components, 300 Hz apart in frequency, change relative phase by about 2 radians, slipping by 1 radian each from the mean. This is a very plausible figure for the sort of changes in direction of the vectors in the random vibration diagram that might alter its resultant appreciably, and in view of the difficulty of providing a rigorous argument we shall adopt it on empirical grounds.

The same process underlies an entirely different phenomenon – the speckle pattern observed when a laser beam is scattered off a rough surface (e.g. white card).[6] It is the surface, whose roughness has a scale many times the wavelength, that provides the random element, but now in the spatial rather than the temporal dimension. To enlarge on this point, the scattered light has the same character as would be produced by a myriad of point sources in the illuminated area, each emitting exactly the same frequency and vibrating with a perfectly definite, but random, phase with respect to all the rest.† When the observer's eye is focussed on the surface,

---

† Such a source of light is *random* but *coherent*. A better way of producing random coherent illumination of an area is by internal scattering in a translucent material. Take a cylinder of polyethylene, about 2 cm in diameter and 2 cm long, wrap it in aluminium foil (leaving the flat ends clear) and let a laser beam fall on one end; the other end glows brilliantly and uniformly, apart from the speckle pattern. The light intensity at any point is the resultant of wavelets that have reached it by multiple scattering, and one has only to move a wavelength or two along the end surface to alter the phases of the wavelets significantly. Thus the phase of the light emitted from different points varies randomly over a very short distance, yet this pattern of phases is coherent in the sense that it retains its form so long as the system is not disturbed.

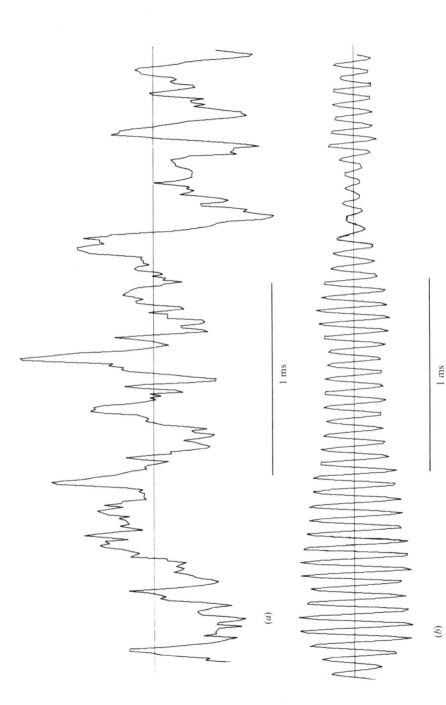

Fig. 13. Random noise (*a*) after low-pass filter, so that only frequency components below 10 kHz appear, (*b*) after band-pass filter, so that only components in the range 13.6–13.9 kHz appear. (Recording by Mr P. Mole.)

the light from each point source focusses to a point (or at any rate a very small area) on the retina and any one receptor receives light from a great many sources that lie close together on the rough surface. The response of the receptor is determined by the vector sum of all the contributions, and some receptors will happen to receive a collection of vectors that add up to a sizeable resultant, while others, though receiving just as many contributions, will experience a much smaller resultant. The distribution of resultants seen by the receptors will follow (40), and this accounts for the granular appearance of the illuminated surface. So long as the observer keeps his eye quite still, the pattern remains steady, for the phase-coherence of the laser beam is remarkably good and the random phase pattern imposed by a rigid rough surface does not alter. But as soon as the eye is moved, the phase relationships at the retina are altered and the pattern flickers. In explaining the phenomenon in terms of the response of individual receptors it has been assumed that each point source focusses to a point on the retina, but this is not necessary. In fact, if a foil, pierced with a very fine pinhole (say $\frac{1}{5}$ mm diameter), is held before the eye, diffraction at the hole causes each point source to illuminate a considerable area of the retina. Any one receptor receives weaker contributions from many more sources, but this does not alter the wide spread of the distribution of intensities at different parts of the retina. It does, however, enlarge the scale of the granular appearance. For neighbouring receptors, indeed all lying within the diffraction circle of the pinhole, receive very nearly the same collection of elementary contributions with the same phase relationships, and the resultant, correspondingly, is very nearly the same. This experiment is strongly recommended to any reader with access to a low-power continuous laser. For those who are without this facility the photographs in fig. 14 must suffice; the legend explains how they were obtained. In this case there was no lens and the pinhole may be taken as an array of randomly phased point sources, each illuminating the whole of the photographic plate. It is easy to appreciate from the previous example that the relative phases of the many contributions at any one point can only be significantly altered by moving far enough on the plate for the difference in path length from the two sides of the hole to approach a third or half a wavelength – anywhere within this radius the resultant illumination must be roughly constant. The spatial changes in phase relationships here are exactly analogous to the temporal changes when vibrations of slightly different frequency are compounded; making the pinhole smaller is equivalent to narrowing the frequency range and thus allowing only slower variations of the amplitude of quasi-sinusoidal vibration. To put the matter on a more quantitative basis, let the pinhole diameter be d, and the distance from pinhole to photographic plate D; then elementary geometry shows that moving $D\lambda/d$ on the plate imposes an extra λ between the rays from the extreme edges of the hole. Making allowance for the hole being circular, not square, and (in accordance with the known effect this has on the width of the diffraction pattern) treating it as a square of side $0.8d$, we

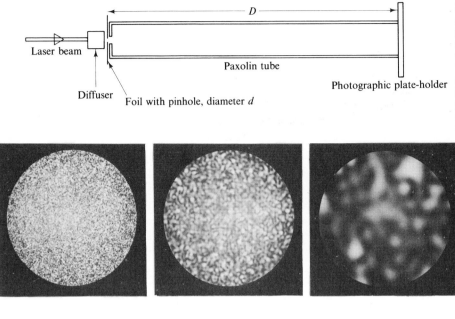

Laser beam

Diffuser

Foil with pinhole, diameter d

Paxolin tube

Photographic plate-holder

D

(*a*)   (*b*)   (*c*)

Fig. 14. Arrangement for photographing speckle pattern. The diffuser is a short rod of polyethylene wrapped in aluminium foil. The three photographs, reduced here by a factor 3/2, were taken with the following pinhole diameters: (*a*) 2.5 mm, (*b*) 0.50 mm, (*c*) 0.13 mm.

arrive at figures of 0.1 mm, 0.5 mm and 2 mm as the distance to move on the plate in order to impose variations in phase of 1 radian on the extreme rays with respect to the central ray. The observed granularity is on a scale perhaps three times as great, but this may be because the eye is caught by major peaks and troughs of intensity, between which there is quite a lot of unnoticed structure, such as shows up more plainly in a record like fig. 13(*b*).

The distribution (40) is quite different from that resulting from a constant-amplitude vibration; for if $\xi = X \cos \omega t$, the fraction of time spent between ξ and $\xi + \mathrm{d}\xi$ is given by

$$P'_\xi \, \mathrm{d}\xi = \mathrm{d}\xi / \pi (X^2 - \xi^2)^{\frac{1}{2}}. \tag{41}$$

This rises to infinite peaks at the two limits $\pm X$, markedly contrasting with the smooth fall to zero of P_ξ. Even when the random components are selected from a narrow frequency range, so that the resultant is a quasi-sinusoidal oscillation with slowly varying amplitude, $X(t)$, the wide variation of X ensures that the peaks in (41) are completely smoothed out. The actual distribution of amplitudes in this case can be derived from fig. 11. Any one choice of phase angles between different components, leading to a single dot in the pattern, remains generally constant over a number of cycles, so that the dot rotates very nearly in a circle, whose radius is the instantaneous value of X. The diagram as a whole is then to be interpreted, after dissection into circular annuli, as showing the relative probability of different values of X:

$$P_X \, dX \propto X \rho(X) \, dX,$$

i.e. in normalized form, and with ρ substituted from (37),

$$P_X \, dX = X \, e^{-\frac{1}{2}X^2/\bar{I}} \, dX/\bar{I}. \tag{42}$$

$P_X = 0$ when $X = 0$ and rises to a maximum when $X = (\bar{I})^{\frac{1}{2}}$. It can now be shown that the distribution (42) does indeed, as asserted above, smooth out the peaks of (41) into Gaussian form. For the probability of ξ lying in a given range, when all possible values of X are weighted according to (42), is given by the expression

$$P_\xi \, d\xi = d\xi \int_{|\xi|}^{\infty} P'_\xi P_X \, dX = \frac{d\xi}{\pi \bar{I}} \int_{|\xi|}^{\infty} \frac{X \, e^{-\frac{1}{2}X^2/\bar{I}} \, dX}{(X^2 - \xi^2)^{\frac{1}{2}}},$$

which is readily integrated when z^2 is substituted for $X^2 - \xi^2$, and yields the expected result (40).

An alternative form of (42) is in terms of $P_I \, dI$, the probability that the quasi-sinusoidal vibration will have an intensity between I and $I + dI$, I being $\frac{1}{2}X^2$. If the intensity is in this range, X must lie between $(2I)^{\frac{1}{2}}$ and $(2I)^{\frac{1}{2}} + dI/(2I)^{\frac{1}{2}}$; substituting $(2I)^{\frac{1}{2}}$ for X and $dI/(2I)^{\frac{1}{2}}$ for dX in (42), we have that P_I is a simple exponential function:

$$P_I \, dI = e^{-I/\bar{I}} \, dI/\bar{I}. \tag{43}$$

This result is peculiar to quasi-sinusoidal vibrations, for which an instantaneous amplitude may be defined.†

The full extent of the variations of intensity of quasi-sinusoidal vibrations expressed by (43) will only be observed by instruments that have adequate time-resolution to react in times small compared with the characteristic time for changes in intensity, which we have empirically determined as $2/\Delta\omega$ when all Fourier components lie in the range $\Delta\omega$. It is not uncommon, however, for a recording instrument to respond only to

---

† It may be noted in passing that statistical mechanics leads one to expect the distribution (43) to apply to a harmonic vibrator subject to random disturbances by, for example, the impact of gas molecules.[7] If the temperature of the gas is T, the fraction of time during which the vibrator has energy between E and $E + dE$ is proportional to $\exp(-E/\bar{E}) \, dE$, and $\bar{E} = k_B T$, k_B being Boltzmann's constant. Each separate molecular collision imparts to the vibrator a little extra oscillatory motion, and the net effect of many collisions at random times is to add many vectors to its vibration diagram, just as in fig. 10. The form of the Boltzmann distribution which, as conventionally derived, possesses something of the mysterious quality of a rabbit emerging from the conjuror's hat, has an immediate molecular interpretation in this simple case. Other examples of this nature will be discussed in a later chapter.

stimulus over longer periods; thus a spectrometer may be able to select a very narrow band $\Delta\omega$, yet the intensity of light in the band may necessitate an exposure time much larger than $2/\Delta\omega$. In this case a resolving power of 10^6 (which means an extremely good spectrometer) implies that $\Delta\omega$ is 10^{-6} times the frequency of the light being studied ($\omega \sim 5 \times 10^{15}\,\text{s}^{-1}$). With $\Delta\omega \sim 5 \times 10^9$ significant intensity fluctuations occur within a time of $10^{-9}\,\text{s}$, and any normal photographic process records an average of the intensity over many such fluctuations. Consequently the recorded intensity lies very close to the mean $\bar{I}$. Just how close is by no means easy to derive rigorously, but a rough estimate can be made. For if the instrumental averaging time (e.g. photographic exposure) is T, we may imagine it divided into n intervals of $2/\Delta\omega$ each, n being $\frac{1}{2}T\,\Delta\omega$. Each interval is short enough for the average intensity to remain fairly constant throughout, but long enough for the mean intensities in successive intervals to be only weakly correlated. Thus what is recorded may be approximately described as n independent values of a quantity I that exhibits the full range of variation given by (43). The standard deviation of I is comparable with $\bar{I}$ itself, and in accordance with the theory of repeated measurements we expect n independent values to have an average which is subject to a standard deviation smaller by $n^{\frac{1}{2}}$. Thus the instrumental averaging should reduce the fluctuation of the recorded signal to about $(\frac{1}{2}T\,\Delta\omega)^{-\frac{1}{2}}$ of the signal itself. With a photographic exposure of 1 second, each grain on the plate will be subjected to the same mean intensity within less than one part in 10^4, so that the statistical fluctuations of intensity will be quite negligible compared with the other sources of unevenness, especially the graininess of the plate.

White noise

The examples just considered have involved the quasi-sinusoidal resultant of random vibrations chosen from a narrow frequency interval, $\Delta\omega$. Let us turn to the opposite extreme in which $\Delta\omega \to \infty$ and the components are spread evenly throughout the entire frequency range with their phases uncorrelated – this is the spectrum known as 'white noise', by analogy with white light which also in its ideal form includes all frequencies. There is no correlation between the displacements at any two instants, however close they may be, since for any choice Δt of time interval the range of frequencies greater than $1/\Delta t$ is infinitely larger than the range less than $1/\Delta t$, and the overwhelming majority of components suffer substantial relative phase changes during Δt. The displacement fluctuates with the distribution (40) but runs through the range of possibilities infinitely rapidly.

It should not be supposed that this is a description of an observable phenomenon; in every physical situation the spectral range is finite, sometimes because there are absorption processes or instrumental deficiencies that eliminate higher frequencies, but in the last resort because of quantization. To give a specific example, an electron accelerated in an X-ray tube, and brought to rest on hitting the target, emits a short pulse of electromagnetic radiation. However hard one tries to shorten the pulse by making the deceleration as sudden as possible, one never finds in the continuous spectrum of emitted X-rays any component with a frequency greater than $E/\hbar$, E being the kinetic energy of the electron and $\hbar$ Planck's constant; the quantum of radiation, $\hbar\omega$, can never carry away more energy than the whole of what the electron has available.

This limit to the frequency reveals itself with so sharp a definition as to have earned a place, for quite a long time, among the precision measurements which, taken together, allow the values of the fundamental constants to be determined. When an X-ray spectrometer is set to detect Bragg-reflected X-rays of a certain wavelength λ, it is found, as illustrated in fig. 15, that none are produced until the accelerating voltage reaches a certain critical value V_c. The electrons then have energy eV_c and emit X-rays whose maximum frequency is $eV_c/\hbar$; correspondingly the minimum wavelength, λ in this case, is $2\pi c\hbar/eV_c$, where c is the velocity of light.

Fig. 15. Intensity of X-rays Bragg-reflected at 14° from NaCl crystal, as a function of the electron energy in the X-ray tube (H. Feder[8]).

Provided the unit cell dimensions of the crystal are well enough known for absolute wavelength measurements to be reliable, the experiment gives a value of $\hbar/e$, which can be combined with the results of other experiments, giving different combinations of fundamental constants, to determine each separately. In recent years the exploitation of the Josephson effect in superconductivity has superseded this particular determination of $\hbar/e$, but the observations are of no less interest for their clear demonstration of the overriding quantal control of the spectrum at the high-frequency end.

It is only very close to the cut-off, however, that this becomes apparent. To see this it is necessary to know what would be observed if a random sequence of pulses, due to electrons striking the target, were resolved by a spectrometer into its Fourier components. We start with a long sequence of NT_0 identical pulses randomly distributed in a time interval T_0, allowing this pattern to repeat indefinitely so that it generates a Fourier series. Later we allow T_0 to increase without limit while the mean frequency of pulses, N, remains constant. A single pulse in the sequence, having the form of a δ-function of strength P, generates a line spectrum with lines of equal amplitude P/T_0 (the pulse strength averaged over the interval) at a spacing

$\delta\omega = 2\pi/T_0$. If the pulse is not so sharp the amplitude of the lines will vary with frequency and we shall represent it by $f(\omega)/T_0$, whose magnitude we suppose to vary relatively slowly. From (16) it is readily seen that for a pulse of form $g(t)$, $f(\omega)$ takes the form $\int_{-\infty}^{\infty} g(t) \, e^{i\omega t} \, dt$, independent of T_0. The lines in a narrow band $\Delta\omega$, $T_0 \Delta\omega/2\pi$ in number, combine to form a quasi-periodic oscillation with mean intensity $(2ff^*/T_0^2) \times (T_0 \Delta\omega/2\pi)$. And when the contributions to the intensity of all the other pulses are added we have for the spectral intensity

$$I_\omega \, \Delta\omega = (2ff^*/T_0^2) \times (T_0 \Delta\omega/2\pi) \times NT_0 = Nff^* \, \Delta\omega/\pi. \qquad (44)$$

Already T_0 has disappeared from the argument, though it is implicit in the sense that it must be large enough for $\Delta\omega$ to contain plenty of lines, i.e. $\Delta\omega \gg 1/T_0$. As $T_0 \to \infty$ this requirement is automatically satisfied and I_ω is the intensity of the continuous spectrum.

For the particular case of a sharp pulse, $g(t) = P\delta(t - t_0)$, $f(\omega)$ becomes $P \, e^{i\omega t_0}$ and

$$I_\omega = NP^2/\pi, \qquad (45)$$

a constant spectral intensity without limit of frequency. It is this that we wish to compare with the observed spectral distribution, but it is important to recognize that I_ω is not necessarily the quantity that emerges most naturally from the measurements. In a crystal spectrometer the angle of Bragg reflection is proportional to X-ray wavelength, and when the X-rays are collected those within a certain narrow angular range, representing a certain wavelength range, $\Delta\lambda$, are recorded together. As the angle of reflection is changed $\Delta\lambda$ stays sensibly constant, and it is the variation of $I_\lambda \, \Delta\lambda$, the intensity in a wavelength range $\Delta\lambda$, that is the nearest approximation to the untreated output of the instrument, and is normally presented as the experimental observation. Although I_ω may be independent of ω, I_λ is not independent of λ. For $\omega = 2\pi c/\lambda$, and $\Delta\omega = -2\pi c \, \Delta\lambda/\lambda^2$; the energy in a narrow band $\Delta\omega$, amounting to $NP^2 \, \Delta\omega/\pi$ according to (45), is spread over the wavelength range $\lambda^2 \, \Delta\omega/2\pi c$, so that

$$I_\lambda = 2cNP^2/\lambda^2,$$

rising rapidly at short wavelengths. When the experiment is performed with very thin targets, so that most of the incident electrons either pass straight through, or radiate only one quantum, all radiation is from electrons of the same initial energy and conditions are optimal for comparison with theory. It is found that I_λ now varies as $1/\lambda^2$ right up to the quantum cut-off, and the classical model gives an extraordinarily accurate description of the mean spectral intensity at the lower frequencies which are not precluded by the necessity of emission in the form of quanta.

In this example the white noise spectrum results from a random sequence of more or less equal pulses, and it is often convenient in analysing the response of an instrument to a white noise input to assume that the input is such a sequence of pulses. Yet this is not strictly justifiable, for if we take the Fourier transform of a pulse sequence and arbitrarily alter the phases, it will certainly not transform back into a sequence of pulses but into a continuous random signal. There is, in fact, a certain residual order in the spectrum of the pulses and it may be that to ignore this is to falsify the analysis. To take a concrete example, consider the response of a lightly damped resonant system to a random pulse sequence (e.g. a pendulum struck at random intervals). At each pulse a damped oscillatory train is started and it is clear that the individual pulses reveal themselves as soon as their separation in time exceeds the decay time of the oscillations. On the other hand if the pulses arrive so frequently that the response at any moment is the combined effect of many, it will be difficult to discover in the fluctuations of the vibratory amplitude any feature that reveals the mean pulse rate. And, to go to extremes, if the pulse rate is faster even than the resonant frequency, so that many pulses arrive during one cycle, the precise character of the input noise certainly will not matter. To express the argument in terms of the Fourier transform of the input, let us replace a continuous noise signal, not consisting of separate pulses, by a pulse sequence constructed as follows: divide the signal into sections of random length, and integrate each section to yield the magnitude of the pulse to be placed at the centre of each section, replacing the original signal. Now take the Fourier transforms of the original signal and of its replacement. There will obviously be no difference for any component whose period takes in many pulses. It is clear, therefore, that in analysing the response of an instrument to white noise it is perfectly valid to imagine the noise as a pulse sequence with mean pulse rate higher than any frequency to which the instrument can respond. Nor will it matter exactly what detailed character is assigned to the pulse sequence. It might be thought that some difference would arise between identical pulses and pulses of random size and sign; at low frequencies, however, this is not so except, of course, at zero frequency where the identical pulses add to give a non-vanishing resultant. If one has any doubt about this, imagine the random pulses sorted into different sizes and treated as a large number of different random sequences of identical pulses. Since each sequence gives a low-frequency transform that is purely noisy, the sum of all will be no less, and can be no more, noisy.

Autocorrelation

Although we have defined I_ω in the context of noise spectra, for which it is particularly valuable, there is no objection to using it to describe spectral functions that are perfectly well phase-correlated, being Fourier transforms of smooth functions of time. Indeed, since many spectrometers and related instruments respond to intensity without regard to phase, the properties of I_ω deserve discussion in a wider context than noise. To start with a specific example, the Lorentzian spectrum described by (28) has a spectral intensity given by:

$$I_\omega = 2ff^* = 1/2\pi^2(\alpha^2 + \Delta^2).\qquad(46)$$

In using I_ω rather than f, all information about phases has been jettisoned, and (46) holds not only for the exponentially decaying oscillation of (27) but for innumerable other time sequences resulting from applying arbitrary phase changes to the components. This is irrelevant when the spectrum is uncorrelated in phase, being the output of a noise source or some other equivalent generator, so that I_ω contains all the information that one could wish to make use of. But one may still ask the question – what can one say about the time sequence of displacement when only I_ω is known? Clearly a complete specification of $g(t)$ is impossible, but there is still much of interest locked up in I_ω, notably the autocorrelation function which we now define and relate to I_ω.

If $g(t)$ is an arbitrary function, which may be complex, the autocorrelation function of $g(t)$ is defined as $\Gamma(\tau)$ where

$$\Gamma(\tau) = \overline{g(t)g^*(t-\tau)}/\overline{gg^*}$$

$$= \int g(t)g^*(t-\tau)\,dt \bigg/ \int g(t)g^*(t)\,dt, \tag{47}$$

the integral being taken over the whole range of t, or a whole period if $g(t)$ is a periodic function. The general significance of $\Gamma(\tau)$ is clear from its definition and its name – it measures the way in which two values of $g(t)$, separated by τ, are related to one another on the average. At one extreme, if $g(t)$ is strictly periodic (not necessarily sinusoidal), so that $g(t-n\tau_1) = g(t)$ for all integral values of n, $\Gamma(n\tau_1) = 1$; more than this, $\Gamma(\tau)$ is periodic in τ with period τ_1, since $g(t-\tau-n\tau_1) = g(t-\tau)$, and therefore $\Gamma(\tau + n\tau_1) = \Gamma(\tau)$. At the other extreme, if $g(t)$ is a white noise sequence, the values of g separated by an interval τ will be quite unrelated, however small τ may be, and $\Gamma(\tau) = 0$ unless $\tau = 0$. Other properties worth noting are the following:

(1) $\Gamma(0) = 1$.

(2) $\Gamma(-\tau) = \Gamma^*(\tau)$. If $g(t)$ and therefore $\Gamma(\tau)$ are real, $\Gamma(-\tau) = \Gamma(\tau)$.

(3) If $g(t) = a\,e^{-i\omega t}$, a being a complex constant,

$$\Gamma(\tau) = \int_0^{2\pi/\omega} e^{-i\omega t}\,e^{i\omega(t-\tau)}\,dt \bigg/ \int_0^{2\pi/\omega} dt = e^{-i\omega\tau}.$$

In this case the amplitude and phase of $g(t)$, as described by a, disappear from $\Gamma(\tau)$ which records only the periodicity. Already one can see why $\Gamma(\tau)$ may be useful when phase information is lacking.

(4) Following the notation of (16), define the Fourier transform

$$\gamma(\omega) = \frac{1}{2\pi}\int_{-\infty}^{\infty} \Gamma(\tau)\,e^{i\omega\tau}\,d\tau = \frac{1}{\pi}\int_0^{\infty} \Gamma(\tau)\cos\omega\tau\,d\omega \quad \text{if } g(t) \text{ is real.} \tag{48}$$

Then $\Gamma(\tau) = \int_{-\infty}^{\infty} \gamma(\omega)\,e^{-i\omega\tau}\,d\omega = 2\int_0^{\infty} \gamma(\omega)\cos\omega\tau\,d\omega \quad \text{if } g(t) \text{ is real.} \tag{49}$

(5) The *Wiener–Khinchin theorem*. Write $g(t)$ as a Fourier sum,

$$g(t) = \sum_i a_i\, e^{-i\omega_i t}.$$

Then, according to (47),

$$\Gamma(\tau) = \overline{\sum_i a_i\, e^{-i\omega_i t} \sum_j a_j^*\, e^{i\omega_j(t-\tau)}} \bigg/ \overline{\sum_i a_i\, e^{-i\omega_i t} \sum_j a_j^*\, e^{i\omega_j t}}.$$

When the sums in numerator and denominator are multiplied, all terms are periodic with frequency $\omega_i - \omega_j$ and vanish on averaging, unless $i = j$.

Hence $\quad \Gamma(\tau) = \sum_i a_i a_i^*\, e^{-i\omega_i \tau} \bigg/ \sum_i a_i a_i^* = \sum_i a_i a_i^* \cos \omega_i \tau \bigg/ \sum_i a_i a_i^*$

if $g(t)$ is real. As we proceed from a sum of discrete frequencies to a continuous spectrum, a given frequency range $d\omega$ contributes to the sum a value of $\Sigma_i a_i a_i^*$ which is simply $\tfrac{1}{2} I_\omega\, d\omega$, ω being restricted to positive values.

Hence $\quad\quad\quad \Gamma(\tau) = \int_0^\infty I_\omega \cos \omega\tau\, d\omega \bigg/ \int_0^\infty I_\omega\, d\omega,$ $\quad\quad\quad$ (50)

and, from (49),

$$\gamma(\omega) = \tfrac{1}{2} I_\omega \bigg/ \int_0^\infty I_\omega\, d\omega.$$

Apart from the normalizing coefficient, which ensures that $\Gamma(0) = 1$, $\gamma(\omega)$, the Fourier transform of $\Gamma(\tau)$, is just the spectral intensity function; conversely, as expressed in (50), $\Gamma(\tau)$ is the Fourier transform of $I(\omega)$. This is the Wiener–Khinchin theorem, which gives precise expression to the information contained in the spectral density when all information about phase is thrown away.

With a white noise signal, I_ω is constant throughout the frequency range and (50) shows, as already noted, that $\Gamma(\tau)$ vanishes for all values of τ except zero. If, however, the constant intensity spectrum is cut off at a certain limiting frequency ω_0:

$$I_\omega = I_0 \quad \text{when } \omega < \omega_0,$$
$$= 0 \quad \text{otherwise,}$$

then $\gamma(\omega) = 1/2\omega_0$ so long as $\omega < \omega_0$ and, from (50),

$$\Gamma(\tau) = \frac{1}{\omega_0} \int_0^{\omega_0} \cos \omega\tau\, d\omega = \frac{\sin \omega_0 \tau}{\omega_0 \tau} \quad \text{(cf. (13)).} \quad\quad (51)$$

The displacements are fairly strongly correlated over time intervals less than π/ω_0, but above that interval are hardly correlated at all – a result compatible, of course, with the Uncertainty Relation.

The top-hat form of $I(\omega)$ leading to the autocorrelation function (51) is perhaps met only rarely. More usual is a nearly random function of time whose autocorrelation decays exponentially,

$$\Gamma(\tau) = e^{-|\tau|/\tau_c}, \tag{52}$$

so that after an interval of τ_c (the *correlation time*) memory of earlier behaviour is largely lost. Then, from (48),

$$\gamma(\omega) = \frac{1}{\pi} \int_0^\infty e^{-\tau/\tau_c} \cos \omega\tau \, d\tau = \tau_c / \pi(1 + \omega^2\tau_c^2).$$

The intensity spectrum I_ω is proportional to $\gamma(\omega)$, but its magnitude is left undetermined by $\Gamma(\tau)$. We have therefore that

$$I_\omega = I_\omega(0)/(1 + \omega^2\tau_c^2), \quad I_\omega(0) \text{ being a constant.} \tag{53}$$

The intensity in the spectrum begins to fall off rapidly at frequencies above $1/\tau_c$, but the cut-off is not sharp as in the previous example. Both examples show that the absence of frequencies beyond ω_0 prevents the displacement changing arbitrarily quickly – not a surprising result, and indeed this analysis adds to our appreciation of what variations are possible with noise of a given spectral character very little that could not be guessed by taking thought. The autocorrelation function, however, has wide application as a quantitative tool in physical problems, and one or two examples will   **103, 233, 364** appear as we proceed.

5 Spectrum analysis

Computation of Fourier transforms[1]

Because of the importance of Fourier transforms in many branches of science, some far removed from vibrations as understood in this book, much effort has been devoted to the development of numerical procedures for converting an input time sequence into a spectrum of frequencies.† One in particular, the Fast Fourier Transform, is ingenious and subtle, but it is not our purpose to dwell on matters of mathematical or computational expertise. Apart from these the physical interest derives from the scale of the operations now possible, which were inconceivable to the manual computer. Very long sequences of numbers may be transformed so as to reveal periodicities, and instruments have been developed to use this facility to the full, with automatic data recording and analysis replacing the laborious personal efforts of earlier generations.

Certain points of interest arise from the character of the input which normally takes the form of a finite sequence of discrete numbers, for example the displacement at equal intervals of time, rather than a continuous function. The finite length of the input tape is of no consequence if the function to be transformed is periodic; one chooses a single period (or an integral number if there are small variations or experimental scatter and one desires to analyse the average waveform), and the Fourier series may then be computed by use of (4.10) directly. In terms of time and frequency, let the period be T_0, so that the fundamental frequency ω_0 is $2\pi/T_0$, and let the input data consist of N values of $F(t)$, a typical value being $F(nT_0/N)$ $(\equiv F_n)$ with n running from 0 to $N-1$. If $F(t)$ were continuous, we should write the amplitude of the mth harmonic (neglecting constant factors) according to (4.10):

$$a_m = \int_0^{2\pi/\omega_0} F(t)\,e^{im\omega_0 t}\,dt.$$

† We shall continue to use time (t) and frequency (ω) as the variables in the transforms, but it will be appreciated that the applications are not restricted to these. Thus a function $f(x)$ of posi- tion on a line may be analysed into a spectrum of wave-numbers $g(k)$ – here x is analogous to t, wave-number k to frequency ω, and wavelength λ to period T_0.

Since, however, only discrete values of $F(t)$ are known, the integral is replaced by a summation: The nth input value is F_n, occurring at a time nT_0/N, i.e. $2\pi n/N\omega_0$, and

$$a_m \propto \sum_{n=0}^{N-1} F_n e^{-im(2\pi n/N)}$$

$$= \sum_{n=0}^{N-1} \left[F_n \cos\left(mn\frac{2\pi}{N}\right) - iF_n \sin\left(mn\frac{2\pi}{N}\right)\right]. \tag{1}$$

This last form is appropriate to the computer's preference for real numbers.

Let us note what is implied by (1); $2\pi/N$ is the angular unit that results from dividing a circle into N equal parts, and any choice of the integers m and n denotes one of the set of N angles. A table of the cosines and sines of these angles, $2N$ numbers in all, forms the basic set of multipliers for the N data points F_n. The computation of the real part of a given a_m involves multiplying each value of F_n by the appropriate cosine multiplier and summing the lot. To compute another coefficient, m is changed to m' (say), and this means changing the order in which the multipliers are employed on successive F_n. Obviously the same process applies to the imaginary part of a_m, with the sine multipliers now brought into play. Each a_m therefore involves $2N$ processes of multiplication and a final addition, and given enough computing time one can build up a long sequence of a_m from a_0, the mean value of F, to very high harmonics.

There is a limit, however, set by the discrete nature of the input data. For if we write the form of (1) appropriate to a_{m+N}, the argument of the cosine and sine is $(m+N)2\pi n/N$, i.e. $2mn\pi/N + 2n\pi$. Since $2n\pi$ is irrelevant in the argument, the multipliers in a_{m+N} occur in the same order as in a_m, and a_m is therefore a periodic function of m, repeating itself exactly after N values. Everything that can be said about the Fourier transform of F, as specified by N discrete data points, is contained in the amplitude of N harmonics. The reason is easily demonstrated by a simple example (fig. 1).

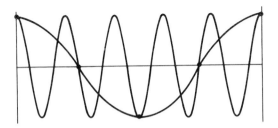

Fig. 1. The displacements represented by the black dots can be interpreted as a sinusoid in any number of ways, of which two are shown.

The points represent a possible set of input values for the case $N=4$, which are equally well fitted, if these points are all that is known, by the line of long wavelength, $\cos(n\pi/2)$, or by the line of short wavelength, cos $(5n\pi/2)$, the arguments of which differ by $2n\pi$. There is no way of telling, without intermediate data points, how irregular the true function F may be, and therefore no criterion for preferring one curve over the other (or for that matter cos $(9n\pi/2)$, cos $(13n\pi/2)$ etc.), except that the longest wavelength performs its limited task with the least fuss. Having recognized, then, that we should make a choice of not more than N Fourier components to describe the N data points, it is sensible to keep m as low as possible, letting it run between $\pm\frac{1}{2}N$. With real input, $a_{-m} = a_m^*$, so that $\frac{1}{2}N$ complex coefficients are all that can be usefully extracted from the data. The highest harmonic that can be determined in a waveform described by N values per cycle is the $\frac{1}{2}N$th, having a period equal to twice the interval between data points. Each harmonic needs both its amplitude and its phase to specify it completely, so that $\frac{1}{2}N$ harmonics need N numbers and use up the whole input material.

Another point of practical importance is the computer time involved. Each Fourier coefficient requires $2N$ multiplications and, with $\frac{1}{2}N$ coefficients accessible, the total number of multiplications needed to extract everything available from the data is N^2. A computer that takes less than a microsecond to perform each multiplication can processs thousands of data points in a minute, and tens of thousands are not ruled out, so that very considerable sequences of observations can be analysed into a correspondingly detailed spectral distribution, provided the computer can store and retrieve rapidly all the input data. With numbers as large as this, however, a more efficient program may bring great rewards, and the Fast Fourier Transform is such a one, requiring only about $N \ln N$ operations to process N data points. The saving in computer time, by something like $\ln N/N$, is a substantial gain – a factor of 100 when $N = 1000$.

These large numbers do not normally arise in the recording and analysis of periodic phenomena, where 1000 Fourier coefficients would provide a much better specification of the waveform than is generally required. It is another matter when a long sequence (e.g. a century of weather records) is to be examined for hidden periodicities, sometimes in very great numbers as in the output of the Michelson interferometer used as a spectrometer in the manner to be described presently. In such circumstances one must recognize that the practical considerations enforcing truncation of what should have been an endless sequence necessarily introduce distortions. A simple procedure is to treat the N data points as one cycle of a periodic variation, and extract N Fourier coefficients on this assumption. But any strong component with a period that does not fit into the cycle length will be altered from a continuous sinusoid to one which suffers a discontinuous phase change every N points. The Fourier transform of such a dislocated sinusoid has satellite components around its principal periodicity, and in interpreting the transform care must be taken not to attach significance to what may be a computational artefact. This is a problem that has received much attention, but beyond noting its existence there is no need here to enter into the matter more deeply.

The ear and the eye; optical spectrometers

There are many processes which allow the recording and analysis (other than by computer alone) of a time sequence – a gramophone disc, for example, or a magnetic tape, both of which convert the electrical signal from a microphone into permanent, retrievable forms; or the signal may be sampled at regular intervals and stored digitally.† We may, if we wish, transfer the recording of a symphony to punched tape; if we are satisfied to reject all frequencies above 10^4 Hz we need to record the amplitude of the signal at intervals of about 10^{-4} s, so that a Haydn symphony will demand over 10^7 entries in the form of numbers of, say, 10 binary digits each ($2^{10} = 1024$, and this number of steps in registering the amplitude at any instant is not excessive if very loud and very soft passages are to be adequately covered). Given this representation of the music we feed it into a computer and extract the Fourier transform, whose highest frequency components tell us more about the violins and flutes than about the double basses and bassoons, while the lowest frequencies (< 20 Hz) are musically irrelevant since they are not perceived by the ear. There will be in the transform, one hopes, a marked periodicity in the amplitude as a function of frequency, with strong preference for those frequencies that correspond to the notes of the scale. Further, to select one note at random, in the vicinity of that frequency (440 Hz) that corresponds to a, the amplitude will vary in a complicated and highly significant fashion, for it is the beating of these close frequencies that determines the precise loudness of this note at all times. For example, if different movements are in different keys a given note may occur more frequently in one than in the others, a fact that is registered in the spectrum by the presence with roughly equal amplitude of two components perhaps only 10^{-3} Hz apart; it will be appreciated that a change in phase of one of these components will bring in this particular note during the wrong movement. Similarly, the most obvious feature of music, pulse or (in a subtler form) rhythm, shows in the Fourier transform as phase correlations between components differing by something around 1 Hz – very hard to pick out from the spectrum yet immediately apparent to even the tone-deaf.

Clearly the Fourier point of view is irrelevant to the music *qua* music. We do not take in half-an-hour's performance and then at leisure appreciate the complexity of its Fourier transform. The ear is a remarkable instrument in its ability to discriminate two notes of nearly the same frequency, but the degree of discrimination is, at its best, a few tenths of 1 Hz, not the 10^{-3} Hz mentioned above. In return for this we are able to appreciate sudden changes in the sound and thus listen to music (and everything else) as a sequence of events in time. The ear, in fact, occupies an intermediate position between the tape recording or disc which responds to the immediate sound pressure and a sound spectrograph

† This rather sophisticated process was used to record the sounds, and draw the diagrams, for figs. 7 and 8.

designed to sort out the sinusoidal components. The recorder ideally takes no account of the frequency spectrum but discriminates one instant from the next – the spectrograph discriminates one frequency from another but does not indicate the instant when a given frequency is present. Indeed, it cannot, for a sinusoidal train of finite length Δt is composed of frequencies in a range $1/\Delta t$ and to obtain fine discrimination in frequency involves analysing a correspondingly long sequence of oscillations. The ear, delighting in its sensitivity to sudden noises such as the percussive attack of each note played on a piano, must pay by a limited frequency discrimination.

Recognizing this, one must be careful not to over-emphasize the spectrum analysis aspect in treating particular cases. As already remarked, the most obvious frequency in a musical performance is the pulse rate – something that would not appear in a Fourier analysis of the time sequence of displacement (e.g. the voltage signal from a microphone) but would show if the instantaneous acoustic power level was recorded, with no attempt at frequency discrimination. Or, to take another example, when a nerve transmits information as a series of impulses, it is the frequency of arrival of impulses during a short interval that indicates the intensity of the signal. The concept of instantaneous frequency is of more value than that frequency we have labelled ω and derived from analysis of a long record. Instantaneous frequency may be defined in relation to the time interval between successive recognizable markers on the input signal – the short electrical impulse in a nerve-ending, or the momentary rise in acoustic power level every time the big drum is struck. Even with sinusoidal oscillations, or any others of well-defined waveform, instantaneous frequency may be defined through the interval between zeros of displacement; indeed many electronic frequency meters work on this principle. A frequency-modulated carrier wave is described in these terms by a graph of instantaneous frequency against time, and this graph is also, in normal f.m. communication, the variation of acoustic displacement with time that is to be extracted from the carrier and fed to the loudspeaker. It has therefore a much more direct significance than a Fourier analysis of the modulated carrier. But of course the concept of instantaneous frequency breaks down when the signal becomes complicated, without recognizable markers. With music or speech, therefore, neither Fourier analysis nor instantaneous frequency gives more than a faint inkling of what meaning is there to be appreciated, and to make progress it is necessary to devise descriptive techniques which are related to the peculiar abilities of the ear – rather good frequency discrimination, rather good time discrimination, and a pattern-recognition capacity in the brain that has so far eluded understanding.

Compared with the ear, the eye is a poor analyser of spectra. It accepts electromagnetic oscillations at frequencies between 4.5 and 8×10^{14} Hz, and only a very sensitive observer could distinguish 1% variations with assurance. Thus we might set $\Delta\omega \sim 3 \times 10^{13}$, while the time discrimination

68

of the eye is perhaps 10^{-2} s, as judged by the rather weak flicker due to mains lighting. The product $\Delta\omega\,\Delta t \sim 3\times10^{11}$, enormously larger than the theoretical lower limit of unity or thereabouts. There are compensations – the nervous system associated with the eyes has the far more important task of analysing directional variations and interpreting them as shapes, a faculty in which the ear is extremely rudimentary.

In optical measurements, as was remarked in chapter 4, limitations to simultaneous discrimination in time and frequency by conventional instruments have until recently made consideration of the Uncertainty Principle irrelevant. The laser has changed this, but here we confine attention to the essential problem of normal spectroscopy – to analyse what is effectively a continuous source of phase-incoherent radiation, the result of many molecular systems emitting independently. The information extracted has been all in the frequency (or wavelength) domain – the frequencies of the various spectral lines and their profiles, i.e. the frequency spread around each discrete line, with hardly a thought for what goes on in any one emitter. The instruments have the property of accepting a continuous time sequence and converting it into its power spectrum, I_ω, or something from which I_ω may be derived without difficulty.

Two instruments deserve discussion as a start, because of the contrast they present in their principles of operation – the diffraction grating, and the Michelson interferometer used as a Fourier transform spectrometer.[2] The grating sorts into a multitude of channels the Fourier components of the input; the Michelson interferometer forms the autocorrelation coefficient which is then subjected to computer processing to yield the power spectrum. On the assumption that the general principles of design and use of these instruments is sufficiently well understood we may proceed to contrast the processing to which they subject an input signal in the form of a time sequence, $F(t)$, representing, for example, the variation of electric field strength in the electromagnetic wave. Each line of the grating scatters the incident wave into a wide fan of angles, and the telescope, set at a particular angle θ to the incident wave direction, collects and superimposes the scattered waves, but with a time lag τ between the waves from successive grating lines (when light is incident normally on a grating with spacing d, the extra distance travelled by successive waves is $d\sin\theta$, so that $\tau = d\sin\theta/c$). The output of the telescope is therefore $\sum_{n=0}^{N-1} F(t+n\tau)$ if the grating has N lines. This is a rapidly varying function if $F(t)$ is made up of components at optical frequencies, and the eye (or a bolometer, or some other detector) only records the mean square value of F, or possibly some other average. If N is large only those Fourier coefficients for which τ is an integral multiple of the period of vibration will survive the process of summation, so that in each direction of viewing, a particular frequency and its harmonics emerge in almost pure form. The degree of purity is determined by N, being proportionately better as N is made larger. This is readily seen by imagining a single pulse falling on the grating, to emerge in the direction θ as an evenly spaced succession of N pulses with an interval

90

τ between each. Now an endless succession of pulses has a spectrum consisting of perfectly sharp frequencies which are multiples of the fundamental $2\pi/\tau$; a finite succession, containing nN cycles of the nth harmonic, is represented by a slightly broadened band of frequencies; with a $\sin\phi/\phi$ type of amplitude distribution round the central maximum, and a width between zeros given by $\Delta\omega = 4\pi/N\tau$, as follows from (4.21). The relative spread, $\Delta\omega/\omega$, is thus $2/Nn$ and very small when $N \sim 10^4$–10^5.

The quantity Nn is taken as a measure of the resolving power of the grating, and structure in the power spectrum closer in frequency than ω/Nn is not revealed by the grating. This may be thought of as a result of the smearing process which turns every sharp line into a $\sin\phi/\phi$ profile or, what comes to the same thing, we may prefer to explain it in terms of the autocorrelation coefficient of $F(t)$. If $F(t)$ has any tendency to repeat itself at time intervals greater than $N\tau$, nothing of this will show in the observed spectrum since the grating cannot superimpose two portions of $F(t)$ separated by so long an interval. But these long-term correlations arise from structure in the power spectrum involving frequency differences less than something like $2\pi/Nn$, and this leads to the same criterion for the resolution of fine structure in the spectrum.

Whereas the grating superimposes N replicas of $F(t)$ with progressive time delays, the Michelson interferometer takes only two replicas, by reflecting the input wave from two mirrors and combining the output in a detector (which we take to have a square-law characteristic). The time delay, τ, between the replicas is varied by moving one of the mirrors parallel to itself by a high-precision slide and lead-screw. If we designate the output of the detector by $S(\tau)$,

$$S(\tau) = \overline{[F(t)+F(t-\tau)]^2}$$
$$= 2\overline{F^2} + 2\overline{F(t)F(t-\tau)}$$
$$= 2\overline{F^2}[1+\Gamma(\tau)].$$

This shows that if the mirror is steadily racked so as to vary τ at a uniform rate, the output of the detector is a record, apart from a constant, of the autocorrelation function of the input. The Fourier transform, carried out automatically from the punched-tape record of $S(\tau)$, therefore displays the power spectrum of the source directly. The example shown in fig. 2 illustrates the power of the method in the sub-millimetre wavelength range where the extreme weakness of available sources, even the Sun, virtually precludes conventional spectroscopy. The resolution here is not especially high – only the first 2 cm of path difference produces significant variations of S; but in principle, especially at shorter wavelengths, very high resolution is attainable. If τ can be varied between 0 and some maximum value τ_0, $\Gamma(\tau)$ is known in this range and, by the same argument as in the last paragraph, fine structure involving frequency differences of the order of $2\pi/\tau_0$ is resolvable. The relative frequency difference resolvable is $2\pi/\omega\tau_0$, which is the reciprocal of the number of wavelengths in the path difference

Fig. 2. Fourier transform spectroscopy in the extreme infrared (sub-millimetre wavelengths). The input to the spectrometer was sunlight as received on a high mountain. with everything except the very long wavelengths filtered out. The variation of detector output with path difference is shown in (*a*) and the spectrum derived by Fourier transformation in (*b*). The conventionally defined 'frequency' means the reciprocal of the wavelength, and the record covers the wavelength range from about 3 mm at the left to 0.3 mm at the right. Some of the absorption lines due to oxygen and water vapour in the atmosphere are labelled. Most of the other discernible features are real, but many are not yet identified. (Diagram provided by Prof. D. H. Martin and Mr E. Puplett.)

introduced by the mirror movement. We may take this number of wavelengths as a measure of the resolving power, and make immediate comparison with the corresponding quantity for the grating, which is the path difference in wavelengths between the wave trains from opposite ends; thus something like the width of the grating is to be compared to twice the mirror movement, and the latter may be tens of centimetres. It requires high skill to make a slide for the mirror that is optically true over tens of centimetres, but this is nothing to what would be involved in ruling and handling a high-quality grating as large as this.

A movement of 10 cm introduces a path difference of 20 cm, about 5×10^5 wavelengths of visible light. If N values of the detector output are recorded at equal intervals during the traverse, the Fourier transform of $\Gamma(\tau)$ has as its maximum frequency the $\frac{1}{2}N$th harmonic of an oscillation of period τ_0, i.e. $\omega_{max} = \pi N/\tau_0$; in other words, to use the instrument up to a certain frequency, the number of data points must be twice the resolving power at that frequency, say 10^6 in the example quoted. If the computer available cannot handle this quantity of information, $\omega_{max}\tau_0$ must be reduced, either by moving the mirror the same distance and recording the output at longer intervals, or by recording at the same intervals while reducing the movement. In the first case the high-frequency end of the spectrum is abandoned, in the second the resolution of fine structure is coarsened.

Linear transducers; response functions

In the acoustic field the classical instrument for spectrum analysis was the Helmholtz resonator described earlier. The air in the cavity is set into resonance if the incident sound contains a Fourier component at or near its characteristic frequency, and the ear hears this note strongly amplified, while other notes are rejected. Nowadays it is easier to pick up the sound with a microphone, converting the pressure variations into voltage variations, and either to subject the record to computer analysis, or to feed the signal through a number of resonant circuits in parallel, recording the output of each separately so as to have a continuous record of the strength of a number of components in the incident sound. The Helmholtz resonator and the resonant circuit are examples of *linear transducers*, by which is meant a device which performs a linear transformation on an input signal and feeds out the result. We have already discussed what constitutes a linear transformation with time-independent parameters. It may involve multiplication by a constant (linear amplifiers and attenuators have this property), differentiation (resistance–capacitance networks), time delay (by generating a wave at the input of a transmission line and picking it up when it reaches the output), and other transformations or mixtures of all these. The restriction to linearity in the transducer ensures that superposition holds – two signals added and fed into the transducer emerge as the sum of what each would have been converted into if it alone had been injected. Conversely, a complicated input may be imagined to be resolved into its Fourier components, each of which is transformed independently; and because of time-independence a sinusoidal input emerges as a sinusoid of the same frequency. It is therefore possible to specify the character of the transducer not by the processes or combinations of processes (multiplication, delay etc.) it performs, but by what it does to each Fourier coefficient. A typical sinusoid entering as $A\,e^{-i\omega t}$, ω being real, emerges with its amplitude changed by a multiplier characteristic of the frequency and having suffered a phase change. That is to say, $A\,e^{-i\omega t}$ is converted to $\chi(\omega)A\,e^{-i\omega t}$, where $\chi(\omega)$, the *transfer function*, is a complex function of frequency. Clearly since every input time sequence can be resolved into its Fourier coefficients, specification of $\chi(\omega)$ fully determines the action of the transducer:

If the input $\qquad g_0(t) \quad$ is $\quad \displaystyle\int_{-\infty}^{\infty} f(\omega)\,e^{-i\omega t}\,d\omega,$

the output $\qquad g_1(t) \quad$ is $\quad \displaystyle\int_{-\infty}^{\infty} f(\omega)\chi(\omega)\,e^{-i\omega t}\,d\omega_0.$

Just as the requirement that the input be real implies that $f(-\omega)=f^*(\omega)$, so for the output to be real $\chi(-\omega)f(-\omega)=\chi^*(\omega)f^*(\omega)$, and therefore

$$\chi(-\omega)=\chi^*(\omega). \tag{2}$$

An important particular case arises when the input $g_0(t)$ is a single impulse P at time $t=0$; then $f(\omega)$ is equal to $P/2\pi$, independent of frequency, and

74

$$g_1(t)=\frac{P}{2\pi}\int_{-\infty}^{\infty}\chi(\omega)\,e^{-i\omega t}\,d\omega.$$

The *impulse response function* $h(t)$ of the transducer, which is the output signal $g_1(t)$ resulting from an input consisting of a unit impulse ($P=1$) at $t=0$, is $1/2\pi$ times the Fourier transform of the transfer function:

$$h(t)=\frac{1}{2\pi}\int_{-\infty}^{\infty}\chi(\omega)\,e^{-i\omega t}\,d\omega. \tag{3}$$

Conversely, by (4.16),

72

$$\chi(\omega)=\int_{-\infty}^{\infty}h(t)\,e^{i\omega t}\,dt. \tag{4}$$

This may be understood as a consequence of superposition if one calculates the response of the transducer to a sinusoidal input $e^{-i\omega t}$ by dissecting the input into a succession of impulses, each of which elicits the appropriate response. Thus in the time interval between t_0 and t_0+dt_0 the input is equivalent to an impulse of magnitude $e^{-i\omega t_0}\,dt_0$, and on integrating over all impulses comprising the sinusoidal input we have that the output is given by

$$g_1(t)=\int_{-\infty}^{\infty}h(t-t_0)\,e^{-i\omega t_0}\,dt_0.$$

When z is written for $t-t_0$, this is converted into the form

$$g_1(t)=e^{-i\omega t}\int_{-\infty}^{\infty}h(z)\,e^{i\omega z}\,dz.$$

But with a sinusoidal input, $g_1(t)=\chi(\omega)\,e^{-i\omega t}$, from which (4) follows directly.

It will be appreciated that since $\chi(\omega)$ and $h(t)$ are uniquely related, the impulse response function is just as good a specification of the properties of the transducer as is the transfer function, and indeed is often more convenient. If one wishes to know the output when the input, $g_0(t)$, is of arbitrary form, the argument just applied, involving breaking it down into a succession of impulses, yields the result

$$g_1(t)=\int_{-\infty}^{\infty}g_0(t_0)h(t-t_0)\,dt_0, \tag{5}$$

and it may be easier to evaluate the integral directly than to find the spectrum of $g_0(t_0)$, multiply by $\chi(\omega)$ and take the Fourier transform of the resultant. The result (5) may be recognized as the convolution of g_0 with h, and application of the convolution theorem provides an alternative method for deriving the other results.

A specific example of an impulse response function is the response of a simple harmonic oscillator to an impulse – a pendulum, for instance, at rest and struck sharply at $t = 0$. The resulting damped vibration is described by (2.28) which, it must be remembered, applies only to positive values of t;

32

$$h(t) = h_0 \sin \omega' t \cdot e^{-t/\tau_a} \qquad (t > 0). \tag{6}$$

Hence, from (4),

$$\chi(\omega) = \int_0^\infty h_0 \sin \omega' t \cdot e^{-t/\tau_a + i\omega t} \, dt$$

$$= \frac{h_0}{2i} \int_0^\infty [e^{-(1/\tau_a - i\omega' - i\omega)t} - e^{-(1/\tau_a + i\omega' - i\omega)t}] \, dt$$

$$= \omega' h_0 / [(\omega'^2 - \omega^2 + 1/\tau_a^2) - 2i\omega/\tau_a]. \tag{7}$$

76 This gives the amplitude of steady oscillation set up by a periodic force, and has the characteristic resonant quality if $\omega' \tau_a \gg 1$. When ω is near ω' the denominator is small and the response correspondingly large, with a maximum $\sim \frac{1}{2} i h_0 \tau_a$ at resonance, when $\omega^2 = \omega'^2 + 1/\tau_a^2$; as τ_a is increased the maximum becomes larger and the frequency range within which the response is comparable to this maximum becomes narrower. This behaviour will be studied from a number of viewpoints in the next chapter, and at this stage it is enough to note that a reason for the ubiquity of resonant response obeying (7) is to be found in the simplicity of the impulse response in the form of a damped sinusoidal oscillation – so many systems behave in this way when struck, and all must of necessity show precisely the same resonant character. This is, of course, only another way of saying that simple linear systems with one natural frequency have very little choice of behaviour.

Causality

A point of considerable interest arises from the relation (3) between the transfer function $\chi(\omega)$ and the impulse response function $h(t)$. If we were to choose some arbitrary form of $\chi(\omega)$, satisfying only the reality condition (2), and evaluate $h(t)$, we should expect in general to arrive at a function occupying the whole range of positive and negative t. This, however, is not physically sensible – a passive transducer does not anticipate the arrival of an impulse, and any realistic $h(t)$ must vanish identically for all negative values of t; the pendulum just discussed is a typical example. There must therefore be certain restrictions on the form of $\chi(\omega)$ that any real transducer can generate. These restrictions are not very severe, for although we demand that $h(t)$ shall vanish in the negative range of t we have no physical arguments suggesting that its behaviour shall be in any way restricted in the positive range,† and a wide variety of functions must therefore still be allowable for $\chi(\omega)$.

Perhaps the easiest way to approach the matter is to ask how a transducer will respond to an oscillatory input that increases or decreases exponentially with time and is described by $A\,e^{-i\omega t}$, with ω complex. There is no problem in extending the definition of $\chi(\omega)$ to include complex frequencies, and we therefore express the output as $\chi(\omega)A\,e^{-i\omega t}$, having the same exponential variation of amplitude with time as the input. Now consider the case when the imaginary part of ω is positive and the input increases with time. At any instant the output may be regarded as the resultant of the impulse responses due to all previous values of the input. Since at earlier times the input was smaller, with an exponential decrease towards the past, and since the impulse response is finite, there are no circumstances in which the combined impulse responses can ever diverge; $\chi(\omega)$ is therefore finite at all values of ω lying in the upper half-plane in which $\text{Im}\,[\omega] > 0$. On the other hand, if $\text{Im}\,[\omega] < 0$ the response of the transducer at any instant is the resultant of a succession of impulse responses generated by an input which was exponentially larger the further back we look into the past. If the impulse response does not die away fast enough there is a very real likelihood that the largest contributions to the resultant originate in those impulses which came earliest; and when we come to form the resultant by integrating backwards towards $-\infty$ we can have no assurance that the integral will converge. In fact it is a very limited class of transducers that lack singularities of $\chi(\omega)$ in the lower half-plane, that is, localized values of ω at which χ becomes infinite. For there is a general theorem (Liouville's theorem) in complex variable theory which shows that the only function that is regular (i.e. continuous and differentiable) everywhere on the complex plane is a constant.[3] A transducer for which χ is constant produces an output that is exactly proportional to the input, so that an impulsive input yields an instantaneous impulsive output, $h(t) = \alpha\delta(t)$. Only the simplest and most idealized transducers such as networks of perfect resistors have this property, and one may therefore expect all real transducers to exhibit singularities in the lower half-plane, but none in the upper half-plane.

We are not, however, concerned at this stage to investigate the pathology of $\chi(\omega)$, since it is its regularity in the upper half-plane that is the characteristic expression of what is conventionally called *causality* – the fact that the effects never precede the impulse which causes them. The argument of the last paragraph showed that the regularity of χ in the upper half-plane was a necessary consequence of causality; and it is readily seen

---

† Except that $h(t)$ should be finite at all times; the analysis does not apply to an unstable transducer which, when excited by however small an impulse, goes out of control and develops an output that increases without limit.

that the converse is true, that if χ is thus regular causality is satisfied. For just as an impulse response function which is not a δ-function, vanishing for all $t > 0$, must generate at least one singularity of χ in the lower half-plane, so one which does not vanish for all $t < 0$ must generate at least one in the upper half-plane. The absence of singularities in the upper half-plane therefore guarantees the vanishing of the impulse response function at all negative times.

The restrictions placed by causality on the real-frequency response of a linear transducer may therefore be expressed in terms of the possible behaviour of $\chi(\omega)$ along the real axis that ensures regularity of χ in the upper half-plane. We shall assume that χ is finite for all real ω and that it tends to zero as $\omega \to \infty$. The first assumption is not necessarily true when $\omega = 0$ (in certain circuits incorporating superconductors, for example) or at a non-zero real value of ω if the transducer incorporates a strictly lossless resonant element. These are such special cases, however, as should not be allowed to lead us into more delicate analysis than is required by the vast majority of applications. As for the second assumption, concerning the behaviour as $\omega \to \infty$, it is something to be borne in mind when applying the theorem we are about to derive, for if it is not satisfied the integrals involved may diverge and make nonsense of the theorem; but again, the exceptional cases are rare and normally arise as artefacts when the real physical system is replaced by a simplified model whose limiting behaviour has not been considered with sufficient care.

If $\chi(\omega)$ is regular throughout the upper half-plane, so also is $\chi(\omega)/(\omega - \omega_0)$, ω_0 being a real constant, except for the simple pole at ω_0 where this function becomes infinite. An integral round the contour C shown in fig. 3, which incorporates an infinitesimal semicircle to avoid the pole, must therefore vanish, according to Cauchy's theorem:

$$\oint_C \frac{\chi(\omega)}{\omega - \omega_0}\, d\omega = 0.$$

Fig. 3. Contour for integration in the complex plane.

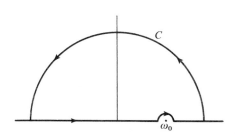

If χ/ω falls to zero more sharply than ω^{-1}, as assumed, the large semicircle may be extended to infinite radius and there make no contribution to the contour integral. The contribution from the real axis may be dissected into that due to the semicircle at ω_0 and the rest, $\int_{-\infty}^{\infty} [\chi(\omega)/(\omega - \omega_0)]\, d\omega$, the stroke through the integral sign indicating that the principal part is to be taken (integrate from $-\infty$ to $\omega_0 - \varepsilon$, and from $\omega_0 + \varepsilon$ to ∞, and then let $\varepsilon \to 0$). Then

$$\int_{-\infty}^{\infty} \frac{\chi(\omega)}{\omega - \omega_0} \, d\omega + \int_{\text{semicircle}} \frac{\chi(\omega)}{\omega - \omega_0} \, d\omega = 0.$$

The integral round the semicircle is carried out with $\chi(\omega)$ constant at its value $\chi(\omega_0)$; putting $\omega = \omega_0 + r\, e^{i\theta}$, r being constant, and integrating θ from π to 0, the result is obtained:

$$\int_{-\infty}^{\infty} \frac{\chi(\omega)}{\omega - \omega_0} \, d\omega = i\pi \chi(\omega_0).$$

Finally, expressing $\chi(\omega)$ in terms of its real and imaginary parts, $\chi = \chi' + i\chi''$, and separating real and imaginary terms in the equation, we find that

$$\chi''(\omega_0) = -\frac{1}{\pi} \int_{-\infty}^{\infty} \frac{\chi'(\omega) \, d\omega}{\omega - \omega_0}$$

and

$$\chi'(\omega_0) = \frac{1}{\pi} \int_{-\infty}^{\infty} \frac{\chi''(\omega) \, d\omega}{\omega - \omega_0}.$$

$$\text{(8)}$$

A perhaps more convenient form, involving integration only over the positive frequency range, results from applying the reality condition (2) in the form $\chi'(-\omega) = \chi'(\omega)$ and $\chi''(-\omega) = -\chi''(\omega)$. Then

$$\chi''(\omega_0) = -\frac{2\omega_0}{\pi} \int_0^{\infty} \frac{\chi'(\omega) \, d\omega}{\omega^2 - \omega_0^2}$$

and

$$\chi'(\omega_0) = \frac{2}{\pi} \int_0^{\infty} \frac{\omega \chi''(\omega) \, d\omega}{\omega^2 - \omega_0^2}.$$

$$\text{(9)}$$

These are the Kramers–Kronig relations, which show that knowledge of the real part of $\chi(\omega)$ for all real frequencies is sufficient to define the imaginary part completely; and vice versa.[4] The two relations are not independent, but alternative forms of the same restriction on χ imposed by causality. This is readily seen by observing that if $\chi(=\chi' + i\chi'')$ is regular in the upper half-plane, so also is $i\chi(= -\chi'' + i\chi')$. If then the first equation of (8) is true, so also is that obtained by replacing χ'' on the left by χ' and χ' on the right by $-\chi''$; and this yields the second equation of (8). It may be asked whether (8) represents a complete description of the limitations imposed by causality, or whether a stronger statement is possible. The answer is provided by the fact that for any reasonable choice of χ' on the real axis (by 'reasonable' is meant continuous and differentiable, and such as to allow the Kramers–Kronig integrals to converge) there exists one function and one only, extending over the whole complex plane, with the properties of having a real part on the real axis equal to χ' and of being regular in the upper half-plane. This means that causality places no restrictions on the form of χ', but allows only one form of χ'' for a given χ', i.e.

364

that given by (8), and this is as strong a statement as we can make. There is something rather satisfying about the way in which, having specified completely one half of the impulse response function (by making it vanish when $t < 0$), we find we have halved our freedom of choice of $\chi(\omega)$, being allowed to specify either the real or the imaginary part[†] but not both.

As an illustration of the Kramers–Kronig relations consider the resonant response (7) whose essential form may be expressed, by redefining the variables,

$$\chi(\omega) = 1/[(1 - \omega^2) - i\beta\omega], \text{ where } \beta \text{ is a positive constant; } \beta \ll 1. \quad (10)$$

Then

$$\chi'(\omega) = \frac{1 - \omega^2}{(1 - \omega^2)^2 + \beta^2 \omega^2}, \qquad \chi''(\omega) = \frac{\beta\omega}{(1 - \omega^2)^2 + \beta^2 \omega^2}. \quad (11)$$

These two functions are shown in fig. 4. First we note that (10) satisfies the basic formulation of the causality principle, in that the poles of $\chi(\omega)$, at which the denominator vanishes, occur when $\omega = \frac{1}{2}[(4 - \beta^2)^{\frac{1}{2}} - i\beta]$, and are both in the lower half-plane. Unfortunately the formal verification of (8) involves a tedious integration, but the qualitative validity of the equations may be understood by drawing, which probably gives a better feel for what is going on. It is convenient in evaluating the second equation of (8) to note that $\int_{-\infty}^{\infty} d\omega/(\omega - \omega_0) = 0$, so that a constant added to $\chi''(\omega)$ in the numerator makes no difference. We therefore write

$$\chi'(\omega_0) = \frac{1}{\pi} \int_{-\infty}^{\infty} \frac{\chi''(\omega) - \chi''(\omega_0)}{\omega - \omega_0} \, d\omega;$$

the requirement of taking the principal value disappears, since the integrand remains finite even when the denominator vanishes. The shift of horizontal axis to pass through the curve of $\chi''(\omega)$ when $\omega = \omega_0$, which is all that this device amounts to, is shown in fig. 4(b). It is clear that just above ω_0 both numerator and denominator are positive, while just below both are negative; either way the integrand is positive throughout the range in which it is largest, and $\chi'(\omega)$ is correspondingly positive. On the other hand, if we were to choose ω_0 above the peak of χ'' the integrand would be predominantly negative. The general form of fig. 4(a) emerges clearly, and of course is substantiated precisely by carrying through the integration.

After this discussion of the Kramers–Kronig relations it must be confessed that they seem to have rather limited utility. There are circumstances in which measurements of either χ' or χ'' prove very troublesome and where information on one may be extended to the other by computer integration of (9). But the integrals are only very slowly convergent, and consequently the information on, say, χ' must extend over a very wide frequency range; moreover experimental errors in χ' can produce anomalies

[†] Or a suitable mixture of both, but we shall not try to delineate the latitude permitted in such circumstances.

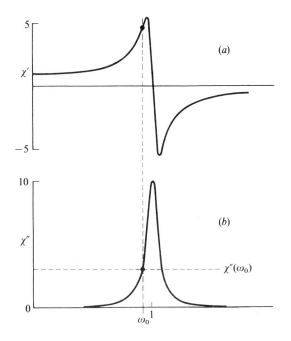

Fig. 4. Real (χ') and imaginary (χ'') parts of the response function of a resonant transducer, to illustrate Kramers–Kronig relations.

in χ'' which tempt the user to infer interesting physical behaviour where nothing special really exists. If both χ' and χ'' can be measured, but with limited accuracy (as for instance in certain studies of the optical properties of solids),[5] the relations may help in testing whether the results are compatible, and so provide a valuable critique of the experiment. But the case for presenting this theory here ultimately rests more on its very existence than on its utility; the qualitative idea of causality, involving a temporal ordering of causes and effects, is so fundamental to the physical universe that one should always try to discover its quantitative corollaries. In their way the Kramers–Kronig relations express a general truth that is independent of specific models, a distinction they share with thermodynamics, and deserve respect even if, unlike thermodynamics, they contribute little of practical value.

One special case is, however, worth noting briefly. If $\omega_0 = 0$ in (8), the second equation relates the real part of χ at zero frequency to the logarithmic integral of the imaginary part:

$$\chi'(0) = \frac{1}{\pi} \int_{-\infty}^{\infty} \chi''(\omega)\, d\omega/\omega = \frac{1}{\pi} \int_{-\infty}^{\infty} \chi''(\omega)\, d(\ln \omega).$$

In a resonant system $\chi''(\omega)$ is confined to a fairly narrow range of ω, and the integral can be computed from experimental measurements of loss. A complex system may have many loss peaks but, provided they are far enough apart in frequency to be resolved, each may be assigned its independent contribution to the zero frequency susceptibility. This idea has been put to good use in determining the contribution of electron spin paramagnetism (*Pauli paramagnetism*) to the magnetic susceptibility of a metal. There are many different contributions from various sources in a typical metal, which cannot be separated solely by measurement of the static susceptibility. This was the cause of much regret for many years since the Pauli contribution is potentially of great importance in testing the theory of electron–electron interaction in metals; if the electrons did not interact at all among themselves, the magnitude would be uniquely related to the thermal capacity of the electrons, a directly measurable quantity, and departures from this relation would give a quantitative estimate of the interaction strength. Now in a magnetic field of one or two tesla the electron spins should exhibit an absorption line in the microwave frequency range, and the logarithmic integral of this absorption gives the Pauli paramagnetism independently of any other mechanism. The line, though sharp in many insulators, is so broad in most metals as to disappear in the background contributed by other loss mechanisms, but in sodium it stands out strongly enough to be measured. It is fortunate, since the absorption line does not have the standard resonance profile, that the Kramers–Kronig theory is independent of any special model and can be applied even if the fine details of the line shape are imperfectly understood. The outcome of some years' devotion to this technically very demanding problem has been a firm value for the Pauli paramagnetic susceptibility in sodium, markedly different from what would be expected if the electrons were non-interacting, yet luckily not so different as make untenable theories of metals that start, as a first approximation, by neglecting this interaction.[6]

Resonant filters; sound spectrograph

Returning to experimental affairs, let us enquire what information can be discovered concerning the structure of a complicated input signal by passing it through a resonant filter, i.e. a transducer whose impulse response function is the decaying oscillation expressed by (6) and whose transfer function is the sharply peaked function (7). The formal solution has already been written down in (5), and applied to a pure sinusoidal input to show how the output is small unless the frequency is fairly close to the resonant frequency ω_0. If $\omega'\tau_a$ is large, the real and imaginary parts of the denominator in (7) are equal when the frequency is about $1/\tau_a$ away from resonance, and the power output, proportional to $\chi\chi^*$, is then reduced to half its maximum. We may therefore take $2/\tau_a$ as a convenient measure of the frequency bandwidth of the transducer, and this is also the reciprocal of the time-constant for decay of intensity in the impulse response.

When an arbitrary input $g_0(t)$ is fed into such a transducer, the input in the time interval t_0 to t_0+dt_0 gives an impulse $g_0(t_0)\,dt_0$ and starts a decaying oscillation which by time t is reduced by a factor $e^{-(t-t_0)/\tau_a}$. The same output at time t would be produced from a lossless resonant transducer if the input were reduced by this factor. The two equivalent processes are illustrated in fig. 5 – the real time sequence fed into the real transducer is replaced by an exponentially attenuated time sequence fed into an ideal transducer with $1/\tau_a = 0$. Now the ideal transducer, stimulated by a pure sinusoid that has been applied to it ever since the start of time, responds infinitely more to an input that is exactly tuned than to any other, as is obvious from (7). The response at time t is therefore to be obtained by forming a Fourier transform of the input (fig. 5(b)) and picking out the amplitude of the component at ω_0, the transducer frequency. From this viewpoint the essential features of the transducer become clear. First, even if the input is a pure sinusoid, the exponential attenuation spreads its frequency spectrum over a range of the order $\omega_0 \pm 1/\tau_a$, as is clear from (4.27) or from the virtually equivalent analysis leading to (7). There is no possibility of making a spectral analysis of a non-sinusoidal input to a higher degree of resolution than this. Secondly, the output at any moment t does not refer to the spectral content at that exact moment (if such a statement has any meaning), but is related to the average over an interval of about τ_a before the time of observation. The process of analysis by a resonant transducer is thus subject to exactly the same limitations as are expressed by the Uncertainty Principle $\Delta\omega\,\Delta t \gtrsim 1$.

75

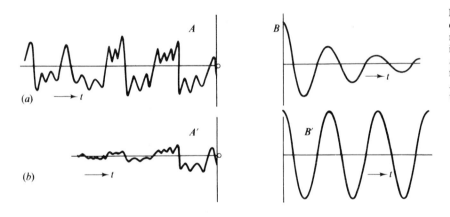

Fig. 5. The signal A entering the damped resonant transducer, with impulse response B, at $t = 0$ emerges in the same form as the damped signal A' entering the ideal transducer B'.

Another way of looking at the same effect is to note that, since the real transducer only responds to frequency components of the input that lie within the range roughly described by $\omega_0 \pm 1/\tau_a$, we may eliminate from the input all but this narrow band of frequencies. Once this is done, what may have started as a jagged function has been reduced to something much smoother, for the resultant amplitude of the components in the range $1/\tau_a$ of frequency cannot significantly change in a time less than τ_a. The 'filtered' input is then approximately sinusoidal with mean frequency ω_0, but with amplitude and phase fluctuating slowly, at the sort of speed, in fact, that the transducer can respond to. It is these amplitude fluctuations that show up as output variations and are interpreted as fluctuations in the amplitude of components at frequency ω_0.

Clearly, then, if we are concerned to know how the spectral content of a signal changes with time we must be reconciled to limiting our frequency resolution to match any requirement to detect sudden changes. In the analysis of speech, for example, where an ordinary word may take $\frac{1}{3}$ second to say, and may contain several different noises as vowels and consonants alternate, we may wish a time resolution not worse than $\frac{1}{50}$ second. We must not then hope for a frequency resolution better than $\Delta\omega \sim 50\,\mathrm{s}^{-1}$, or $\Delta\nu \sim 8$ Hz. Even so we are in a position to obtain a very detailed analysis; neglecting all but the frequency range up to 3000 Hz, in which most of the interest of human speech lies, we may still divide it into over 300 separate frequency bands and record the variations of amplitude in each with the desired time resolution. Or if we are appalled at the thought of handling so many frequency channels we may cut them down to 30 and allows the luxury of a time resolution of 4 ms which is probably less than the ear can discern. One sound spectrograph, used for analysing speech, divides the range from 80 to 8000 Hz into as many as 500 channels, and records them one at a time. A short record of speech is speeded up and fed into a filter circuit whose bandwidth is adjustable down to about 45 Hz, and the time variation of the output recorded as a horizontal line of varying blackness; the same record is fed through again and again with the pass band slowly swept to higher frequencies, and the line displaced upwards as the frequency is increased. A typical result is reproduced in fig. 6, which shows how the title of this book appears when spoken in Southern English. The most prominent features are the wide band hissing noise, with the energy concentrated at higher frequencies, characteristic of the s's in 'physics', and the broader sh-sound towards the end of 'vibration'. The pure vowels are all short in duration, but the two diphthongs (the first i and the a in 'vibration') show clearly the variation of quality as the sound proceeds. Many other examples are to be found in the book of Potter, Kopp and Green.[7]

The pure vowels are sounded by concentrating the energy in certain bands characteristic of the vowel (*formants*). In normal speech the vocal chords emit a rather irregular sequence of impulses, at a mean frequency that is lower in men than in women, and fluctuates as the voice is inflected for emphasis, becoming higher in pitch at moments of rhetorical fervour or emotional stress. The cavities of nose, mouth and pharynx are tuned by manipulation of lips, tongue, etc. so as to respond selectively to certain frequencies in the rich spectrum of this output from the vocal chords, and it is these that form the vowel sound. In a well-sung note the irregular series of impulses is locked in frequency so that the output of the vocal chords is a fairly pure tone rich in harmonics; the appropriate harmonics are selected by the vocal cavities to give the note its vowel character. Sopranos and basses singing the same vowel will impose it on quite different fundamental frequencies by selecting different harmonics. For example, the first vowel in 'father' is distinguished by a strong formant around 900 Hz; a bass singing this word on the A♭ a tenth below middle C (about 100 Hz) will

Fig. 6. Sound spectrograph, obtained with a Kay Sona-Graph, model 7029A, spectrum analyser. For explanation see text (Cambridge University Department of Linguistics.)

contrive to emphasize the ninth harmonic, a contralto singing an octave above will emphasize the fourth and/or fifth, while a soprano still another octave above, at 400 Hz, will have to make do with the second harmonic, though the higher components may emerge hardly shifted and the vowel will be perfectly recognizable. Fig. 7 illustrates this point. The fundamental periods differ by a factor of four, but the very strongly marked harmonics (the ninth for the baritone and the second for the falsetto) are not very different in frequency. The few cycles recorded here tell one little about such highly significant features as the attack (the first stages in producing the steady note) and variations of pitch (vibrato) and tone colour (harmonic content). These necessitate a sound spectrograph, and examples will be found, including analysis of a Caruso top note, in the book already cited.[7]

The analysis of speech by sound spectrograph is a fascinating adjunct to more general phonetic studies, but one that cannot be followed here. Let us turn to the rather simpler matter of the waveform produced by musical instruments, for which the sound spectrograph is usually unnecessarily elaborate. For we are primarily interested in a note of steady pitch with a periodic, nor normally sinusoidal, waveform. The oscillograph records in fig. 8 are typical of the many studies that have been made to relate the timbre of instruments to their waveform; in comparison with speech, their Fourier analysis presents no significant problem.

In the next volume we shall discuss the mechanism of tone production in certain instruments, and for the present a very general account is sufficient to make the particular point that is relevant here, and which may be introduced by comparing the waveforms produced by a bassoon, a guitar and a bell. The first two show a periodicity that is absent from the last, and it is no accident that only the first two are used in Western music that depends on the harmony of combined notes.† For the basic intervals on which the diatonic scale is constructed are related by simple numerical ratios (2 for the octave, 3/2 for the fifth, 5/4 for the major third), and the overtones (harmonics) contained in a periodic waveform supply octaves, twelfths etc. above the fundamental that do not clash disturbingly with the notes of other instruments playing in a concord; it is possible for the ear to recognize without ambiguity the harmonic meaning of the sound. Most listeners in fact are unaware of the overtones when a wind-instrument plays a single note – their ears and minds interpret what they receive as a single note, but a note of a certain quality (timbre). Three oboes playing a major triad (C, E, G) are normally heard as these three notes, not as an extremely complex mixture of the fundamentals with ten or twelve audible

† When bells, timpani and other percussion instruments, even xylophones, are used, it is either in a nonharmonic context or with such support from conventional instruments as will clarify the harmonic meaning beyond their powers of disruption. English ears, accustomed to the austere splendour of change-ringing, find the sound of a carillon playing harmonized music grotesque in the incompatibility of the bell timbre with the logic of diatonic harmony.

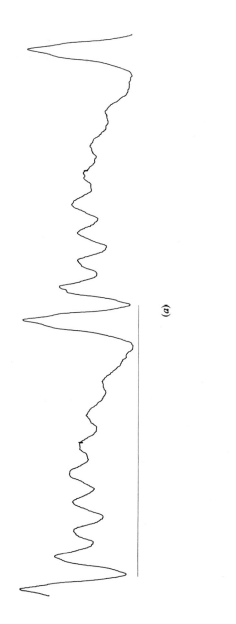

(a)

(b)

Fig. 7. Waveforms of the vowel *a* in *father* sung (*a*) on A♭ (104 Hz) in natural baritone and (*b*) two octaves higher (416 Hz) in falsetto. The strongly marked frequency giving the vowel its characteristic quality is the ninth harmonic in (*a*), i.e. 936 Hz, and the best approximation to this in (*b*), i.e. the second harmonic 832 Hz. Both traces are 20 ms long; here and in fig. 8 the fundamental period in each is indicated by a straight line under the trace.

Fig. 8. Waveforms of
(*a*) bassoon playing G
(100 Hz), (*b*) bassoon
playing g′ (400 Hz),
(*c*) guitar about $\frac{1}{2}$ s after
plucking e′ (338 Hz),
(*d*) handbell about $\frac{1}{2}$ s after
striking. The first two
show highly reproducible
waveforms from one cycle
to the next, indicating a
pure harmonic series; the
guitar is slightly less
perfect, indicating small
departures from the
harmonic series; the bell
obviously suffers
progressive changes of
waveform – Fourier
analysis gives the principal
components the
frequencies 2.120 and
4.355 kHz, in a ratio that
is significantly different
from 2 (cf. fig. 4.2(*f*)).

(*a*)

(*b*)

Fig. 8—*contd.*

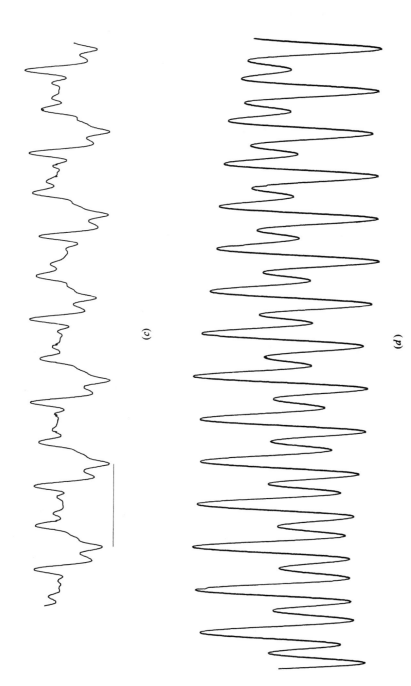

(c)

(d)

overtones attached to each. A trained listener, however, can hear the separate constituents, and it is worth remembering that much progress in the analysis of tone quality, including the discovery of the nature of vowel sounds, was made before oscillographs were invented; Helmholtz used acoustic resonators, but Bell and Paget relied on their ears.

We must now note an important distinction between different types of musical instrument. Although both the guitar (and similarly the piano) and a wind-instrument produce waveforms that maintain their character over many cycles and therefore consist of a strict harmonic series of frequencies, the former has this property somewhat accidentally while in the latter it is intrinsic. The string, like the bell, is capable of executing independent harmonic vibrations in a large number of different modes, and when it is struck or plucked many of these are excited. As a consequence of the wave velocity on a string being independent of frequency, and the one-dimensional character of the wave, the modes have frequencies that are very nearly exact multiples of the fundamental; the bell, on the other hand, not having the simplicity of a uniform string, is far richer in non-integral multiples. The periodicity of the guitar or piano note arises therefore because the only vibrations it can maintain happen to be such as combine to give a periodic resultant. The wind-instrument, however, has the acoustic vibrations of the air in its tube excited by periodic puffs from the reed or lips; the back-reaction of the acoustic vibrations locks the reed frequency to a resonant frequency of the tube. If locking is complete the motion of the reed is now strictly periodic, though far from sinusoidal, and it may excite higher frequency acoustic vibrations if one or more of the higher resonant models of the tube lies close to a harmonic of the reed frequency. But if the instrument should be so badly designed that its higher modes are seriously out of tune with multiples of the reed frequency, they will not be appreciably excited and the emergent sound will still be periodic but with a lesser admixture of the higher harmonics. Nothing the untuned modes can do will break the periodicity which is controlled by the reed, though they may make it more difficult to lock the reed vibration to any acoustic mode; as an extreme example, the 'tacet horn' designed by Benade[8] is actually unplayable.

The reed is an example of a non-linear oscillator whose intrinsic waveform is not sinusoidal. This does not prevent one from carrying through a formal Fourier analysis of its motion into fundamental frequency and harmonics, for this is merely a mathematical procedure applicable to any periodic function. But the analysis is irrelevant to the physical processes at work; one could not excite any one of these sinusoidal vibrations alone, in the way that one can with a piano string by singing loudly when the dampers are raised.

Mixing and combination tones

With a wind-instrument the non-linear process occurs at the point where the sound is generated, and the resulting vibrations are modified by the tube acting as a linear transducer. The reverse process, in which independent vibrations in a linear system are passed through a non-linear transducer, distorts the waveform so as to generate new Fourier components whose frequencies are combinations of those originally present. The simplest example of practical importance is the rectifier which, in its ideal form, transmits unchanged all positive inputs and blocks everything negative. A sinusoidal voltage applied across a rectifier and resistance in series passes a current waveform consisting of sine loops interspersed with blank regions, as shown in fig. 9; the mean current is no longer zero and Fourier

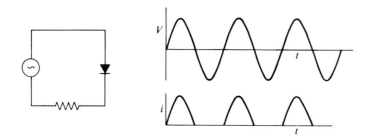

Fig. 9. Simple half-wave rectifier, without smoothing. The sinusoidal generator voltage V results in the intermittent current i.

analysis shows, as is obvious from the diagram, that a complete series of harmonics has been generated. If the output waveform is given by

$$y = \cos \omega t \quad \text{when} \quad -\pi/2 < y < \pi/2,$$

$$= 0 \text{ otherwise,}$$

then according to (4.10) the mth Fourier amplitude is given by

67

$$a_m = \frac{1}{2\pi} \int_{-\pi/2}^{\pi/2} \cos x \, e^{-imx} \, dx.$$

This leads to the following expansion:

$$y = \frac{1}{\pi}\left[1 + \tfrac{1}{2}\pi \cos \omega t + \tfrac{2}{3} \cos 2\omega t - \tfrac{2}{15} \cos 4\omega t \right.$$

$$\left. - 2 \sum_{m \text{ even}} (-1)^{\frac{1}{2}m} \cos m\omega t/(m^2 - 1) \right]. \quad (12)$$

A more usual rectifier circuit incorporates a capacitor as in fig. 10, which also shows the current through the resistor. On the rising positive part of

Fig. 10. Smoothed half-wave rectification.

the cycle the rectifier passes current through the resistor and simultaneously charges the capacitor to the point that the voltage across it equals
the input voltage (this assumes zero resistance in the rectifier). A little time
after the peak, at P, the applied voltage is falling at the same rate as the
combination of resistor and capacitor can discharge itself if cut off from the
input; the discharge curve PQ is of exponential form, $e^{-t/\tau}$, with $\tau = CR$,
and is drawn to be tangential with the sine curve at P. Until P is reached,
forward current through the rectifier is needed to hold the voltage to the
sine curve. But beyond P the sine falls off more sharply than the exponential and the condition for isolation from the input is satisfied, the rectifier
being back-biased. At Q forward biasing starts again, so that the heavy
curve represents the variation of voltage across, and therefore current
through, the resistor. If the time-constant τ is made much greater than the
oscillation period, by choosing a large enough capacitor, the output is
almost flat and the strong harmonic content expressed by (12) is greatly
reduced. Still better is to use the full-wave rectifier circuit of fig. 11. When

Fig. 11. Smoothed full-
wave rectification.

X is positive, current cannot flow through A and C and passes through R
in the direction of the arrow; when Y is positive, the current is blocked by
B and D and passes through R in the same direction. The voltage input is
thus two half-waves, and with the capacitor present the output is correspondingly flatter than with the half-wave rectifier of fig. 10.

When heavy current is required of a rectifier capacitive smoothing may
be impracticable, and the use of a 3-phase a.c. supply is advantageous in
reducing the ripple. The supply consists of a neutral line and three others
with alternating voltages phased at 120° to each other. If each is rectified in
a half-wave circuit and the outputs combined, the result is the superposition of three current waveforms like that in fig. 9, and gives the
waveform shown in fig. 12(b). Still better is to use a full-wave rectifier
circuit on each phase so that six outputs at 60° are added, and with suitable
phase-changing networks the process may be multiplied further to give 12
or more outputs evenly spaced in phase. If a single-phase waveform is
expressed as a Fourier series

$$y = \sum_{-\infty}^{\infty} a_n\, e^{-in\omega t},$$

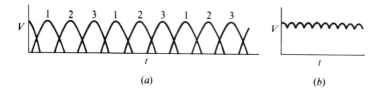

(a)   (b)

Fig. 12. Unsmoothed three-phase rectification. The intermittent outputs, labelled 1, 2 and 3, interlock to give a steady current with ripple superimposed.

the addition of N identical waveforms at phase spacing of $2\pi/N$ eliminates all those terms for which n is not a multiple of N. This is immediately obvious from the vibration diagram for any Fourier component, which is a closed polygon of N sides, having zero resultant, if $n \neq mN$, but a straight line if $n = mN$. Thus the 12-phase rectifier just mentioned, feeding straight into a resistive load with no capacitive smoothing, has an output waveform whose Fourier components can be written down from (12),

$$y \propto 1 + \tfrac{2}{143}\cos(12\omega t) + \tfrac{2}{575}\cos(24\omega t) + \cdots,$$

which is very close to the scalloped waveform from a full-wave rectifier, at 12 times the supply frequency, superimposed on a much larger steady current.

These are ideal rectifiers that we have been discussing, with output waveforms whose structure is independent of input amplitude and is proportional to the input. In this sense they share some of the properties of a linear circuit, but the resemblance is spurious since when more than one sinusoid is injected the output is not the sum of what each would give if present alone. It is unnecessary to labour this point, but something more valuable can be learnt from an analysis of a less ideal rectifier, one in fact that is only slightly non-linear. A real rectifier (e.g. a p–n semiconductor junction) has a characteristic like that shown in fig. 13. When the voltage is

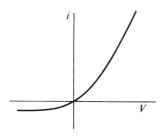

Fig. 13. Rectifier characteristic.

large and positive it passes current easily, when large and negative it passes so little as to be virtually blocked. For large voltages, then, it approximates to an ideal rectifier. For small voltages, however, the characteristic is only slightly curved, and now it is easily seen that the mean d.c. current is proportional to the square of the applied voltage. For this condition of square-law rectification to obtain, the voltage must be small enough for the characteristic to be adequately represented by the quadratic form

$$i = \alpha V + \beta V^2. \tag{13}$$

Substituting $V = V_0 \cos \omega t$ (note that since the system is non-linear we must beware of using the complex representation), we find that

$$i = \alpha V_0 \cos \omega t + \beta V_0^2 \cos^2 \omega t$$

$$= \tfrac{1}{2}\beta V_0^2 + \alpha V_0 \cos \omega t + \tfrac{1}{2}\beta V_0^2 \cos 2\omega t, \tag{14}$$

The d.c. current being $\tfrac{1}{2}\beta V_0^2$ and the waveform being distorted by the production of a second harmonic whose relative magnitude depends on the strength of the applied voltage.

Rectification depends, of course, on the i–V characteristic having a different form for positive and negative voltages. A symmetrical characteristic, for example

$$i = \alpha V + \gamma V^3, \tag{15}$$

gives for an input $V_0 \cos \omega t$ an output

$$i = \alpha V_0 \cos \omega t + \gamma V_0^3 \cos^3 \omega t$$

$$= (\alpha V_0 + \tfrac{3}{4}\gamma V_0^3) \cos \omega t + \tfrac{1}{4}\gamma V_0^3 \cos 3\omega t,$$

with odd harmonics only and no steady term.

When two different frequencies are fed into a non-linear circuit, mixing occurs and other frequencies are generated. Thus with $V = V_1 \cos \omega_1 t + V_2 \cos \omega_2 t$ substituted in (13) the output is given by

$$i = \tfrac{1}{2}\beta(V_1^2 + V_2^2) + \alpha V_1 \cos \omega_1 t + \alpha V_2 \cos \omega_2 t + \beta V_1^2 \cos 2\omega_1 t$$

$$+ \beta V_2^2 \cos 2\omega_2 t + \tfrac{1}{2}\beta V_1 V_2 \cos(\omega_1 - \omega_2)t + \tfrac{1}{2}\beta V_1 V_2 \cos(\omega_1 + \omega_2)t. \tag{16}$$

In addition to terms of the form (14) arising from each input component separately the output contains, in the last two terms, combination frequencies generated by the mixing of the inputs, with amplitudes determined by $V_1 V_2$ and frequencies equal to the sum and difference of the input frequencies. The same input applied to (15) yields no d.c. term, of course, but a rich collection of frequencies – ω_1, ω_2, $3\omega_1$ and $3\omega_2$ from each component alone, $\omega_1 \pm 2\omega_2$ and $2\omega_1 \pm \omega_2$ by mixing. There are no plain sum and differences tones, which require a rectifying characteristic for their production.

In general, if the characteristic can be expressed as a power series:

$$i = \alpha V + \beta V^2 + \gamma V^3 + \cdots \tag{17}$$

and the input takes the form $V_1 \cos \omega_1 t + V_2 \cos \omega_2 t + \cdots$, the output can contain only such combination frequencies as take the form $n_1 \omega_1 + n_2 \omega_2 + \cdots$, where the ns are integers and $|n_1 + n_2 + \cdots|$ is one of the exponents occurring in the power series (17), or is less than one of the exponents by an even number.

A fairly simple practical demonstration of mixing, which may also have useful applications, involves feeding three signals into a semiconductor diode whose output is amplified to excite a loudspeaker; two signals are the outputs of triode oscillators at about 10^9 Hz, which can be made to give very pure sinusoids, stable in frequency; and the third is the output of a square-wave oscillator at 10^5 Hz.[9] If one triode oscillator is kept at constant frequency while the other is slowly tuned, an audible note is heard when the difference in frequency, $\omega_1 - \omega_2$, between the two is close to one of the harmonics of ω_3, the harmonic-rich square wave. What is heard is $\omega_1 - \omega_2 + n\omega_3$, and as ω_2 is varied the note is first heard very high and then descends below the audible range, rising again to a high and finally inaudible frequency. The successive tuning positions at which the note goes through zero frequency provide calibration points for the tunable oscillator at exactly equal spacings of ω_3. As many as 50 points can be obtained, corresponding to the difference $\omega_1 - \omega_2$ running between $\pm 25\omega_3$.

The difference tone can be experienced directly if two audiofrequency signals at a high level of intensity are heard together. To demonstrate the effect convincingly as a property of the ear it is advisable to feed the amplified output of two signal generators into separate loudspeakers, so as to give no chance for non-linear effects in the loudspeakers to produce the difference tone. With the head between the loudspeakers two notes rather close together at a high audible frequency (several kHz) are accompanied by a low note which rises and falls rapidly as one of the signal generators is tuned through a relatively small range. The sound level required to hear the difference tone may prove unpleasantly loud unless one has an especially good (or unusual) ear. Not only the simple difference tone may be heard – other combinations like $2\omega_1 - \omega_2$ are also produced and it is clear that the mechanism of the ear has rectifying powers and higher order terms in its response which are significant at a high intensity.

A somewhat obscure, and possibly related, phenomenon is made use of (though deprecated by purists) to produce the effect of very low notes from church organs that have not the space to accommodate 32 ft pipes. A 16 ft pipe at an octave above the desired note and a $10\frac{2}{3}$ ft pipe at an octave and a fifth are sounded together; that is, if ω is desired, 2ω and 3ω are sounded and the listener is persuaded that he has heard the difference frequency ω. This is called an *acoustic bass* and is sometimes explained as arising by the above mechanism of non-linearity in the ear. It is, however, hard to believe this explanation since the amplitude would depend, according to (16), on the product of the two component amplitudes, and listeners in different parts of the building would get a very different intensity of difference tone relative to the two generating tones. One should not rule out a psychological origin – possibly the brain hearing two low notes (the device only works for very low notes apparently) at 3ω and 2ω interprets them as arising from a still lower fundamental at ω. To carry the conjecture further, the audible illusion may be fostered by perfect tuning of the two constituents so that their waveform is strictly repetitive with a basic repetition frequency of ω, and this may provide a reason for the normal practice of strapping the two pipes together. For it is known that two organ pipes, like other maintained oscillators, in close proximity may influence one another (this is a non-linear effect which will be examined in some detail in chapter 12), and may pull their frequencies into a simple numerical ratio even if they are initially not quite perfectly tuned. But there is yet another explanation possible, which is that the mechanical interactions between the two pipes clamped together and set into vibration by the air within may, through some unidentified rectifying effect, generate a real difference tone on the spot.

The non-linear generation of combination tones is the mechanism underlying the superheterodyne principle of radio detection. Audio-frequency information transmitted as amplitude modulation of a carrier wave having a high frequency, 1 MHz say, is amplified and detected in two stages; first the incoming signal, presumed to be very weak, is mixed in a non-linear device with a much stronger local oscillator signal of constant amplitude and differing in frequency from the carrier. With a square-law detector such as is described by the second term of (13) (the linear term being irrelevant to the mixing process), an input consisting of the incoming signal $f(t) \cos \omega_1 t$ and the local oscillator $A \cos \omega_2 t$ is transformed into $\beta[f(t) \cos \omega_1 t + A \cos \omega_2 t]^2$, which contains among others the term $\beta f(t) A \cos (\omega_1 - \omega_2)t$. The information, expressed by $f(t)$, has been transferred to the lower difference frequency (above the audible limit) which can be chosen for convenience of amplification; at the same time its amplitude has been multiplied by βA, which can be made much greater than unity by suitable choice of local oscillator amplitude. In this way a very weak signal is enhanced before ever entering a conventional amplifier, and with a good local oscillator, producing an output almost free of random variations, the quality of the information is maintained in the process. After amplification the output is rectified, and the information appears as signal variations imposed on no carrier frequency but capable of actuating a loudspeaker or recorder directly.

Mixing of modulated radio waves sometimes occurs when they are reflected by the ionosphere.[10] Above the neighbourhood of a powerful transmitter the intensity may be so great as to impart extra kinetic energy to the electrons to an extent that is not negligible in comparison with their thermal energy. The reflecting properties of the ionosphere are thus slightly modified, and the radio waves from a second, distant, transmitter may be modulated, as it is reflected from this region, in synchronism with the rise and fall of the intensity of the nearby transmitter. Since this variation of intensity is the signal carried on an amplitude-modulated wave, a listener tuned to the second transmitter may be surprised to hear the program of the first breaking in on that of the second, even though the two transmitters have quite different carrier frequencies. This is the wave-interaction effect, sometimes called the Luxembourg effect after the first transmitting station powerful enough to produce a discernible non-linearity.

6 The driven harmonic vibrator

The simple harmonic oscillator, driven by a sinusoidally varying force, is central to the discussion of vibrating systems, being a model for so many real systems and therefore serving to unify the description of very diverse physical problems. In view of this it is worth spending some time examining it from several different aspects, even though it might be thought that a formal solution of the equation of motion contained everything useful to be said on the matter. Indeed, if one were concerned only with physical systems that could be modelled exactly in these terms a single comprehensive treatment would suffice for all. Real systems, however, normally only approximate to this idealization, and alternative approaches may then prove their worth in allowing the behaviour to be apprehended semi-intuitively, often enough with sufficient exactitude to make mathematical analysis unnecessary. The reader who has progressed to this point will be familiar enough with the most elementary analyses not to be worried that we approach the problem indirectly, picking up an argument that has already been partially developed. More familiar treatments will be introduced in due course.

Transfer function, compliance, susceptibility, admittance, impedance

The essential framework for this approach has already been laid down in chapter 5, where the concept of the transfer function $\chi(\omega)$ was introduced. **105** We had in mind there a linear transducer into which a sinusoidal signal $A\,e^{-i\omega t}$ was fed, and from which emerged an output signal $\chi(\omega)A\,e^{-i\omega t}$. Precisely the same idea may be applied to a mechanical system or electrical circuit by reinterpreting the signal $A\,e^{-i\omega t}$ as the oscillating driving force and the output signal as the response of some part of the system, for example the displacement or velocity of one mechanical component, the charge on a capacitor, or the current in some part of the circuit. If $\chi(\omega)$ relates mechanical displacement to force it is called *compliance*, if it relates electrical or magnetic polarization to field strength it is *susceptibility*, if it relates current to e.m.f. it is *admittance*. Alternatively, the disturbing input signal may be a current and the output the resulting e.m.f. in which case the frequency-dependent relation between the two is the *impedance*, the electrical analogue of what in elastic systems is an *elastic modulus*. We do not attempt to define each of these quantities with high precision for it is not their exact meaning that matters – only the close analogy between them that allows the same discussion of $\chi(\omega)$ to be transferred to new situations with only minor modifications.

For a simple resonant system, with only one coordinate available to describe the response to an input force, the character of $\chi(\omega)$ is defined by the impulse response function – normally a damped sinusoid as in (5.6), leading to (5.7) for $\chi(\omega)$. Let us look at a number of illustrations, taking care to choose as input some form of disturbance that does not lead to an infinite response. Thus with a simple pendulum $\chi(\omega)$, as in (5.7), is appropriate to describe the relation between displacement as output and force as input, rather than the other way round. For a force $F(t)$ of finite magnitude and applied over a short interval δt results in a finite initial velocity and virtually zero displacement; on the other hand, an impulsive displacement $\xi(t)$ lasting for an interval δt would require infinite forces (an infinite forward impulse to displace the pendulum to ξ instantaneously, followed by an opposite impulse to bring it to rest there; after δt the same process in reverse). Immediately after the impulsive force assumed to be of unit strength, the pendulum bob, of mass m, is still at the origin but moving with velocity $1/m$. The subsequent motion is then described by (5.6), h_0 being assigned such a value as to ensure that the initial velocity is correct:

$$h(t) = (m\omega')^{-1} \sin \omega't \cdot e^{-\omega''t} \qquad (t>0) \tag{1}$$

ω'' being written for $1/\tau_a$. Hence, from (5.7),

$$\chi(\omega) = -1/mY, \tag{2}$$

where Y is written for the resonant denominator that will recur continually:

$$Y = \omega^2 - \omega'^2 - \omega''^2 + 2i\omega''\omega = \omega^2 - \omega_0^2 + 2i\omega''\omega, \quad \text{from (2.24)}. \tag{3}$$

Y may also be expressed in terms of the two complex frequencies that make up the damped oscillation (1). If $\Omega_1 = \omega' - i\omega''$ and $\Omega_2 = -\Omega_1^* = -\omega' - i\omega''$,

$$Y = (\omega - \Omega_1)(\omega - \Omega_2). \tag{4}$$

The response of the pendulum to an oscillatory force $F e^{-i\omega t}$ is given by the expression

$$\xi(t) = \chi(\omega)F e^{-i\omega t} = -F e^{-i\omega t}/mY. \tag{5}$$

If ω'' is small (light damping), $\chi(\omega)$ has a high peak of magnitude $i/2\omega''\omega m$ when ω is very close to the natural frequency ω_0. We shall not discuss the form of $\chi(\omega)$ further at this point, but first present other illustrations.

The series resonant circuit of fig. 1 may be excited impulsively by attaching a cell across AB for an interval δt and straightway short-circuiting these terminals. The finite current excited by the cell can then oscillate in the closed circuit, dying out exponentially as the electrical

Fig. 1. Series resonant circuit.

energy is dissipated in the resistor. It is not appropriate to regard the current as the input driving force and the e.m.f. across AB as the resulting output, since it requires an infinite e.m.f. to establish a current suddenly through the inductor. Before jumping to the conclusion that $\chi(\omega)$ in (2) describes the relation of oscillatory current to oscillatory e.m.f., we must satisfy ourselves that the impulse response function has the form (1); but in fact it has not. In the initial stages after the cell is connected, the whole e.m.f., V, is taken up in forcing the current through the inductor to rise at a rate V/L. After time δt the current is $V\delta t/L$, and if $V\delta t$ is a unit impulse, the initial value of the current response, $h(0)$, is $1/L$. There is no significant potential drop across the capacitor since the charge that has so far flowed, $h(0)\,\delta t$ or $\delta t/L$, can be made infinitesimal by allowing δt to go to zero while increasing V to maintain $V\delta t$ at the value unity. Immediately after short-circuiting the terminals AB there is no e.m.f. round the circuit, so that if the resistance is small the e.m.f. across it, R/L, and the equal and opposite e.m.f. across the inductor, are both small, tending to zero as R is diminished. Initially, then, the current response $h(t)$ is nearly unchanging at its value $1/L$, consistent with a cosine rather than a sine form, as in (1), for $h(t)$. The precise form of $h(t)$ is most readily derived by noting that the charge $q(t)$ on the capacitor is completely analogous to the displacement of the pendulum. In particular, starting at zero it must be a sine function, and the initial rate of change $1/L$, determined by the initial value of the current, fixes the coefficient. Hence

$$q(t) = (\omega' L)^{-1}\, e^{-\omega'' t} \sin \omega' t, \qquad (6)$$

and for the current response we have that

$$h(t) = \dot{q}(t) = (\omega' L)^{-1}\, e^{-\omega'' t}(\omega' \cos \omega' t - \omega'' \sin \omega' t), \qquad (7)$$

a decaying sinusoidal oscillation, but one that starts at a different phase from (1).

Corresponding to (7) the admittance of the circuit follows by use of (5.4):   **106**

$$\chi(\omega) = i\omega/LY, \qquad (8)$$

differing in phase from (2), through multiplication of the numerator by $-i\omega$, but otherwise very similar in form and, when ω'' is small, with a sharp peak at the natural frequency ω_0. The reason for the appearance of $-i\omega$ is clearly that this represents the operation of time-differentiation by which the response (7) is derived from that in (6).

The parallel resonant circuit of fig. 2 may be treated in the same way by reversing the significance of e.m.f. and current. It is now inappropriate to treat e.m.f. as the driving force, since applying a cell to the terminals immediately sends an infinite current through the capacitor. On the other hand, an impulsive current of unit strength deposits unit charge on the capacitor plates and establishes an e.m.f. $V = 1/C$ between the terminals.

Fig. 2. Parallel resonant circuit.

When these are open-circuited after the charge has been injected, V oscillates as the capacitor discharges through the inductor, R being large enough to cause only slow damping, and the impulse response is roughly of cosine form. Once more, like (7), it is the derivative of a sine-like response function as may be seen from the fact that V is also given by $L\, \mathrm{d}i/\mathrm{d}T$, and that after the impulse there is no current through the inductor. In this case, then, with the required modifications made in (8),

$$\chi(\omega) = \mathrm{i}\omega/CY, \qquad (9)$$

and $\chi(\omega)$ is now the impedance of the circuit. The close relationship between series and parallel resonant circuits is most obvious, as we shall see shortly, when they are analysed by the conventional circuit procedure.

Before proceeding to this, however, let us examine a slightly modified parallel resonant circuit, as drawn in fig. 3, the lossy element being provided by a small resistance in series with the inductor rather than a large resistance in parallel. The impulse response function, relating e.m.f. output

Fig. 3. Parallel resonant circuit (alternative arrangement to fig. 2).

to current impulse input, is almost the same as for fig. 2, but with a small phase shift. This is because V is not $L\, \mathrm{d}i/\mathrm{d}t$, being supplemented by a contribution Ri from the resistor. To see the exact form it is easier to note that after the impulse there is no current through L to discharge the capacitor, so that initially $dV/dt = 0$. In this case, then, V is the integral, rather than the derivative, of a sine-like response function; the form of V is easily constructed to have this property and the correct initial value:

$$V(t) = (\omega' C)^{-1} e^{-\omega'' t}(\omega' \cos \omega' t + \omega'' \sin \omega' t)$$

leading to

$$\chi(\omega) = (-2\omega'' + i\omega')/CY, \tag{10}$$

which differs from (9) only through the presence of $-2\omega''$ in the numerator, a small change when the resonance is sharp ($\omega'' \ll \omega'$). As with the circuit of fig. 2, $\chi(\omega)$ is the impedance.

The expressions (8) to (10) may be derived much more readily by circuit analysis. For example, the impedance of the series resonant circuit (fig. 1) is $R + j\omega L - j/\omega C$, so that its admittance takes the form

$$\chi = (R + j\omega L - j/\omega C)^{-1} \tag{11}$$

To compare this with (8) it is necessary to identify ω_0 and ω'' with the circuit parameters. The free vibration of a circuit has been analysed in chapter 2, and reference to (2.24) shows that

$$\omega'^2 = 1/LC - \tfrac{1}{4}R^2/L^2,$$

and

$$\omega'' = \tfrac{1}{2}R/L.$$

When these values are substituted in (8), (11) follows immediately (remember $j = -i$). So also with the parallel resonant circuits. The close relationship between the formulae for χ for the circuits of figs. 1 and 2 is explained by the circuit analysis. For in the first case the impedance is that of three components added together – one real (R), one positive imaginary ($j\omega L$) and one negative imaginary ($-j/\omega C$); in the second case it is the admittance that is the sum of three analogous terms, $1/R$, $j\omega C$ and $-j/\omega L$. There is, from the point of view of the impulse response function, no essential difference between the series and parallel circuits. Each shows the characteristic resonance peak in $\chi(\omega)$, and they only differ in the meaning to be attached to χ, that is, in what we have called the appropriate choice of input and output parameters.

More generally, the small difference between (9) and (10) and the larger difference between these and (2) and again rather accidental than essential. A family of curves for $\chi(\omega)$, embracing these as special cases, can be generated from the exponentially damped oscillation that is characteristic of a simple linear vibratory system, by allowing the oscillation to begin at an arbitrary phase in its cycle. For the general form of $h(t)$ in simple systems we write

$$h(t) = A e^{-\omega'' t} \cos (\omega' t - \theta), \tag{12}$$

yielding

$$\chi(\omega) = A[-\omega' \sin \theta + i(\omega + i\omega'') \cos \theta]/Y. \tag{13}$$

The special cases we have analysed fall into this general form when A and θ are chosen correctly. For (2), $A = 1/m\omega'$ and $\theta = \pi/2$; for (8), $A = (\omega'^2 + \omega''^2)^{\frac{1}{2}}/\omega' L$ and $\theta = -\tan^{-1}(\omega''/\omega')$; for (10), $A = (\omega'^2 + \omega''^2)^{\frac{1}{2}}/\omega' C$ and $\theta = \tan^{-1}(\omega''/\omega')$. There is no difficulty devising a circuit to fit any other value of θ. In fig. 4, for instance, the output, V_{out}, across the terminals CD is determined not solely by the current through r, nor by the rate of change of current through l, but by a mixture of both. By varying the ratio l/r the phase of the output may be changed relative to that of the oscillatory current established by the impulsive input across AB.

Fig. 4. Series resonant circuit tapped across inductance l and resistance r.

It is worth analysing this circuit in detail to bring out a point which has hitherto been irrelevant to the examples chosen. The most general response of a simple system to an impulse consists not only of the damped oscillatory term (12) but may contain an instantaneous response to the impulse itself. Thus when an impulsive e.m.f. is applied across AB, it appears in the circuit entirely across the inductor L, and a fraction l/L is incorporated in the output response, which therefore takes the form, for unit input,

$$h(t) = (l/L)\,\delta(t) + r\dot{q} + l\ddot{q},$$

where $\dot{q}$ is the current response of the circuit as given in (7). Hence

$$h(t) = (l/L)\delta(t) + (\omega' L)^{-1}\,e^{-\omega'' t}[(r - 2\omega'' l)\omega'\cos\omega' t$$
$$- (\omega'' r + \omega'^2 l - \omega''^2 l)\sin\omega' t].$$

The Fourier transform of $h(t)$ now includes a frequency-independent term, l/L, from the impulse, and gives after evaluation the transfer function

$$\chi(\omega) = \frac{\omega}{LY}(\omega l + \mathrm{i}r),$$

which is the same as the form that emerges immediately from circuit analysis

$$\chi(\omega) = \frac{r + \mathrm{j}\omega l}{R + \mathrm{j}\omega L - \mathrm{j}/\omega C}.$$

It will be appreciated that even with the additional complication of the direct response to the impulse, the most general form of response of a simple linear system is still determined by a very small number of parameters – the amplitudes of the direct and delayed (oscillatory) responses, and the phase, frequency and decay time of the latter – and these can be modelled by a simple circuit such as that of fig. 4. All examples in this general category, then, whether they be electrical, magnetic, mechanical, acoustic etc., may be replaced by their circuit analogue for the purpose of analysis. The advantage of doing this is that familiarity with the manipulation of impedances allows results to be derived economically, once the analogue has been set up, with the circuit parameters correctly related to those of the real system. It is easy enough to look at the equation of motion of a mass on a spring:

$$m\ddot{z} + \lambda \dot{z} + \mu z = F \cos \omega t, \tag{14}$$

and that of a series resonant circuit

$$L\ddot{q} + R\dot{q} + q/C = V \cos \omega t, \tag{15}$$

and to recognize that mass and inductance form a pair of equivalent parameters, as do the damping coefficient λ and resistance, spring constant μ and reciprocal of capacitance, force and e.m.f. It is not quite so easy, unless one is practised in the art, to do the same with, say, a Helmholtz resonator. For this reason we do not propose to lay any stress on the technique of modelling a non-electrical system by means of circuits, useful though this is in analysing complex systems. But it is essentially a practical tool, not a means of gaining insight into the physics of a system; as such it should be acquired when professional necessity drives, not elevated into an important principle. The far more important physical principle is that the similarities of behaviour of simple systems are not accidental, but spring from the necessary restrictions to the form of the impulse-response function.

The form of the resonance curve

Although some emphasis has been placed on the analogies between different systems, considerable variation is possible through the influence of the phase θ introduced in (12). This becomes clear when we examine the shape of the resonant response, $\chi(\omega)$; to begin with we shall assume the damping parameter ω'' to be small. This leads to a narrow, high-resonance peak and exhibits the essence of the phenomenon in its clearest form. If ω'' is small we are mainly concerned with the behaviour when ω is very near ω_0, so that $\omega - \Omega_1$ (see (4)) is small, and $\omega - \Omega_2$ large enough to be adequately represented by $2\omega_0$. To the same degree of approximation we may replace ω by ω_0 at all points except where the difference $\omega - \omega_0$ appears. Then the various forms of $\chi(\omega)$ that we have met may be summed up in the simplified form of (13):

$$\chi(\omega) \sim \tfrac{1}{2}A[\mathrm{i} \cos \theta - (\sin \theta + \tfrac{1}{2}Q^{-1} \cos \theta)]/(\omega - \Omega_1), \tag{16}$$

in which, following (2.2), Q replaces $\tfrac{1}{2}\omega'/\omega''$. When Q is large enough to make negligible the term in which it appears, the following result emerges:†

if $h(t) = \mathrm{Re}\,[a\,e^{-i\Omega_1 t}]$; $t > 0$, a being in general complex,

$$\chi(\omega) \sim \tfrac{1}{2}ia/(\omega - \Omega_1),$$

or

$$\tfrac{1}{2}A\,e^{i(\theta + \pi/2)}/(\omega - \Omega_1), \tag{17}$$

if a is written as $A\,e^{i\theta}$. All the most important features of resonant response are contained in (17), which is the Lorentzian function already encountered in (4.28) as the Fourier transform of a decaying exponential. The extra phase factor in (17) obliges us to extend the discussion that was given there.

76

The form of χ is perhaps best appreciated graphically; $1/\chi$ varies as $(\omega - \Omega_1)e^{-i(\theta + \pi/2)}$, and if $\theta = \pi/2$ (as for the pendulum), $1/\chi \propto \Omega_1 - \omega$ and is represented, as in fig. 5, by the vector LN between ω, a point on the real axis, and $\Omega_1(=\omega' - i\omega'')$. As the applied frequency is changed, L, representing ω, moves evenly along the real axis and the denominator

Fig. 5. Graphical construction for response of resonant system, for example displacement of a pendulum.

varies symmetrically about its minimum value when $\omega = \omega'$. The line NP has the right magnitude (apart from a constant) to represent $1/(\Omega_1 - \omega)$, but the wrong phase, being in fact $1/(\Omega_1^* - \omega)$. If, however, one draws a circle on the diameter joining ω_0 and Ω_1^*, the vector P^*N^* correctly represents the variation of $1/(\Omega_1 - \omega)$ in magnitude and phase angle. The variation of the real and imaginary parts of $\chi(\omega)$ are shown in fig. 6(a), which is virtually the same as fig. 5.4; χ and χ'' are equal in magnitude when $|\omega - \omega'| = \omega''$, and at that point the intensity (or power) of response, proportional to the square of the amplitude P^*N^*, has fallen to half its value at the peak. The width of the resonance, $\Delta\omega$, is defined as the

366

† A useful simplification of the impedance of a series resonant circuit involves the same degree of approximation. Instead of $R + j(\omega L - 1/\omega C)$ one writes ω as $\Omega_1 + \delta$, and to first order in δ the impedance is $2jL\delta$, i.e. $2jL(\omega - \Omega_1)$. Similar linear approximations can of course be applied to other resonant systems.

difference in frequency between these points, i.e. $2\omega''$ or ω'/Q. The quality factor, Q, is just as useful in describing the forced vibration as in defining the decay of free vibration. Moreover this particular result does not depend on the value of the phase angle, θ, since the power response $P \propto \chi\chi^*$; i.e. from (17),

$$P(\omega) = P_0/(\omega - \Omega_1)(\omega - \Omega_1^*) = P_0/[(\omega - \omega')^2 + \omega''^2]. \qquad (18)$$

The real and imaginary parts of $\chi(\omega)$ are, however, much affected by the phase angle θ; with $\theta = 0$, as for the current response (8) of a series resonant circuit, the real and imaginary parts in fig. 6 are interchanged

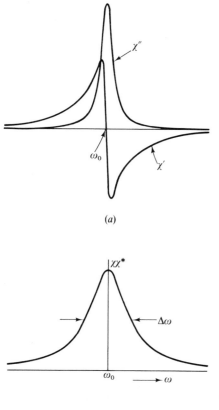

(a)

(b)

Fig. 6. (a) Real and imaginary parts of χ derived from fig. 5; (b) $\chi\chi^*$ for a resonant peak.

(apart from sign changes). In general, the horizontal line in fig. 5 should be tilted up at an angle $\theta - \pi/2$, as illustrated in fig. 7, but Re $[\chi]$ and Im $[\chi]$ are determined with respect to a horizontal real axis. The curves are now markedly asymmetric, as in fig. 8; Re $[\chi]$ goes through zero at ω_1 and Im $[\chi]$ has its maximum at ω_2, both below ω'. It is obvious from the diagram that the phase angle between input and output varies with

Fig. 7. The form of fig. 5 when the phase of resonant response is changed; $\chi' = 0$ at ω_1, χ'' is maximal at ω_2, χ' is maximal at ω_3.

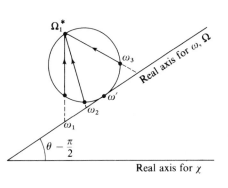

Fig. 8. Real and imaginary parts of χ derived from fig. 7; the horizontal scale is widened compared to fig. 6, to show asymmetries.

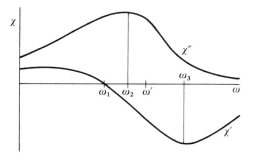

frequency in the same way for all θ, except that the overall range, which is always π, runs between $\theta \mp \pi/2$ according to the equation derived from (17),

$$\arg(\chi) = \theta + \pi/2 - \tan^{-1}\left(\frac{\omega''}{\omega - \omega'}\right). \tag{19}$$

This function, shown in fig. 9, together with (18) completely specify the form of $\chi(\omega)$.

The diagram of fig. 5 may be interpreted in physical terms so as to illuminate the phase and amplitude variations across the resonant peak. For the sake of clarity let us consider the specific case of a mass on a spring, damped by a viscous liquid. By turning the problem round and asking what force is needed to maintain a certain amplitude of motion, rather than what amplitude results from applying a certain force, we can divide the driving force into constituent parts, each performing a separate function. First consider the undamped vibrator which, left to itself, would oscillate at constant amplitude and at its natural frequency, $\omega_0 = (\mu/m)^{\frac{1}{2}}$. The vibration diagram in fig. 10(a) illustrates this situation; the vector A, rotating at angular velocity ω_0, represents the motion of the mass, the force required for its acceleration being supplied by the spring, and represented by the oppositely directed vector B. If, however, we wish the vector A to rotate at a different speed, corresponding to the mass being forced to vibrate at a frequency other than its natural resonant frequency, the spring force must

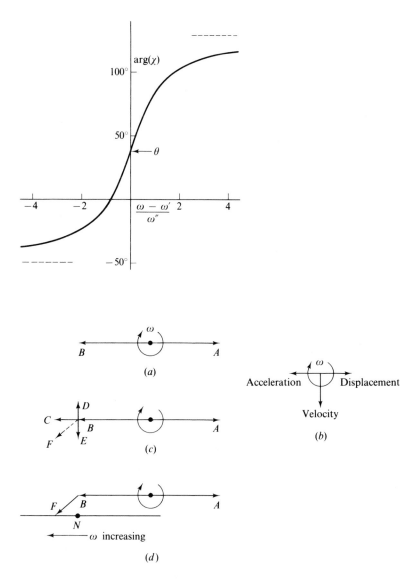

Fig. 9. Variation of the phase χ through resonance; the broken lines are the asymptotes, 180° apart.

Fig. 10. Vector diagrams for forces needed to maintain vibration. The instantaneous displacement A and acceleration B for a free-running lossless vibrator are shown in (a); (b) shows the directions of displacement, velocity and acceleration in (a). For (c) and (d) see text.

be supplemented by an additional force in the same phase, increasing the magnitude of A if the frequency is to be raised. The required force is shown as C in fig. 10(c). If the viscous damping is now introduced, the mass experiences another force, directed opposite to the velocity and represented by D. For the vibration to be maintained at constant amplitude this force must be countered by the driving force, which must therefore contain a component E $(=-D)$. The required driving force is then the vector sum of C, which changes the frequency and alters in magnitude as the resonance is traversed, and E which is sensibly independent of frequency across the narrow resonance. The final result is shown in fig.

$10(d)$, with the end of the resultant force vector F moving along the horizontal line, and the force required to maintain a given amplitude being least at resonance, represented by N. At the low-frequency end F is parallel to the displacement A, being needed to stretch the spring, while at the high-frequency end it is parallel to the acceleration, being needed to supplement the spring which alone could not provide for so rapid a vibration. It will be recognized that this discussion has only constructed by physical reasoning the form of the denominator in (17), i.e. the line LN in fig. 5.

It is also instructive to examine the energy balance. The force C, supplementing the spring, can be considered to make a contribution to the potential energy of the vibrator, which is exchanged with kinetic energy twice per cycle; thus C is responsible for the oscillatory power described by the first term in (3.10). The force E, on the other hand, being in phase with the velocity, steadily transfers energy to the vibrator, to be dissipated as heat, the process described by the second term in (3.10) or by (3.13). From this point of view it is easy to see how a lightly damped resonant system will behave under non-linear damping, by frictional forces, for example. In order to maintain strictly sinusoidal motion the force C is needed as before; but the damping force is no longer represented by the real part of a rotating vector of constant length. Such a driving force E, supplying enough power to provide for the mean dissipation, will compensate the mean drag force but will be too large at some phases of the vibration and too small at others. This implies that if we drive the vibrator with a strictly sinusoidal force it cannot execute strictly sinusoidal vibrations, for this uncompensated part of the force will superimpose vibratory motion at a harmonic of the fundamental frequency. In a lightly damped system, however, the drag forces are small compared to the spring force, so that the uncompensated part cannot excite a large amplitude, especially at a frequency so far removed from the natural resonance. To a good approximation, then, we may ignore the harmonic generation and assume that sinusoidal vibrations require sinusoidal forces determined by the frequency shift from resonance and by the mean dissipation.

To treat in more detail the specific example of a constant frictional force, f, we note that when the amplitude is ξ_0 the work done per cycle is $4f\xi_0$, which is to be matched by a viscous force 'constant' λ giving $\pi\lambda\omega_0\xi_0^2$. The appropriate amplitude-dependent form of λ is therefore $4f/\pi\omega_0\xi_0$, inversely proportional to ξ_0. As ξ_0 is changed, the force C in phase with the acceleration changes in proportion, but the force E remains constant. The magnitude of F, instead of being given as for viscous damping by the equation

$$F^2 = 4m^2\omega_0^2\xi_0^2[(\omega_0-\omega)^2 + \omega''^2], \quad \text{where } \omega'' = \tfrac{1}{2}\lambda/m,$$

is now determined by the equation

$$F^2 = 4m^2\omega_0^2\xi_0^2[(\omega_0-\omega)^2 + 4f^2/\pi^2m^2\omega_0^2\xi_0^2].$$

Hence for frictional forces the response to a force of constant magnitude but varying frequency is like that of an undamped vibrator:

$$\xi_0 = (F^2 - 16f^2/\pi^2)^{\frac{1}{2}}/2m\omega_0(\omega_0 - \omega).$$

The only essential difference is that F must be large enough to overcome the friction; also it should not be forgotten that if F is only a little larger than required for this purpose the assumption of light damping is hardly valid. It may be found instructive to compare this approximate result, obtained so simply, with the exact analysis, a considerably more troublesome calculation. The result of the exact treatment is

$$\xi_0 = \frac{1}{m(\omega_0^2 - \omega^2)}\left[F^2 - \frac{f^2(\omega_0^2 - \omega^2)^2}{\omega^4}\tan^2\left(\frac{\omega_0\pi}{2\omega}\right)\right]^{\frac{1}{2}},$$

which reduces to the approximate solution when ω is close to ω_0.

Frictionally damped vibrators may not be commonly encountered, but aerodynamic damping is certainly non-linear in many real systems, particularly vibrating strings. At low speeds the damping force is viscous, proportional to speed, but at higher speeds it varies as the square of the speed. In the latter case we have that

$$F^2 \propto \xi_0^2(\omega - \omega_0)^2 + \beta\xi_0^4, \quad \beta \text{ being a constant.}$$

The amplitude at resonance varies as $F^{\frac{1}{2}}$, but far from resonance, where it is controlled by spring constant or inertia rather than by losses, it varies as F. With increasing driving force the effective Q-value of the resonance falls.

Low-Q resonant peaks

If the damping term is not small, the approximation (17) is inadequate and the complete resonant denominator, Y, involving both natural frequencies, Ω_1 and Ω_2, must be used. To draw attention to but one point ignored in the high-Q approximation, the response at zero frequency is controlled equally by Ω_1 and Ω_2 which lie equidistant from the origin. The true response is twice as great as is suggested by the high-Q approximation. We shall consider only two special cases of low-Q resonances, the displacement response $\chi_d(\omega)$ as given by (2), and its 'time-derivative' $\chi_v(\omega)$ as given by (8), the velocity or current response. By means of (3) these two responses are expressed in terms of the displacement response at zero frequency, $\chi_d(0)$, which we write as χ_0:

$$\chi_d = -\omega_0^2\chi_0/(\omega^2 - \omega_0^2 + 2i\omega''\omega), \tag{20}$$

$$\chi_v = -i\omega\chi_d = i\omega\omega_0^2\chi_0/(\omega^2 - \omega_0^2 + 2i\omega''\omega). \tag{21}$$

The power responses take the form

$$P_d = \chi_d\chi_d^* = \omega_0^4\chi_0^2/[(\omega^2 - \omega_0^2)^2 + 4\omega''^2\omega^2], \tag{22}$$

$$P_v = \chi_v\chi_v^* = \omega^2\omega_0^4\chi_0^2/[(\omega^2 - \omega_0^2)^2 + 4\omega''^2\omega^2]. \tag{23}$$

These expressions for the power response, although derived for an input expressed in complex terms, are also correct when it is expressed in real terms; whether the input is expressed as $a_0 e^{-i\omega t}$ or $a_0 \cos \omega t$, and the response as $a_1 e^{-i(\omega t+\phi)}$ or $a_1 \cos (\omega t+\phi)$, P_d and P_v represent $(a_1/a_0)^2$.

The maximum of P_d occurs at a frequency ω_d (*displacement resonance*) given by $\omega_d^2 = \omega_0^2 - 2\omega''^2$, which may be manipulated by use of (2.27) into the form

$$\omega_d = \omega_0 \left(\frac{4Q^2-1}{4Q^2+1}\right)^{\frac{1}{2}} \sim \omega_0(1-1/Q^2+\cdots). \tag{24}$$

On the other hand, *velocity resonance*, the point at which P_v reaches its peak, is always at the undamped natural frequency, ω_0, no matter how low Q may be. The reason can be seen from the series resonant circuit of fig. 1; the amplitude of the impedance is smallest when L and C cancel each other, at ω_0, leaving only R. It is clear from (24) that the separation between the frequencies of displacement and velocity resonance is normally much less than the width of resonance, ω_0/Q, and it is not often that the distinction is worth making.

Substituting ω_d for ω in (22) yields the peak value of P_d:

$$(P_d)_{max} = \chi_0^2(Q+1/4Q)^2. \tag{25}$$

The displacement is enhanced by a factor very nearly equal to Q, compared with the response to a steady force. Similarly, substituting ω_0 for ω in (2) we find the peak velocity response to be nearly proportional to Q (not Q times the zero frequency response since $\chi_v(0)$ vanishes):

$$(P_v)_{max}\omega_0^2\chi_0^2(Q^2+\tfrac{1}{4}).$$

It is, of course, this property that makes tuned circuits useful in radio receivers, etc. They respond to a narrow band of frequencies as if the inductance and capacitance which normally limit the high-frequency current were absent. In themselves, they have no amplifying power – they simply prevent the signal being blocked by unfavourable impedances.

This point is brought out clearly when one considers the practical problem of generating an oscillatory magnetic field by passing an alternating current through a solenoid, represented by L and R in series. If the mains frequency is suitable there is available a source of comparatively low voltage capable of delivering an extremely large current, since the generator and feeder line resistance may be very small. It may then be convenient to put a capacitor in series with the coil to convert it into a series-resonant circuit (we ignore economic considerations – at mains frequency the capacitor may in fact be too large and expensive to be worth while). The impedance of the coil is reduced from $j\omega L + R$ to R, and the current increased by $(1+\omega^2L^2/R^2)^{\frac{1}{2}}$, which is approximately Q. The voltage across the coil is correspondingly large, about Q times as great as the supply voltage, and there is an equal, but oppositely phased, voltage across the capacitor which must be robust enough to stand this.

On the other hand, one may have as source a power oscillator delivering a high voltage but with a significant output resistance so that even short-circuited it can pass only a small current. The solenoid must then have many turns, with correspondingly high L and R, and the performance may be optimized by the use of a parallel resonant circuit as in fig. 3. At resonance the impedance is resistive and equal to L/CR, i.e. about Q^2 times greater than R; the oscillator delivers very little current when Q is high, and therefore its own resistance is less deleterious. But a current i entering the resonant circuit requires a voltage $Q^2 Ri$ across its terminals, and this produces a current $Q^2 Ri/(R+j\omega L)$ through the coil, approximately Q times as great as the input current from the oscillator. If the internal resistance of the oscillator is the controlling agent, limiting the current output, a parallel resonant circuit improves things by a factor Q, which may well be more than 100 for a copper coil at room temperature.

One last point is worth making here; when the losses in a resonant system are reduced so as to make the peak narrower, narrowing takes place through raising the peak to a greater extent than the wings, not by cutting down the wings. This is obvious from (22) and (23), whose denominators can only be decreased by reducing the value of ω''. Fig. 11 brings out this point by showing the resonant peaks for the same system but with Q varied by changing the losses. The diagram also shows how the difference between the displacement response at zero and infinite frequencies is χ_0, determined by the force constant (or its equivalent) and in no way by the losses. This is relevant to the theory of the optical properties of materials (especially anomalous dispersion) which we shall discuss in chapter 7.   **204**

Transient response

By building up the theory of the driven vibrator from the impulse response we have avoided formally solving the equation of motion, and indeed could continue in this vein since, as we have demonstrated fully, the impulse response contains the complete solution. Nevertheless it is desirable to give the formal solution also, for it often provides the most economical analysis of a problem. Typical equations of motion for resonant systems driven by sinusoidal forces are given in (14) and (15), and we may take as representative of all such systems the form

$$\ddot{\xi} + k\dot{\xi} + \omega_0^2 \xi = f_0 e^{-i\omega t}. \tag{26}$$

It is a standard theorem that the general solution of an equation of this type, in which the left-hand side is a linear function of ξ and the right-hand side a function of time alone, with ξ absent, is the sum of *any* solution of (26) plus the general solution of the equation formed by replacing the right-hand side by zero, i.e. (2.23). We know how to meet these requirements from what has gone before. To find some solution of (26), substitute $\xi = A e^{-i\omega t}$ which will generate the steady-state solution analogous to (5):

$$\xi = f_0 e^{-i\omega t}/(-\omega^2 - i\omega k + \omega_0^2) = -f_0 e^{-i\omega t}/Y.$$

Fig. 11. Magnitudes of χ_d and χ_v for a given vibrator whose Q-value is varied by changing the damping term; the three values $Q = 0.87$, 3 and 10 are illustrated.

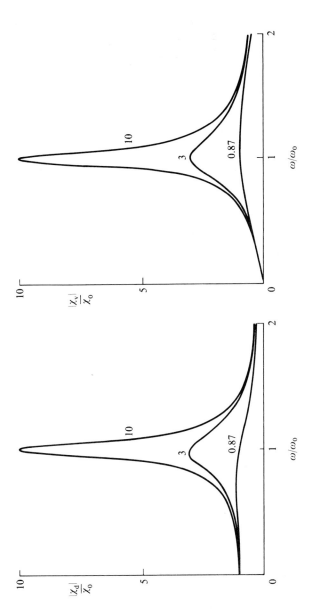

This must be supplemented by the general solution of (2.23) which can be written as the sum of two exponentials:

$$\xi = B\,e^{-i\Omega_1 t} + C\,e^{-i\Omega_2 t}$$
$$= e^{-\frac{1}{2}kt}(B\,e^{-i\omega' t} + C\,e^{i\omega' t}),$$

where, as in (2.24), if damping is subcritical, $\omega' = (\omega_0^2 - \frac{1}{4}k^2)^{\frac{1}{2}}$. The general solution of (26) then takes the form

$$\xi = -f_0\,e^{-i\omega t}/Y + e^{-\frac{1}{2}kt}(B\,e^{-i\omega' t} + C\,e^{i\omega' t}). \tag{27}$$

If the origin of t is chosen so that the actual force is $f_0 \cos \omega t$, the real part of (27) is the physically meaningful solution.

The first term of (27) is the steady-state solution, and the second term is a decaying transient generated when the oscillatory force is first applied or is changed in magnitude. The coexistence of the steady and transient oscillations is to be thought of as a consequence of the linearity of the system, which responds additively to two independent applied forces as if each were alone present. As a special case, one of the forces may be zero, and the transient is the general form of the free oscillation, the response to zero force. Since the transient has the frequency of free vibration, ω', which is not normally the same as the applied frequency, ω, the two terms beat together to produce a periodic modulation of the output which, however, gets less pronounced as the transient decays, and ultimately resolves itself into the steady forced vibration. If the force is represented by $f_0 \cos \omega t$ and is applied at time $t = 0$ to the oscillator at rest ($\xi(0) = 0$, $\dot{\xi}(0) = 0$ immediately after the force is applied), we must find the form of ξ that is real and satisfies these initial conditions. Resolving all exponentials into their trigonometrical components, and redefining the integration constants, we have as a general form for real ξ,

$$\xi = -f_0 \frac{(\omega^2 - \omega_0^2)\cos \omega t - \omega k \sin \omega t}{(\omega^2 - \omega_0^2)^2 + \omega^2 k^2} + \beta\,e^{-\frac{1}{2}kt}\cos \omega' t + \gamma\,e^{-\frac{1}{2}kt}\sin \omega' t.$$

It is now easy to find β and γ so that ξ and $\dot{\xi}$ vanish at the outset, giving as the solution

$$\xi = -f_0 \frac{(\omega^2 - \omega_0^2)\cos \omega t - \omega k \sin \omega t}{(\omega^2 - \omega_0^2)^2 + \omega^2 k^2}$$

$$+ f_0\,e^{-\frac{1}{2}kt} \frac{(\omega^2 - \omega_0^2)\cos \omega' t - \dfrac{\omega^2 + \omega_0^2}{2\omega'}k \sin \omega' t}{(\omega^2 - \omega_0^2)^2 + \omega^2 k^2} \tag{28}$$

$$\doteq -\xi_0[\cos(\omega t + \varepsilon) - e^{\frac{1}{2}kt}\cos(\omega' t + \varepsilon)], \tag{29}$$

where $\tan \varepsilon = \omega k/(\omega^2 - \omega_0^2)$ and $\xi_0 = f_0/[(\omega^2 - \omega_0^2) + \omega^2 k^2]^{\frac{1}{2}}$, the approximation being valid for a reasonably sharp peak, excited at a frequency near resonance so that $\frac{1}{2}(\omega^2 + \omega_0^2)/\omega'$ can be replaced by ω. As ω increases through ω_0, ε varies from π to 0; ξ_0 is to be taken as positive.

Fig. 12. Record of rigid pendulum displacement under the influence of an oscillatory force (*a*) some way off resonance, (*b*) near resonance, (*c*) tuned to resonance. The magnitude of the force is the same in all diagrams; note that the initial rate of growth is the same. For (*d*) the pendulum is more heavily damped, with an oil dashpot, to show the decay of the transient and the approach to the steady state. The scale in (*d*) is not the same as in the others.

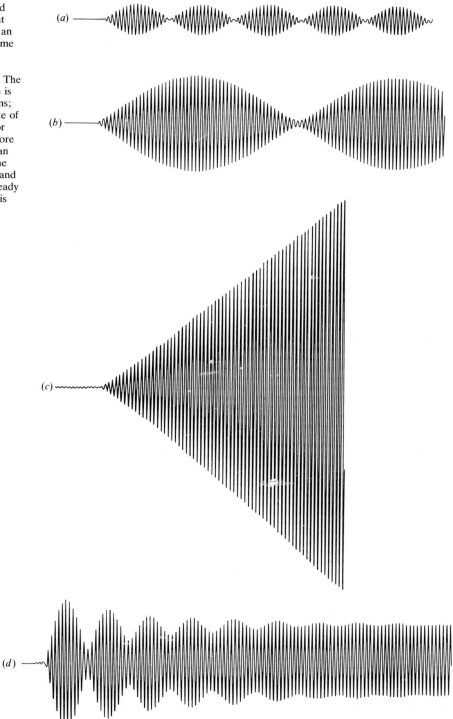

The form of (29) is clearly shown in the recordings, in fig. 12, of the displacement of a very lightly damped pendulum driven by a constant oscillatory force. When ω and ω' differ considerably, the two terms beat rapidly and the resultant oscillation has an amplitude lying between $\omega_0(1 \pm e^{-\frac{1}{2}kt})$, though only in fig. 12(d) is the damping evident. As resonance is approached the envelopes converge at the same rate, so that fewer beats are visible. Exactly at resonance when $\omega = \omega'$, the amplitude follows the lower envelope, tending exponentially to its steady-state value. In these circumstances, $\tan \varepsilon = 4Q$, so that with a sharp resonance $\varepsilon \sim \frac{1}{2}\pi$ and the build-up is closely approximated by

$$\xi = \frac{f_0}{\omega_0 k}(1 - e^{-\frac{1}{2}kt}) \sin \omega_0 t \sim \frac{f_0 t}{2\omega_0} \sin \omega_0 t, \quad \text{when } \tfrac{1}{2}kt \ll 1. \tag{30}$$

The amplitude initially grows as $\frac{1}{2}f_0 t/\omega_0$, independent of k. Damping does not affect the initial stages in the growth of the resonant amplitude, but simply determines how long the growth continues. If there is no damping, the amplitude exactly on resonance grows linearly without limit. Off resonance, the amplitude oscillates sinusoidally, returning to zero periodically since, when $k = 0$ in (29),

$$\xi = 2\xi_0 \sin \bar{\omega}t \sin \Delta t, \tag{31}$$

where $\bar{\omega} = \frac{1}{2}(\omega + \omega')$ and $\Delta = \frac{1}{2}|\omega' - \omega|$. This is simply the beating of two vibrations of equal amplitude.

All these results are, of course, obtainable from the impulse response function, as may be illustrated for the specially interesting case of a very lightly damped system, the same approximations being applied as in the derivation of (17). Let the force $\text{Re}\,[f_0\, e^{-i\omega t}]$, f_0 being complex, be applied when $t = 0$, and let the response function be $\text{Re}\,[a\, e^{-i\Omega_1 t}]$. Then according **106** to (5.5), at time t the response takes the form:

$$\xi(t) = \tfrac{1}{4} \int_0^t (f_0\, e^{-i\omega t_0} + f_0^*\, e^{i\omega t_0})(a\, e^{-i\Omega_1(t-t_0)} + a^*\, e^{i\Omega_1^*(t-t_0)})\, dt_0,$$

the limits of integration being set by the start of the excitation and the vanishing of $h(t)$ when t is negative. The rapidly oscillatory terms in $\omega + \Omega_1$ and $\omega - \Omega_1^*$ may be neglected in the approximation employed, and the result is

$$\xi(t) \sim \tfrac{1}{2}\,\text{Re}\,[if_0 a\,(e^{-i\omega t} - e^{-i\Omega_1 t})/(\omega - \Omega_1)].$$

In particular, for a short while after starting, so long as $(\omega - \Omega_1)t \ll 1$,

$$e^{i(\omega - \Omega_1)t} \sim 1 + i(\omega - \Omega_1)t,$$

and

$$\xi(t) \sim \tfrac{1}{2}\,\text{Re}\,[f_0 a t\, e^{-i\omega t}], \tag{32}$$

which is equivalent to the result in (30), when a is given its correct value i/ω_0.

Let us look at the transient problem from the point of view of the vibration diagram, building up by vector summation the resultant of all the impulse responses initiated by successive increments of the applied force $f_0 e^{-i\omega t}$. First we neglect the resonator losses, so that the solution of the equation,

$$\ddot{\xi} + \omega_0^2 \xi = f_0 e^{-i\omega t},$$

is built up out of increments, each being the vibration that would be excited in the system at rest if it were allowed to experience the applied force for a single element of time δt_0. We therefore look for a solution of the equation,

$$\ddot{\xi} + \omega_0^2 \xi = f_0 e^{-i\omega t_0} \delta t_0,$$

for $t > t_0$, given that $\xi = \dot{\xi} = 0$ immediately before t_0. Integrating through the element of time δt_0 we see that ξ remains zero while $\dot{\xi}$ jumps to $f_0 e^{-i\omega t_0} \delta t_0$. The general form of the subsequent oscillations is given by

$$\delta \xi = A e^{-i\omega_0 t} + B e^{i\omega_0 t},$$

and A and B must be chosen to satisfy the initial conditions. Hence

$$\delta \xi = \frac{if_0 \, \delta t_0}{2\omega_0} [e^{i(\omega_0 - \omega)t_0 - i\omega_0 t} - e^{-i(\omega_0 + \omega)t_0 + i\omega_0 t}].$$

The first term, containing $e^{i(\omega_0 - \omega)t_0}$ in its amplitude, changes phase only slowly with t_0 if ω and ω_0 are very close; subsequent increments can therefore build up to a substantial resultant. The second term, by contrast, is governed by $e^{-i(\omega_0 + \omega)t_0}$ and is rapidly oscillatory. For excitation near resonance it never amounts to much and can be dropped in comparison with the first term. We therefore write

$$\delta \xi = \frac{f_0 \, \delta t_0}{2\omega_0} e^{i(\omega_0 - \omega)t_0 + \frac{1}{2}i\pi} e^{-i\omega_0 t},$$

which is represented on a vibration diagram by a short line element whose length is given by

$$\delta s = \tfrac{1}{2} f_0 \, \delta t_0 / \omega_0, \tag{33}$$

and which lies at an angle ϕ to the real axis, where

$$\phi = (\omega_0 - \omega)t_0 + \tfrac{1}{2}\pi. \tag{34}$$

As usual, the overall rotation of the curve represented by $e^{-i\omega_0 t}$ is set aside in constructing the vibration diagram. Different elements δt_0 at different times t_0 add extra elements to the curve, whose shape $\phi(s)$ is parametrically defined in terms of t_0 by (33) and (34). In this case the curve is a circle, since

$$d\phi/ds = 2\omega_0(\omega_0 - \omega)/f_0,$$

which is constant; $d\phi/ds$ is the curvature, whose reciprocal, $f_0/2\omega_0(\omega_0 - \omega)$, is the radius of the circle. As time proceeds, the length of the arc grows steadily in an anticlockwise sense if $\omega < \omega_0$, starting from the point A in fig. 13(a) so that the first vector has phase $\pi/2$, in accordance with (34) when $t_0 = 0$. The resultant, AB, follows the behaviour of the amplitude expressed by (31), $2\xi_0 \sin \Delta t$, the maximum value being $2\xi_0$, i.e. $f_0/\omega_0(\omega_0 - \omega)$ if $\omega_0 + \omega$ is replaced by $2\omega_0$. This is just the diameter of the circle, and the time between successive zeros, π/Δ, is the period of revolution of B round the circle, $2\pi/\dot{\phi}$.

Fig. 13. Vibration diagrams for a driven resonant system; for explanation see text.

(a)   (b)

If there is damping we must imagine each vector element to start at the length given by (33) but to diminish exponentially as time proceeds. The resulting vibration diagram is then an equiangular spiral. For example, fig. 13(b) shows how the diagram might look after B has accomplished just over $2\frac{1}{2}$ revolutions, while after many revolutions the start of the curve has shrunk into the origin, and thenceforth the resultant amplitude AB stays constant as B goes round. Such diagrams represent the decay of the beat pattern due to the transient, and could be made with little extra difficulty to yield the correct form (28). It would be a real enthusiast for vibration diagrams, however, who would contend that this is an easier approach than the analytical solution of the equation. If, however, the problem is such as to make an analytical solution laborious, the vibration diagram can prove its worth by exhibiting in pictorial form the essential features of the solution.

To illustrate this point, let us examine the response of a resonant system to a gliding tone, for example an oscillatory force of constant amplitude but steadily varying frequency, such as might be expressed in the form

$$f = f_0 \, e^{-i(\frac{1}{2}\beta t^2 + \omega_0 t)}.$$

The phase of the applied force is $\frac{1}{2}\beta t^2 + \omega_0 t$, and if one defines instantaneous frequency as the rate of change of phase, in this case it is $\beta t + \omega_0$, increasing linearly with time and matching the natural frequency of the system when $t = 0$. When this force is applied to a lossless resonant system, the vibration diagram is no longer circular, for although (33) remains true (since the amplitude is constant) ωt_0 in (34) must be replaced by $\frac{1}{2}\beta t_0^2 + \omega_0 t_0$, so that

$$\phi = -\tfrac{1}{2}\beta t_0^2 + \tfrac{1}{2}\pi. \tag{35}$$

Hence $$\mathrm{d}\phi/\mathrm{d}s = \beta t_0 = 2\omega_0 \beta s/f_0. \tag{36}$$

The vibration diagram now has a curvature proportional to distance measured along it. Passing through the origin ($s = 0$) the curvature becomes zero as it changes sign, so that there is an inflexion here; and as one proceeds along the curve the steady increase of curvature causes it to spiral in to a limiting point at each end. In fact (36) defines the Cornu spiral, which we have already met and seen illustrated in fig. 4.14. The following discussion refers to this diagram.

78

In this example of the use of the spiral, the origin is reached when $t = 0$, all parts to the left being contributed by earlier responses to the gliding tone. The amplitude of the resonse is therefore the length of the vector from P to some point on the spiral, and this point moves along the spiral towards Q at a constant rate. At early times the amplitude grows smoothly, and the phase changes rather rapidly. This reflects the fact that the instantaneous frequency of response is not ω_0; at a point on the spiral near P, the tangent is always nearly normal to the vector from P and therefore the phase of the response changes at the same rate as that of the applied force. Before resonance is reached, then, the system follows the oscillations of the force, but with a small amplitude and a phase lag of $\pi/2$. As the gliding tone passes through resonance the response rises sharply and thereafter oscillates in amplitude and phase, but now remaining close in phase to the ultimate phase represented by PQ; i.e. the system is set ringing at its own natural frequency and subsequent excitation, off resonance, can only create small additional oscillations to beat with the principal vibration. The amplitude-variation with time is shown in fig. 14.

The ultimate amplitude is the same as if the force $f_0 e^{-i\omega_0 t}$, tuned to resonance, had been applied for as long as it takes the point to move from R to S on the spiral, i.e. for the gliding tone to slip in phase, relative to the free vibrator, from $\pi/4$ to zero and back to $\pi/4$. Since the phase error is $\frac{1}{2}\beta t^2$, the time needed is $(2\pi/\beta)^{\frac{1}{2}}$ and in this time, according to (30), the amplitude will rise from zero to $f_0(\frac{1}{2}\pi/\beta)^{\frac{1}{2}}/\omega_0$; this is the amplitude at which the lossless resonator is set ringing.

If the resonator is lossy, the left-hand end of the spiral must be imagined to shrink as B progresses along the curve, and the end-point A follows B at a suitable distance. The mathematical details involve Fresnel integrals with complex arguments, which are accessible in tabulated form but hardly enlightening. The physical character of the behaviour is fairly obvious without numerical work. If the gliding tone shifts frequency rapidly enough to pass through the middle region RS in much less than the decay time, the build-up follows fig. 14 very closely. After resonance the response oscillates in amplitude but gradually decays with the natural decay time of the

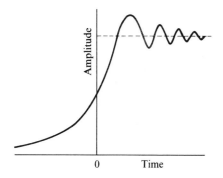

Fig. 14. Time variation of amplitude of response of a lossless resonator to a gliding force which passes through the resonant frequency when $t = 0$.

resonator. On the other hand, if the tone glides so slowly that at each frequency there is time enough for transients to decay, the amplitude follows a resonance pattern as in fig. 11. It is only in between these limits that numerical computation may be helpful, but even then one can probably learn more, and more quickly, by sketching vibration diagrams than by searching through books of tables.

This discussion of transient phenomena may have conveyed the impression that there is more information locked up in the impulse response function than in, say, the steady-state response to an oscillatory disturbance. In fact, as is made clear by (5.3) and (5.4), they are equivalent descriptions; the steady-state response is in no sense an approximation which has to be corrected by the addition of the transient. To take a specific example, when a signal $f_0 e^{-i\omega t}$ is switched on at time $t = 0$ the fact that it has not existed for all time inevitably means it is not a pure sinusoid, but has a certain narrow Fourier spectrum. Each Fourier component, at frequency ω_1, say, is indeed pure and excites a response whose magnitude is determined by $\chi(\omega_1)$, the steady-state compliance at the same frequency. It is now a matter of routine calculation to add the responses over the whole range of the spectrum and to show that the resultant response is a gradual build-up to an ultimately steady oscillation. Because the Fourier transform of the applied signal is singular if it is allowed to continue indefinitely, at constant amplitude, we shall assume it to take the form:

106

$$g_0(t) = f_0 e^{-(i\omega + \lambda)t}, \quad \text{when } t > 0,$$

where λ is small and will eventually be allowed to vanish. Then the Fourier transform follows from (4.16):

$$f(\omega_1) = \frac{f_0}{2\pi} \int_0^\infty e^{-[i(\omega - \omega_1) + \lambda]t} \, dt = \frac{f_0}{2\pi} \Big/ [i(\omega - \omega_1) + \lambda].$$

When this is applied to a transducer whose compliance is $\chi(\omega_1)$, the output is the Fourier transform of $\chi(\omega_1)f(\omega_1)$,

i.e.
$$g_1(t) = \frac{f_0}{2\pi} \int_{-\infty}^{\infty} \chi(\omega_1) e^{-i\omega_1 t} \, d\omega_1 / [i(\omega - \omega_1) + \lambda].$$

For a system governed by (26), $\chi(\omega_1)$ takes the form $-1/(\omega_1-\Omega_1)\times(\omega_1-\Omega_2)$, in accordance with (2) and (4) when $m=1$; $\Omega_1=\omega'-i\omega''$, $\Omega_2=-\omega'-i\omega''$. Hence

$$g_1(t) = \frac{-if_0}{2\pi}\int_{-\infty}^{\infty} e^{-i\omega_1 t}\,d\omega_1/(\omega_1-\Omega)(\omega_1-\Omega_1)(\omega_1-\Omega_2),$$

where Ω is written for $\omega-i\lambda$. This integral is most readily evaluated by contour integration, along a contour that runs the length of the real axis and returns via a semicircle at infinity. There are three simple poles at Ω, Ω_1 and Ω_2, all lying below the real axis. When $t<0$ the semicircle must lie in the positive half-plane to avoid divergence of $e^{-i\omega_1 t}$ at $-i\infty$. There is no contribution to the integral from the semicircle and the integrand is regular within the contour; hence $g_1(t)=0$ for all negative t. For positive t, on the other hand, the contour must lie in the negative half-plane and enclose the three poles, whose contributions to the integral lead to a non-vanishing value:

$$g_1(t) = -f_0[e^{-i\Omega t}/(\Omega-\Omega_1)(\Omega-\Omega_2)+e^{-i\Omega_1 t}/(\Omega_1-\Omega)(\Omega_1-\Omega_2)$$
$$+e^{-i\Omega_2 t}/(\Omega_2-\Omega)(\Omega_2-\Omega_1)].$$

For a high-Q system excited near resonance $\Omega-\Omega_1$ is much smaller than $\Omega-\Omega_2$ or $\Omega_1-\Omega_2$, and the last term can be neglected without serious error. When λ is allowed to vanish, Ω becomes ω, the applied frequency and the first term is the steady-state response in the same form as (5); the second term is the transient, having the frequency and decrement of the free oscillation, and such amplitude as to cancel the first term (with the help of the neglected third term) at $t=0$. We have thus seen that the same result is obtained in three ways – by this approach, by the use of the impulse response function and by direct solution of the differential equation; it could not, of course, have been otherwise.

One last point is worth making before leaving the question of transient response. In discussing the idea of causality in the last chapter we noted that a resonant system excited by a decaying oscillatory force, whose frequency matched one of the natural (complex) frequencies Ω, would **108** show an infinite amplitude of response. At any other frequency, including more rapidly decaying oscillations, the steady-state response, as expressed by $\chi(\omega)$, was finite. It might be thought, therefore, that except at the resonant frequencies $\chi(\omega)$ provides a satisfactory description of the ultimate steady-state response. In a sense this is true, but it is meaningless unless one can achieve the steady state. In particular $\chi(\omega)$ is a quite inadequate description of behaviour at all ω whose negative imaginary part is greater than that of Ω. For if we imagine a decaying oscillatory force switched on at some instant, exciting transients as well as the decaying 'steady-state' response expressed by $\chi(\omega)$, the transients decay less rapidly than the steady-state response when $\text{Im}[-\omega]>\text{Im}[-\Omega]$; the longer we wait, the further we find ourselves from the steady state – the behaviour is ultimately controlled by the precise way in which the force was first switched on and the transients then excited. Neglect of the transients (to which physicists are prone, being so often concerned with steady-state response) can obviously lead to totally erroneous conclusions in this situation. This point, however, has no bearing on the validity of the Kramers–Kronig relations, which (apart from the fact that they are a mathematical theorem and valid as such even if they do not apply to any physical problem) are concerned with the behaviour of $\chi(\omega)$ only in that domain of ω in which transients can be neglected if one is prepared to wait long enough; the relations are not only mathematically sound but physically useful.

110

Response of a resonant system to noise

Let a series resonant circuit of high Q be excited by a randomly fluctuating e.m.f. from a source of negligible resistance and let the fluctuations occur at a much faster rate than the resonance frequency. Then, as discussed in chapter 4, we may replace the input by a closely spaced random succession of pulses which may be taken as equal, but have equal chance of being positive or negative if, as we suppose, the mean e.m.f. vanishes. The advantage of working with pulses is that the response of the circuit is already known. Each pulse excites a decaying oscillation, and the randomness of their times of arrival allows the mean effect of a sequence to be written by summing intensities. If each pulse has magnitude $\pm P (= \int V \, dt$ integrated through the impulse), the intensity of the current response at a time t later is half the square of the amplitude of oscillation in (7):

92

$$\overline{i^2} = \tfrac{1}{2}P^2 \, e^{-2\omega''t}/L^2, \tag{37}$$

if $\omega'' \ll \omega'$ and the second term is neglected. The mean intensity is found by adding all contributions like (37) from earlier pulses. With an average of N pulses arriving in unit time, the resultant is given by

$$\overline{i^2} = \frac{NP^2}{2L^2} \int_0^\infty e^{-2\omega''t} \, dt = \tfrac{1}{2}NP^2 \tau_e/L^2, \tag{38}$$

where τ_e is $1/2\omega''$, the energy decay time; it is as if all pulses arriving in an interval τ_e are to be considered as contributing with their initial intensity $\tfrac{1}{2}P^2/L^2$, the rest being ignored.

One can, of course, derive the same result by determining the response of the resonant circuit to the Fourier components of a white noise source, making use of the power response function P_v, as expressed in (23). This is the appropriate current response, in power form, for a series resonant circuit whose zero frequency susceptibility $\chi_0 = C$, the capacitance. According to (4.45) the spectral intensity of the voltage noise due to a sequence of pulses is given by

92

$$I_\omega \text{ (voltage)} = NP^2/\pi.$$

The circuit response is therefore given by

$$I_\omega \text{ (current)} = NP^2/\pi \times P_v,$$

and when this is integrated over all positive frequencies the result is

$$\overline{i^2} = \frac{NP^2}{\pi} \int_0^\infty \frac{\omega^2 \omega_0^4 C^2 \, d\omega}{(\omega^2 - \omega_0^2)^2 + 4\omega''^2 \omega^2}$$

$$= \frac{NP^2 \omega_0^2 C^2}{\pi} \int_0^\infty \frac{x^2 \, dx}{(x^2 - 1)^2 + \varepsilon^2 x^2}, \quad \text{where } \varepsilon = 2\omega''/\omega_0.$$

Since the integral takes the value $\frac{1}{2}\pi/\varepsilon$, as shown in the following square brackets, we have that

$$\overline{i^2} = \tfrac{1}{4}NP^2 \omega_0^4 C^2/\omega'' = \tfrac{1}{2}NP^2 \tau_e/L^2, \text{ as in (38),}$$

since $\tau_e = 1/2\omega''$ and $\omega_0^2 = 1/LC$.

$$\left[\text{Note that } \int_0^\infty \frac{x^2 \, dx}{(x^2 - 1)^2 + \varepsilon^2 x^2} = \int_0^\infty \frac{dx}{(x - 1/x)^2 + \varepsilon^2}, \right. \tag{39}$$

and $$\int_0^\infty \frac{dx}{(x^2 - 1)^2 + \varepsilon^2 x^2} = -\int_{x=0}^\infty \frac{d(1/x)}{(x - 1/x)^2 + \varepsilon^2} = \int_0^\infty \frac{dz}{(z - 1/z)^2 + \varepsilon^2}, \tag{40}$$

if z is written for $1/x$. Since the two integrals on the left of (39) and (40) are thus shown to be equal, each is equal to half their sum:

i.e. $$\int_0^\infty \frac{x^2 \, dx}{(x^2 - 1)^2 + \varepsilon^2 x^2} = \int_0^\infty \frac{dx}{(x^2 - 1)^2 + \varepsilon^2 x^2} = \tfrac{1}{2} \int_{x=0}^\infty \frac{d(x - 1/x)}{(x - 1/x)^2 + \varepsilon^2}$$

$$= \tfrac{1}{2} \int_{-\infty}^\infty \frac{dy}{y^2 + \varepsilon^2} = \tfrac{1}{2}\pi/\varepsilon. \tag{41}$$

In verifying the result of the simple derivation of (38) we have in fact gone beyond it, since there is no restriction in the second treatment to high-Q resonant circuits; such a restriction was apparently necessitated in the impulse treatment by the requirement that all phases of oscillatory response were equally likely to be superposed at any instant. This is not true if the damping is large during one cycle, yet the same mean square current is excited. Naturally the character of the response will differ. With a sharply tuned circuit only those noise components very close to ω_0 are effective, and the response will be quasi-sinusoidal; with a low-Q system, however, or one that is overdamped, the wide spectrum contributing to the response will ensure that all traces of periodicity are eliminated. Nevertheless $\overline{i^2}$ is the same, a result which seems very surprising at first sight, but one that is well verified by experiment.

The classic experiments in this field were carried out, not with circuits, but with light vanes and galvanometer coils suspended on torsion wires, and undergoing Brownian motion as a result of the random bombardment of gas molecules. We have here an explicit realization of a resonant system subject to noise in the form of a sequence of impulses, and it is worth working through a specific case in some detail. The example we choose is that in which the gas pressure is so low that each molecule behaves independently, arriving at the vane from deep within the gas and leaving again after striking it. There are then no collective motions of the gas contributing to the forces on the vane – only random individual bombardment at an easily calculated mean rate. Taking a flat vane for convenience, consider an element of area dA at a distance r from the axis of suspension (torsion wire). If there are n molecules per unit volume, of which a fraction $f(v)\,dv$ have a velocity component, normal to the vane, lying between v and $v+dv$, each side of the element will be struck by $nvf(v)\,dv\,dA$ molecules in unit time, and we suppose them to rebound elastically. Thus each collision communicates $2mvr$ of angular momentum, such as would set the vane, initially at rest, oscillating with energy $2m^2v^2r^2/I$, I being its moment of inertia. Different molecules striking the vane at different points communicate different increments of energy, and we may define the mean increment, $\overline{\Delta E}$, as the weighted average:

$$\overline{\Delta E} = 2m^2\overline{v^3}\,\overline{r^2}/I\bar{v}, \tag{42}$$

in which $\overline{v^3}/\bar{v} = \int_0^\infty v^3 f(v)\,dv / \int_0^\infty vf(v)\,dv$ and expresses the weighted mean of v^2 when allowance is made for the faster molecules striking the vane proportionately more frequently; $\overline{r^2}$ is a geometrical property of the vane, $\int r^2\,dA/A$, A being its area (one side only).

Following the lines of the argument that led to (38) we now assert that the mean oscillatory energy $\bar{E}$ of the vane is $\overline{\Delta E}$ multiplied by the number of collisions occurring in τ_e, the natural decay time for the energy of oscillation of the vane. Since the number of impacts in unit time on both sides of a vane of area A is $n\bar{v}A$, we have that

$$\bar{E} = 2nm^2\overline{v^3}A\overline{r^2}\tau_e/I. \tag{43}$$

It is now necessary to calculate τ_e, which we do on the assumption that the oscillations are only damped by the gas. When the vane is turning with angular velocity $\dot{\theta} = \omega$, the forward face receives more impacts at a higher relative velocity and a retarding force arises from this differential effect. The element dA, now moving with velocity $r\omega$, receives $n(v+r\omega)f(v)\,dv\,dA$ impacts on its leading side, each of which transmits $2m(v+r\omega)r$ of angular momentum. The trailing side behaves similarly, but with r reversed in sign. The total angular momentum communicated by these molecules in unit time is therefore given by

$$dG = 2nmrf(v)\,dv\,dA[(v+r\omega)^2-(v-r\omega)^2],$$
$$= 8nmr^2\omega vf(v)\,dv\,dA,$$

and this is the retarding torque due to collision. Integrating over the whole vane and over all collisions we must remember that only molecules moving towards the vane strike it, so that if $\bar{v}$ is defined as $\int_0^\infty vf(v)\,dv/\int_0^\infty f(v)\,dv$, then $\int_0^\infty vf(v)\,dv = \frac{1}{2}\bar{v}$. Hence

$$\Lambda \equiv G/\omega = 4nm\bar{v}A\overline{r^2}. \tag{44}$$

Now Λ is the damping parameter appearing in the equation of free motion when it is written in the form $I\ddot{\theta} + \Lambda\dot{\theta} + \mu\theta = 0$, and by analogy with similar equations we see that $\tau_e = I/\Lambda$. Hence, from (43) and (44),

$$\bar{E} = \frac{1}{2}m\overline{v^3}/\bar{v}. \tag{45}$$

Since $f(v)$ is the distribution function for one component of molecular velocity, $f(v) \propto e^{-mv^2/2k_BT}$, from which it follows that $m\overline{v^3}/\bar{v} = 2k_BT$ and

$$\bar{E} = k_BT. \tag{46}$$

89   This is a result which is required by Boltzmann's equipartition law, for the oscillating vane has two degrees of freedom with quadratic expressions for the energy: $E = \frac{1}{2}I\dot{\theta}^2 + \frac{1}{2}\mu\theta^2$, and the mean energy associated with each should be $\frac{1}{2}k_BT$. It is the overriding power of Boltzmann's law that confirms the necessity for the mean energy to be independent of the strength of damping, as we found for $\overline{i^2}$ in the series resonant circuit with which we began this section. It further follows that (46) is not a special result applicable only when molecular bombardments are independent of one another. Indeed the corollary is that the relationship between random bombardment and viscous drag discovered in this explicit calculation must in fact be more general, applying, for example, to denser gases and fluids even though interaction between the molecules precludes the elementary assumption that the random forces arise solely from uncorrelated impacts.

To see what the general relation is let us recast the preceding argument. If angular momentum δL is communicated to a vane at rest by an impulse, the kinetic energy δE communicated thereby is $\frac{1}{2}(\delta L)^2/I$. Hence we may use (42) and (44) to specify the mean strength of an impulsive torque:

$$\overline{(\delta L)^2} = 4m^2\overline{v^3}\,\overline{r^2}/\bar{v} = 2k_BT\Lambda/N,$$

in which $N = n\bar{v}A$ and is the number of impacts on the vane in unit time.

92   But (4.44) provides the link between the spectral intensity of a white noise source and the strength and frequency of the random impulses that simulate it. We may therefore write for the random torque due to the medium:

$$I_\omega(\text{torque}) = N(\delta L)^2/\pi = 2k_BT\Lambda/\pi. \tag{47}$$

In this form, as expected after the discussion of Boltzmann's law, all molecular details have vanished, only the damping parameter and the temperature remaining to define the noise intensity.

The expression (47) for the random torque can be applied to the suspended vane whatever its Q may be, and there is no need to go into the details which run exactly parallel to our analysis of the series resonant circuit. Naturally, for we have just made sure this will be so, the mean energy of the vane is always $k_B T$, but we find as well that this is divisible into potential and kinetic energy, each $\frac{1}{2}k_B T$ on the average. Thus,

$$I_\omega(\text{angular displacement}) = I_\omega(\text{torque}) \times P_d$$

and

$$I_\omega(\text{angular velocity}) \quad = I_\omega(\text{torque}) \times P_v,$$

$I_\omega(\text{torque})$ being given by (47), P_d and P_v by (22) and (23) with $1/\mu$ playing the role of χ_0. Integration over frequency yields the mean-square angular displacement and angular velocity, and for this the integrals evaluated in (41) are required. In the end the law of equipartition is verified in detail,

$$\tfrac{1}{2}\mu\overline{\theta^2} = \tfrac{1}{2}I\overline{\omega^2} = \tfrac{1}{2}k_B T.$$

Experimental study of these matters was first undertaken by Gerlach and Kappler,[1] using thin silvered mica vanes on very fine silica torsion suspensions, and making photographic records of the slow twisting Brownian motion. In spite of difficulties arising from wandering of the zero of the suspension, Kappler was able to measure the mean-square angular displacement, and hence the mean potential energy, over a period of 100 hours. The measurement yielded a value of Boltzmann's constant, k, within about 1% of the accepted value. Later, Jones and McCombie[2] used a moving-coil galvanometer with photocell amplification of the very small angular displacements. They could vary the damping by means of the resistance in the coil circuit (Kappler changed the air pressure), and could follow the changes in character of the motion as the Q-value was altered. Fig. 15 shows some of their traces, and the variation from quasi-periodic to aperiodic motion as Q is lowered is clear (cf. fig. 4.13, depicting entirely **86** analogous behaviour). Like Kappler, they found the amplitude to be in accord with Boltzmann's law.

A point of some interest arising from this investigation is worth a brief digression. As the external resistance is changed, the damping of the galvanometer is shared in differing proportions by the air and by the resistor; but the Brownian motion retains the same mean energy. This means that the resistor must supply white noise related to its damping effect by the same formula (47) as applies to the air. To make the argument exact we need to introduce the parameters of the moving coil – if it has n turns, each of area A, and is hanging in a magnetic field $\mathcal{B}$, it experiences a couple αi when a current flows in it, α being $nA\mathcal{B}$. The damping effect arises from induction – when turning at a rate $\dot{\theta}$, the flux through the coil changes at a rate $\alpha\dot{\theta}$, inducing a current $\alpha\dot{\theta}/R$, if R is the resistance of the coil circuit, consisting of the coil itself and any external resistor connected across it. The induced current exerts a couple $\alpha^2\dot{\theta}/R$ on the coil, and this is the damping term already mentioned in chapter 2. Thus α^2/R acts as a **33** damping parameter in the same way as Λ, and from (47) we can assert that the power spectrum of the injected random torque is given by

$$I_\omega(\text{torque}) = 2k_B T\alpha^2/\pi R.$$

Fig. 15. Records of Brownian motion of a galvanometer coil (R. V. Jones and C. W. McCombie[2]; (a) coil short-circuited and heavily overdamped, (b) coil open-circuited and underdamped, (c) coil critically damped. These curves may be compared with fig. 4.13 showing how the appearance of noise depends on the bandwidth.

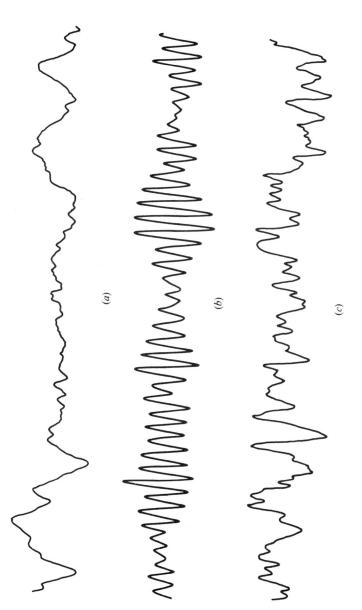

(a)

(b)

(c)

Since this torque results from a current in the circuit, we divide by α^2 to obtain the spectral intensity of the noise current,

$$I_\omega(\text{current}) = 2k_B T/\pi R; \tag{48}$$

and if we like to think of this current as generated by a random voltage source in series with the galvanometer and external resistance, we multiply by R^2:

$$I_\omega(\text{voltage}) = 2R k_B T/\pi. \tag{49}$$

To cast this in conventional form, expressing the mean square noise voltage $\overline{V^2}$ in a band of width β Hz, i.e. $\Delta\omega = 2\pi\beta$, we write

$$\overline{V^2}/R = 4k_B T\beta. \tag{50}$$

Alternatively we may speak of the noise current in a closed circuit,

$$R\overline{i^2} = 4k_B T\beta. \tag{51}$$

This, the *Johnson noise*,[3] is a consequence of the Brownian motion of the electrons in a conductor which, as they collide with the thermally excited ionic lattice (the mechanism of resistance) receive random impulses. The net result in a closed loop is that sometimes there is an average electronic drift in one direction, sometimes in the other, a random current resulting in accordance with (48). It is easy enough to derive these formulae directly from Brownian motion theory, but we shall not digress to this extent. It is sufficient for immediate purposes to recognize that the random collisions simulate the effect of inserting in series with the resistive circuit an ideal random voltage source governed by (49). This remains true for all circuits in which the resistor may be incorporated, and a circuit containing many resistors may be treated as having the appropriate noise voltage inserted in series with each. Since they are uncorrelated in their noise spectra the mean square fluctuations of current or potential at any point in the circuit are to be found by adding the separate mean square contributions from each noise source.

One might imagine that the Brownian analysis of the random electronic motions would reveal a dependence of the noise voltage on details of the model – the dynamics of electron motion, the extent to which their motions are correlated by Coulomb interactions etc. But (50) dispels any such hopes; just like the effect of the surrounding medium on a vane, as expressed by (47), the Johnson noise is an equilibrium effect and, like so many equilibrium properties, has its magnitude determined by gross macroscopic parameters so as to reveal no details, not even whether the carriers are electrons or ions. Microscopic considerations such as the examples mentioned all play their part in determining the actual magnitude of the fluctuations and of the resistivity, but in such a way that the one is uniquely linked to the other.

There is, however, another situation – well away from thermodynamic equilibrium – where microscopic information is conveyed by the magnitude of the noise; this is the *Shot effect*. A thermionic emitter (e.g. a hot tungsten filament) in a simple vacuum diode can pass no more current than the filament emits. With a low voltage between collector (anode) and emitter (cathode), space charge builds up in the intervening vacuum, and the electrons in this gaseous cloud may re-enter the filament on striking it. A high enough voltage, however, prevents space charge accumulating, so that every electron as it leaves the cathode is inexorably drawn to the anode. The saturation current consists of a random stream of electrons, and it is of interest to see how these will excite a resonant circuit connected to the anode (fig. 16). One is tempted to ask whether the impulse due to the electron occurs at the moment it leaves the cathode or at the moment it arrives at the anode; analysis of the potential variations as a charged particle passes from one plate of a capacitor to the other shows that the impulse is not sharp but is spread over the transit time, which typically might be 10^{-8} s. If this is much shorter than the period of the resonant circuit the question becomes unimportant and each electron can be considered to generate a sharp impulse.

Fig. 16. Resonant circuit excited by shot noise of a saturated thermionic diode.

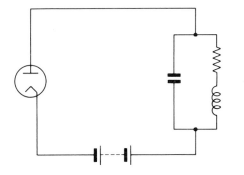

It is now straightforward to apply the impulse response method. An electron leaving the cathode and reaching the anode establishes virtually instantaneously a potential difference e/C across the capacitor, but no current in the circuit. If the Q-value is high the oscillations of voltage start with this amplitude and decay with energy decay time τ_e equal to L/R, during which time Li/Re electrons arrive if i is the mean current. Hence the mean square voltage across the capacitor may be written

$$\overline{V^2} = \frac{Li}{Re} \times \frac{1}{2} \frac{e^2}{C^2} = \frac{1}{2} \frac{Lei}{RC^2} = \frac{Qei}{2\omega_0 C^2}. \tag{52}$$

Hull and Williams,[4] who in 1925 made a careful study of the shot effect, used a resonant circuit of Q-value 100 at 725 kHz, the value of C being 600 pF. With a thermionic current of 20 mA the expected value of $\overline{V^2}$ was about 10^{-7} V^2, giving oscillatory voltages of about $\frac{1}{3}$ mV, quite large enough to measure accurately, even with the rather primitive electronics of the day. It is a tribute to their skill and patience over technical detail that their estimated value of the electronic charge was only in error by 1%.

Let us compare the shot noise (52) with that produced by Johnson noise in the circuit itself. A voltage source, whose spectral intensity is given by (49), in series with L, C and R produces a random voltage across the capacitor of magnitude

$$I_\omega(\text{capacitor voltage}) = \frac{2RkT/\pi}{(1 - \omega^2 LC)^2 + \omega^2 C^2 R^2}.$$

On integrating over frequency, making use once more of (41), we find, not unexpectedly, that the mean square voltage is $k_B T/C$ – a further example of Boltzmann's law; the energy in the capacitor, $\frac{1}{2}CV^2$, is a quadratic function and takes a mean value $\frac{1}{2}k_B T$. With the capacitor used by Hull and Williams, the Johnson noise at room temperature had $\overline{V^2}$ equal to about 10^{-11} V^2, a power level 10^4 less than that of the shot noise.

Many points of physical interest arise from these considerations – why, for example, does shot noise, so large here, play no role in the Johnson noise, and how does the noise level change in the experiment of Hull and Williams as the voltage across the diode is increased to eliminate space charge? But these are matters outside our scope, and we must return to topics that are central to the study of vibrations.

7 Waves and resonators

The simple resonators that have been the principal theme up to this point possessed one resonant mode only and were, moreover, excited by external influences whose reaction to the effect they produced could be ignored. We turn now to extensions of these ideas in two directions. First, the resonator may consist of a string, a tube, a transmission line, a waveguide – all media for the propagation of waves, whether mechanical, acoustic or electromagnetic – and may therefore be capable of excitation in a number of different modes (see the discussion of the vibrations of a string in chapter 2). Secondly, the resonator, which may be one of the above or something simpler, may be embedded in a medium for wave-propagation, being excited by the incidence of a wave and re-radiating a wave as it responds to the excitation. For the most part we shall confine the discussion to one-dimensional systems, leaving the analysis of the resonant modes of two- and three-dimensional systems to the next volume. We shall, however, touch on one three-dimensional problem in treating the response of a small resonant system to a plane wave.

26

Preliminary remarks about one-dimensional waves; characteristic impedance and admittance

A well-developed calculus exists for treating the reflection and transmission of waves on one-dimensional transmission lines, and this is unquestionably the right approach to adopt in engineering design and in the analysis of all but the most elementary problems.[1] Since, however, this is fully explained in numerous textbooks and since we are concerned primarily with building up a physical picture of the behaviour of elementary systems, we shall go no more deeply into this than is necessary. It will soon become clear that very little formal technique is in fact needed at this stage. The concept of *characteristic wave impedance* is valuable in two ways; it greatly aids describing the response of the wave to an obstacle – how much of the energy is transmitted, how much reflected and how much absorbed – and it enables one to bring a variety of different systems into close analogy, so that the same procedures apply to all. It can be introduced most readily by means of the parallel-wire transmission line, carrying an electromagnetic wave of the form shown in fig. 1. From the point of

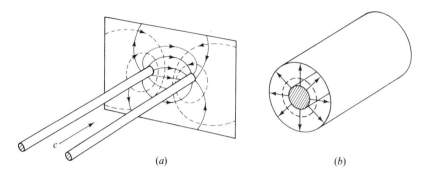

Fig. 1. Electric (full line) and magnetic (broken line) field configurations in the transverse plane of (*a*) twin-wire transmission line and (*b*) coaxial line. In both cases the wave is propagating into the paper.

(*a*)   (*b*)

view of Maxwell's equations we have here a wave travelling in the *z*-direction and described by the field patterns

$$\left. \begin{aligned} \mathscr{E} &= \mathscr{E}_0(x, y)\, e^{i(kz-\omega t)}, \\ \mathscr{B} &= \mathscr{B}_0(x, y)\, e^{i(kz-\omega t)}, \end{aligned} \right\} \tag{1}$$

for which the velocity of propagation, ω/k, is c, the velocity of light in free space, if the conductors are surrounded by empty space, not dielectrics. The conductors control the field pattern, since $\mathscr{E}$ must be normal and $\mathscr{B}$ tangential to them everywhere. In fact, solution of Maxwell's equations shows that for any system of lossless conductors of constant cross-section, of which this is only one example, $\mathscr{E}$ and $\mathscr{B}$ are everywhere normal to and in phase with each other, with the constant ratio $\mathscr{E}/\mathscr{B} = c$. Thus the same holds for the coaxial line of fig. 1(*b*). Looking along the direction of propagation one must turn clockwise to go from the direction of $\mathscr{E}$ to that of $\mathscr{B}$ (i.e. $\mathscr{B} \wedge \mathbf{c} = \mathscr{E}$). This means that when there are waves travelling in both directions on the line, perhaps because of partial or complete reflection at an obstacle, at such points where the electric fields of each add in phase the magnetic fields are in antiphase, and vice versa (see fig. 2). Thus, to use real notation, if the strength of the electric field varies along some line, parallel to the conductors, according to the equation

$$\mathscr{E} = \mathscr{E}_0 \cos(\omega t - kz) + \mathscr{E}_1 \cos(\omega t + kz), \tag{2}$$

the magnetic field varies as

$$\mathscr{B} = \frac{\mathscr{E}_0}{c} \cos(\omega t - kz) - \frac{\mathscr{E}_1}{c} \cos(\omega t + kz). \tag{3}$$

Fig. 2. Electric and magnetic fields in a partial standing wave, $\mathscr{E}_0$ and $\mathscr{B}_0$ referring to a wave propagating into the paper, and $\mathscr{E}_1$ and $\mathscr{B}_1$ to a wave coming out.

61   Both fields form a partial standing-wave pattern, as described in chapter 3, but with maxima and minima interchanged. The maxima of $\mathscr{E}$ are found where $kz = n\pi$, and at these points $\mathscr{E} = \pm(\mathscr{E}_0 + \mathscr{E}_1)\cos\omega t$ while $c\mathscr{B} = \pm(\mathscr{E}_0 - \mathscr{E}_1)\cos\omega t$; where $kz = (n + \tfrac{1}{2})\pi$, however, it is $c\mathscr{B}$ that has the amplitude $\mathscr{E}_0 + \mathscr{E}_1$ and $\mathscr{E}$ the amplitude $\mathscr{E}_0 - \mathscr{E}_1$. At the maxima and minima the oscillations of $\mathscr{E}$ and $\mathscr{B}$ are in phase, but elsewhere there is a phase difference. The ratio $\mathscr{E}/\mathscr{B}$ is in general complex and varies along the line; its value at any point is frequently referred to as the impedance.

An alternative point of view, which approaches more closely the idea of impedance as used in a.c. circuits, starts from the currents carried by the lines and the e.m.f. between them. The electric field lines terminating at the surface of the conductors demand a surface charge of density $\varepsilon_0\mathscr{E}$, and the magnetic field can only be prevented from penetrating the conductors by a surface current of density $\mathscr{B}/\mu_0$. If the conductors carry charge $\pm q$ per unit length and have an e.m.f. V between them ($V = \mathscr{E} \cdot d\boldsymbol{l}$ taken between the conductors along a line lying in a plane normal to the conductors), the capacitance per unit length, C, is defined by the equation $q = CV$; thus between the planes A and B in fig. 3(a) the lines carry charge $\pm CV\,\delta z$.

Fig. 3. Diagrammatic representation of current and electric field on a transmission line. In (a) the planes A and B are δz apart, and $\delta\mathscr{E} = (\partial\mathscr{E}/\partial z)\,\delta z$, $\delta i = (\partial i/\partial z)\,\delta z$. In ($b$) is shown the sign relationship between voltages and currents in the waves incident on, and reflected from, the terminating impedance Z.

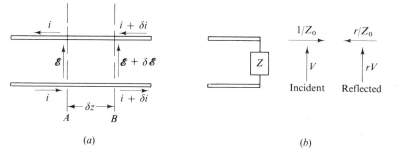

With currents $\pm i$ flowing as shown, magnetic flux passes through the circuit defined by the conductors and the planes A and B. If this flux is written as $Li\,\delta z$, L is the inductance per unit length. It now follows from Faraday's law of induction, applied to this elementary circuit, that

$$L\,\partial i/\partial t = -\partial V/\partial z, \tag{4}$$

and from the conservation of charge that

$$C\,\partial V/\partial t = -\partial i/\partial z. \tag{5}$$

Hence
$$\partial^2 V/\partial z^2 - LC\,\partial^2 V/\partial t^2 = 0,$$

the equation of wave-motion with velocity $c = (LC)^{-\frac{1}{2}}$. Direct calculation of L and C shows that c is $(\varepsilon_0\mu_0)^{-\frac{1}{2}}$, the velocity of light.

We are, however, more interested at the moment in the relation between V and i, which is obvious from (4) and (5); if, as in a travelling wave, the ratio V/i is the same everywhere,

$$Z_0 \equiv V/i = (L/C)^{\frac{1}{2}} = Lc. \tag{6}$$

Because $\mathscr{E}$ and $\mathscr{B}$ in the field pattern are in phase, so too must V and i be, so that Z_0 is a pure resistance if, as assumed here, there is no dissipative process present. The interpretation of Z_0, the characteristic impedance of the line, is very straightforward in terms of the energy flux associated with the wave. When, for example, a wave travels from left to right in fig. 3(a), an observer at the plane A sees current i entering along one conductor and leaving along the other, while an e.m.f. V exists between them. If the line to the right of A were replaced by a resistor Z_0 between the conductors the product $\frac{1}{2}Vi$ would represent the mean rate of heat production therein. Instead, with the line extending to infinity there is no heat production but still a mean flux of energy $\frac{1}{2}Vi$ crossing the plane A and travelling down the line with the wave.

If the line is severed at A and the portion to the right replaced by a resistor Z_0, no reflection of the wave occurs, since voltage and current in the travelling wave are just right to supply the resistor. The line is then said to be correctly terminated. In general, however, an impedance Z terminating a line causes reflection. Let a wave of unit amplitude ($V = 1$) be incident on Z, and let there be a reflected wave of amplitude r. The voltage and current configurations in each wave at Z itself are shown schematically in fig. 3(b); the vertical lines represent the voltages, and the horizontal lines the corresponding currents on one of the conductors (it does not matter which). As with $\mathscr{E}$ and $\mathscr{B}$ in the first treatment, if the voltages add, the currents subtract; but we do not assume r to be positive, or indeed real, and the diagram serves only to remind us of the sign change. The resultant e.m.f. at Z is $1 + r$ and the current is $(1 - r)/Z_0$. Hence r takes such a value that

$$Z = \frac{1+r}{1-r} Z_0$$

i.e.
$$r = \frac{Z - Z_0}{Z + Z_0}. \tag{7}$$

Unless Z is purely resistive, r is complex, indicating a phase change on reflection. Two special cases follow from (7): if $Z = 0$ (short circuit) $r = -1$ and the e.m.f.s in the two waves cancel at the termination; if $Z = \infty$ (open circuit) $r = 1$ and the currents cancel.

As one moves down the line away from the termination, the magnitudes and phases of V and i alter so that the apparent impedance between the conductors varies with a period of half a wavelength. We shall not be concerned to apply the general formula, but it is so easy to derive that we shall do so, and then draw attention to the more immediately relevant special cases. On moving from the termination towards the source through a distance z, one finds the current and e.m.f. of the incident wave advanced in phase by kz, while those of the reflected wave are equally retarded. Instead of an e.m.f. $1 + r$ we now have $e^{-ikz} + r\,e^{ikz}$, and for current we have

$(e^{-ikz} - r\,e^{ikz})/Z_0$. Hence the length z of line terminated by Z, behaves as an impedance Z', where

$$\frac{Z'}{Z_0} = \frac{e^{-ikz} + r\,e^{ikz}}{e^{-ikz} - r\,e^{ikz}}$$

$$= \frac{Z - iZ_0 \tan kz}{Z_0 - iZ \tan kz}, \quad \text{from (7),} \tag{8}$$

$$= \frac{Z + jZ_0 \tan kz}{Z_0 + jZ \tan kz}, \tag{9}$$

to express it in conventional electrical circuit terms. In particular, if the line is short-circuited ($Z = 0$),

$$Z' = jZ_0 \tan kz, \tag{10}$$

while if it is open-circuited ($Z = \infty$),

$$Z' = -jZ_0 \cot kz. \tag{11}$$

In both cases the apparent impedance is imaginary, as it must be when no mechanisms for loss are present. A short length ($kz < \pi/2$) of short-circuited line is inductive, while a short length of open-circuited line is capacitive. When the line is one-quarter wavelength long ($kz = \pi/2$) the short-circuited line behaves as if open-circuited, and vice versa.

When an impedance is connected either in series with or across an otherwise unobstructed infinite line, as in fig. 4, an incident wave is partially reflected and partially transmitted. To take the series connection, Z

Fig. 4. (a) An impedance Z in series with an infinite line appears to an observer at the left as an impedance $Z + Z_0$; (b) an admittance A in parallel appears as an admittance $A + A_0$.

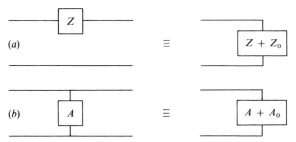

and the line to its right behave as Z_0 in series with Z, so that a wave incident from the left has a reflection coefficient given by substituting $Z_0 + Z$ for Z in (7):

$$r = Z/(Z + 2Z_0). \tag{12}$$

For a wave of unit amplitude incident on the obstruction, the current through $Z + Z_0$ is $(1 - r)/Z_0$, so that the e.m.f. across Z_0 is given by $1 - r$:

$$t = 1 - r = 2Z_0/(Z + 2Z_0). \tag{13}$$

This is the amplitude of the transmitted wave. If the obstacle is non-dissipative, Z is purely imaginary, jX say, and the reflected and transmitted fractions of the incident power are given by

$$rr^* = X^2/(4Z_0^2 + X^2), \qquad tt^* = 4Z_0^2/(4Z_0^2 + X^2), \qquad rr^* + tt^* = 1.$$
(14)

Energy is of course conserved, and the reflected and transmitted waves are in phase quadrature at the obstacle, as follows from (12) and (13), r having an imaginary numerator while that of t is real.

For an obstacle of admittance A placed across the line very similar results obtain:

$$r = -A/(A + 2A_0), \qquad t = 1 + r = 2A_0/(A + 2A_0).$$
(15)

The relationships $t = 1 \pm r$ in (13) and (15) admit of a simple physical interpretation for which we take the case of the obstacle in parallel for the purpose of illustration. To the left of the obstacle there are both incident and reflected waves, and to the right a transmitted wave only, whose amplitudes and phases must be adjusted to give no discontinuity in e.m.f. at the obstacle. Now there are two progressive wave patterns that are continuous in the required sense, (1) a wave that passes the obstacle unchanged and unreflected; (2) equal waves radiating out in both directions from the obstacle and in phase at the obstacle. Neither satisfies all the boundary conditions separately, but a combination of the two does: a progressive wave moving from left to right, having amplitude unity at the obstacle, and a 're-radiated' wave of amplitude r at the obstacle. Then the incident and reflected waves are correctly described, while the transmitted wave is the sum of the two and has amplitude $1 + r$, as required by (15). This is a general result for any lumped obstacle, i.e. one that occupies so small a fraction of a wavelength that the field cannot change significantly across it (or, for a series obstacle, one so short that the currents entering and leaving are the same).† In the special case of a lossless obstacle, the requirement that $rr^* + tt^* = 1$ in addition to $t = 1 + r$ means that a vector

---

† There is considerable local distortion of the field pattern around an obstacle; a capacitor connected by wires between the conductors of a transmission line concentrates the electric field between its plates instead of having it distributed throughout the region between the conductors. The distortion is normally confined to a region round the obstacle of dimensions comparable with the lateral dimensions of the line. If this is much less than the wavelength it is possible to find planes on either side of the obstacle far enough apart for the fields to be undistorted, yet near enough for the space between to be regarded as small for the purpose of treating the obstacle as a lumped impedance. Even when this criterion is not satisfied the concept of a lumped impedance may still be applied to an obstacle whose physical dimensions are small, even though the field distortion is relatively extended. It ultimately breaks down when more than one mode of wave propagation is possible (as in waveguides), for the obstacle irradiated by one mode may generate and re-radiate another. Such a state of affairs obviously demands a more elaborate treatment than anything considered here.

diagram of r and t forms two sides of a right-angled triangle with unit hypotenuse.

185   Later in this chapter we shall refer to obstacles in the form of transducers that are not small in relation to the wavelength and which may connect transmission lines of different Z_0. If these are lossless we may derive all we need to know about them by energy conservation. Consider a non-dissipative transducer to which are connected two lines, with characteristic impedances Z_0 and Z_0', as shown in fig. 5 which also defines the two

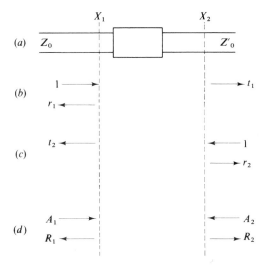

Fig. 5. Location of planes X_1 and X_2 so that all reflection and transmission coefficients are real. In (b) a wave of unit amplitude is incident from the left, in (c) from the right. In (d) the incident waves have amplitudes A_1 and A_2.

reflection coefficients and the two transmission coefficients in terms of the e.m.f. associated with the waves. The two planes X_1 and X_2 are chosen so that the incident and reflected waves are in phase on them; r_1 and r_2 are real and positive, but as yet we know nothing of the phases of t_1 and t_2. Since V^2/Z_0 represents the energy flux in a travelling wave, the absence of dissipation in the transducer implies that

$$1 - r_1^2 = a t_1 t_1^* \left.\right\}$$

and $$1 - r_2^2 = t_2 t_2^*/a,$$ (16)

where $a = Z_0/Z_0'$. Now let waves of amplitude A_1 and A_2 be incident as shown in fig. 5(d), and lead to outgoing waves R_1 and R_2. Then, since R_1 may be regarded as a reflected part of A_1 plus transmitted part of A_2, we have

$$R_1 = r_1 A_1 + t_2 A_2.$$

Similarly $$R_2 = r_2 A_2 + t_1 A_1.$$

But energy conservation demands that

$$R_1 R_1^*/Z_0 + R_2 R_2^*/Z_0' = A_1 A_1^*/Z_0 + A_2 A_2^*/Z_0', \quad \text{for all } A_1 \text{ and } A_2.$$

Hence, substituting for R_1 and R_2 in terms of A_1 and A_2, and equating the coefficients of $A_1A_1^*$, $A_2A_2^*$ and $A_1A_2^*$ to zero, we reproduce (16) and in addition have that

$$r_1t_2^* = -ar_2t_1.$$

These equations are only satisfied if

$$r_1 = r_2 \quad \text{and} \quad t_2^* = -at_1. \tag{17}$$

This is all that energy conservation demands, and it is in fact enough for most of our purposes. But one can go further by invoking the Rayleigh–Carson[(2)] reciprocity theorem:

> If two wires in an electrical network (which need not be small compared with the wavelength, and which may include radiative links) are chosen arbitrarily, and the insertion of an oscillatory voltage generator in one produces a current in the other, then the insertion of the same generator in the other produces the same current in the one; i.e. if $V\,e^{-i\omega t}$ in one produces current $A_{12}V\,e^{-i\omega t}$ in the other, A_{12} being complex in general, then $A_{12} = A_{21}$.

This is a very powerful consequence of the reciprocity of the fundamental laws (e.g. Coulomb's and Ampère's) which can be traced back to the basic properties of space–time that are responsible for Newton's third law, the earliest reciprocity law of all. The Rayleigh–Carson theorem applies, of course, only to linear systems for which the admittances A_{12} and A_{21} can be defined.

Its application to the present problem is illustrated in fig. 6. First the line on the right is short-circuited at X_2 and an e.m.f. is applied at X_1 by

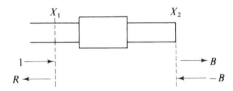

Fig. 6. As in fig. 5, but with the line short-circuited at X_2.

injecting a wave of unit amplitude. At X_2 let the amplitudes of incident and reflected waves be $\pm B$ as shown. Then R is made up of reflected and transmitted waves:

$$R = r_1 - t_2B$$

and

$$B = -r_2B + t_1.$$

Hence $B = t_1/(1 + r_2)$ and $R = r_1 - t_1t_2/(1 + r_2).$

The e.m.f. at X_1 is $1 + R$ and the current at X_2 is $2B/Z_0'$.

Hence

$$A_{12} = \frac{2t_1/Z_0'}{(1 + r_1)(1 + r_2) - t_1t_2}.$$

We now repeat the calculation with the short circuit at X_1 and the input at X_2, which yields A_{21} simply by interchanging subscripts 1 and 2 and replacing Z'_0 by Z_0. The denominator being unchanged, the Rayleigh–Carson theorem then demands that $t_1/Z'_0 = t_2/Z_0$,

i.e. $$t_2 = at_1. \tag{18}$$

It should be noted that this derivation does not make use of energy conservation, and (18) is therefore true for lossy as well as lossless transducers. In general, however, $r_1 \neq r_2$ – there is, for instance, no difficulty in arranging that a wave incident from the right shall be reflected, while one incident from the left shall be absorbed. But if there is no dissipation, (17) and (18) are both true, and t_1 and t_2 must be pure imaginaries, as has already been proved for the lumped obstacle. It follows that a lossless transducer may be replaced, for the purpose of calculation, by a lumped obstacle, a capacitor shunted across the line for example, and flanked by suitable lengths of line to produce the right phases of reflection and transmission. This simulation does not in general work over a range of frequencies, the form of the equivalent obstacle being a function of frequency.

Analogues of characteristic impedance

The characteristic impedance is valuable, as already remarked, as a unifying concept bringing many types of wave-motion into the same formalism. In order to calculate the behaviour of any given one-dimensional wave we need to know its dispersion law, i.e. how k and ω are related, and what reflection is caused by a given obstacle (there are further problems but they all yield to the same treatment and may be considered as examples arise). In each case we can find analogues of e.m.f. and current which enable results such as (9) to be taken over unchanged. In seeking these analogues we look for field variables (*a*) whose product, like Vi, represents power, (*b*) which are in phase for an unattenuated travelling wave and (*c*) which reverse their relative signs when the direction of propagation is reversed. We tabulate below a few examples.

(1) *String carrying transverse waves with velocity* $c_t = (T/\rho)^{\frac{1}{2}}$, *where* $T = tension$ *and* $\rho = mass$ *per unit length*.
If the transverse displacement is ξ, the transverse component of tension = $T\,\partial\xi/\partial z = \mathrm{i}kT\xi$. Taking this as the analogue of V, the analogue of i must be the transverse component of velocity, $-(\partial\xi/\partial t)$ or $\mathrm{i}\omega\xi$ (the sign is fixed by the requirement that Z_0 shall be positive and Vi represent the flux of energy in the direction of wave propagation).

Hence $$Z_0 = \mathrm{i}kT\xi/\mathrm{i}\omega\xi = T/c_t = \rho c_t. \tag{19}$$

(2) *Sound waves with velocity* $c_l = (K/\rho)^{\frac{1}{2}}$, *where* $K = bulk\ modulus$ *and* $\rho = density$.
First we consider waves in a cylindrical pipe of area S, and neglect drag at the walls. If ξ now represents the longitudinal displacement of the medium

and p the local excess pressure, the force due to the wave acting across any cross-section is pS, and the velocity of the medium is $\partial \xi / \partial t$. The rate at which energy is transmitted in the positive direction is $pS \, \partial \xi / \partial t$, and this is the analogue of Vi. Clearly there are different ways of assigning the variables to V and i, but the convenient way is to regard p as equivalent to V and $S \, \partial \xi / \partial t$ as equivalent to i. Then, since the pressure gradient $\partial p / \partial z$ is responsible for acceleration of the medium according to the equation:

$$\partial p / \partial z = -\rho \, \partial^2 \xi / \partial t^2, \quad \text{or} \quad ikp = i\omega \rho \, \partial \xi / \partial t,$$

we have that

$$Z_0 = p/(S \, \partial \xi / \partial t) = \rho c_1 / S. \tag{20}$$

The cross-section of the pipe may be imagined to be made up of any number of smaller elements, as if one had a number of smaller pipes running along together and each carrying a wave of the same amplitude and phase. Then an element of unit area defines a system with characteristic impedance

$$Z_0 = \rho c_1. \tag{21}$$

(3) *Plane electromagnetic waves in free space.*

We can introduce here, as with the sound wave, a characteristic impedance or admittance per unit area, taking $\mathscr{E}$ as the analogue of V. Then since Poynting's vector, $\frac{1}{2}\mathscr{E} \wedge \mathscr{B} / \mu_0$, represents the mean energy flux per unit area, and since $\mathscr{E}$ and $\mathscr{B}$ are at right angles, $\mathscr{B} / \mu_0$ is the consequential choice for i.

Hence $\qquad\qquad Z_0 = \mu_0 \mathscr{E} / \mathscr{B} = \mu_0 c = 377 \; \Omega. \tag{22}$

When the wave is confined by parallel conductors, as in the transmission line of fig. 1, the various elements of plane wave systems must be regarded as being fed in series or in parallel according as they are stacked along $\mathscr{E}$ or along $\mathscr{B}$. Since, however, in such a system $\mu_0 \mathscr{E} / \mathscr{B}$ takes the same value everywhere, we can avoid thinking in these terms by imagining electrodes, shaped exactly like the cross-section of the line, attached to a sheet of uniform resistive material of which any square has resistance between opposite edges of 377 Ω. The resistance between the electrodes is then the value of Z_0 for that line. It is clear that the line can be correctly terminated by attaching such a sheet across its end, for the current that flows is consistent, everywhere on the sheet, with the values of $\mathscr{E}$ and $\mathscr{B}$ in the travelling wave.

A marked similarity can be observed in the expressions for Z_0, in each case the wave velocity being multiplied by a density or its analogue. We have already seen how mass and inductance are analogous for vibrating systems, and the appearance of μ_0, being the specific inductance of free space, underlines this parallelism. There is no need to go into details of the use of Z_0 for other than electrical systems; one elementary example will suffice. Suppose a string hangs vertically under tension T provided by a mass M immersed in a viscous liquid; we ask what fraction of the energy in a transverse wave on the string is reflected by the mass. It is necessary to determine the impedance of the mass in the same terms as were used to define Z_0, i.e. the ratio of applied force to velocity when the mass is vibrating with frequency ω. If its displacement has amplitude ξ_0, and we write v for the velocity $-i\omega\xi_0$, the acceleration is $-i\omega v$, so that

$$F = \lambda v - i\omega M v,$$

where λ is the viscous drag constant. Hence

$$Z = F/v = \lambda - i\omega M,$$

and from (7) and (19),

$$r = \frac{\lambda - \rho c_t - i\omega M}{\lambda + \rho c_t - i\omega M}.$$

If the tension is maintained by a light spring, and the mass M replaced by a light vane so that $i\omega M$ can be neglected, it becomes possible by choosing $\lambda = \rho c_t$ to terminate the line correctly, all wave-motion being absorbed in the liquid without reflection.

Resonant lines

A transmission line short-circuited at both ends can resonate at any frequency that allows an integral number of half-wavelength standing-wave loops to fit in, with zeros (nodes of e.m.f.) at the ends. If there is no dispersion the resonant frequencies form a harmonic series. The same is true for an open-circuited line, but now the ends are antinodes. And for a line that is open-circuited at one end and short-circuited at the other there must be a half-integral number of loops, to allow a node at one end and an antinode at the other. If the length of the line is l, the resonant frequencies are thus determined:

Both ends open-circuited:	$kl = n\pi,$	$\omega = n\pi c/l.$
Both ends short-circuited:	$kl = n\pi,$	$\omega = n\pi c/l.$
One end short-circuited, One end open-circuited	$kl = (n - \tfrac{1}{2})\pi,$	$\omega = (n - \tfrac{1}{2})\pi c/l.$

Similar rules apply to acoustic resonators in the form of uniform pipes. If both ends are closed or open, $\omega = n\pi c/l$, with pressure antinodes at the ends in the first case, pressure nodes in the second; while with one end closed and the other open, as with the transmission line $\omega = (n - \tfrac{1}{2})\pi c/l$. The fundamental is at half the frequency of the fundamentals in the other cases, a fact made use of in designing deep-toned organ pipes for confined lofts. But only odd multiples of the fundamental now appear and the tone lacks richness.

These general remarks somewhat oversimplify the matter. It may not be easy to obtain a perfect short circuit or open circuit and as a result there are usually end-corrections, the resonator behaving as though it were a little longer than its actual length. With an acoustic resonator in the form of a uniform open-ended pipe, the air just outside the ends is in vibration with something like the same amplitude as just inside, and one has to go one or two radii beyond the end before the flow is spread out enough for the velocity to have fallen to a low value. The pressure must oscillate to some small degree at the end to provide the required acceleration of the air outside, and the apparent position of the pressure node therefore lies beyond the end. But so long as the end-correction is simply such as to give a constant extension, Δl, to l, the harmonic series is only displaced down in frequency without losing its simple numerical ratio of frequencies. This is commonly enough what happens when the lateral dimensions are much less than the length. Only when the wavelength is so small as to begin to compare with the lateral dimensions does the end-correction begin to change significantly. For many purposes, then, it can be ignored or treated as an empirical adjustment to l, and for the moment we shall adopt this attitude; but shall return to the matter in due course to discuss the radia-   **179** tion of power from the open end.

This type of resonant line may be thought of either as a series-resonant or parallel-resonant circuit according to the way it is excited. In fig. 7 the two sections of line to either side of the point of excitation can be replaced by reactances, whereupon the character becomes obvious. The circuit (a)

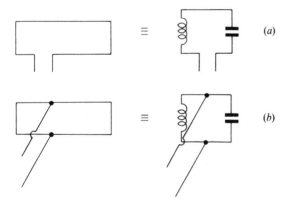

Fig. 7. A closed transmission line fed (a) in series, (b) in parallel, and the equivalent circuits.

fed by an oscillatory current source reveals a resonance at such frequencies as allow the e.m.f. to fall to zero; and in (b) an oscillatory e.m.f. can force no current through the circuit at resonance. The condition for resonance is that the reactance of each section must be equal and opposite. If their lengths are l_1 and l_2, according to (10)

$$\tan kl_1 = -\tan kl_2,$$

i.e.

$$k(l_1 + l_2) = n\pi,$$

the same resonance condition as before, of course.

The essential simplicity of the resonant response is revealed most clearly by the standing-wave pattern produced when a line resonator is excited by a current source at an open end, as in fig. 8. The e.m.f. variation is shown for frequencies around the fundamental resonant frequency, ω_0. The current pattern, with antinodes at the ends, remains virtually unchanged, but as the node of e.m.f. passes through the point of excitation, so the input impedance goes through zero as a linear function of frequency; and that is the essence of a high-Q resonance. If the wave suffers attenuation, so that when ω is real (steady excitation) k is complex, the standing wave may be thought of as a travelling wave totally reflected at the far end, but having lost some of its initial amplitude by the time it returns. As a result the input e.m.f. at resonance does not quite go through zero. In order to make it do so the input wave must have an amplitude that decreases with time so as to match the attenuation of the reflected wave. The true resonance frequency is thus seen to be complex to the extent required to describe the natural decrement of the resonator vibrating freely.

Fig. 8. A current source attached to a closed line, showing how the node of e.m.f. shifts as the resonance is traversed.

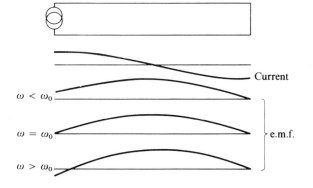

$\omega < \omega_0$

$\omega = \omega_0$

$\omega > \omega_0$

Current

e.m.f.

The regularity of the harmonic series is destroyed by dispersion, for an arithmetic series of wave-numbers, as demanded by the boundary conditions, does not imply an arithmetic series of frequencies unless the velocity is constant. To discuss this problem properly demands an investigation of boundary conditions as well as of the dispersion law – the law expressed by (3.20), for instance, arises from the fourth-order equation (3.19) and the resonances of a finite rod must be found by applying four boundary conditions, two at each end, rather than one at each end as with a string or transmission line. We shall not undertake this investigation here. An easier problem is the non-harmonic series that results when a resonant line is loaded at one point. The effect is easily demonstrated on a piano by threading a fine short nail in and out through one set of three strings such as are used for each of the treble notes. The fundamental is lowered (by perhaps a semitone), but quite as obvious is the strange timbre that results, and the way in which this itself alters as the nail is moved to different positions along the strings. Qualitatively it is easy to see why the harmonic

59

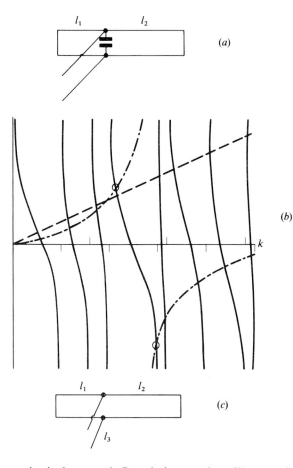

(a)

(b)

(c)

Fig. 9. (a) A closed transmission line with a lumped capacitor across it and excited at that point; (b) diagram for determining the resonances; (c) capacitor replaced by a length of transmission line (exciting line not shown).

series is destroyed. Certain harmonics will cause the string to vibrate with a node at or near the nail, and insertion of the nail hardly changes the pattern of vibration; but those harmonics with an antinode at the nail will be loaded thereby and caused to drop in pitch. For a more detailed analysis consider the transmission line resonator loaded by a capacitor, as in fig. 9, and imagine it excited at the point of loading. The input admittance follows from (10):

$$A = -\mathrm{j}A_0(\cot kl_1 + \cot kl_2) + \mathrm{j}\omega C,$$

and the resonance condition, $A = 0$, implies

$$\cot kl_1 + \cot kl_2 = \omega C Z_0. \tag{23}$$

The left-hand side is plotted in fig. 9, which shows the infinite discontinuities that occur whenever kl_1 or kl_2 is a multiple of π. Two discontinuities are close together if the two sides are almost in resonance together, in which case the capacitor lies near a node. The zeros are evenly spaced and mark the values of k at which the unloaded line is resonant. But the loaded line is resonant where the curves are intersected by the line

$\omega C Z_0$, and immediately one sees how some resonances are shifted more than others, the least affected being those closely flanked by discontinuities.

Also shown on the diagram is the effect of loading the line not with a capacitor but with a short length of open-circuited line (fig. 9(c)), itself capable of resonances. The left-hand side of (23) must now equal $-Z_0$ times the admittance of the load, i.e. $-\cot kl_3$. The infinite discontinuities lead to extra resonances for, as shown by the ringed intersections in fig. 9(b), wherever $\cot kl_3$ has a discontinuity there will be two intersections with the same branch of the left-hand side. In this way the total number of resonant modes is conserved, those originally present in the unloaded line being shifted and supplemented by the resonances of the load. The complexity of even this simple example is enough to discourage a general analysis of lines loaded in more than one point, but fortunately the main characteristics of the behaviour are already plain.

Finger-holes in woodwind instruments

A special case of loading is the drilling of finger-holes in woodwind instruments, to change the pitch. Any real instrument, even the simplest recorder or tin-whistle, is to the physicist a complicated device by reason of the action of the mouthpiece in maintaining vibration. Moreover the bore is rarely uniform; in each instrument special variations of cross-section have been introduced empirically to realise a desirable tone-quality. To bring out the basic properties of finger-holes it is better to present measurements on a uniform pipe; what follows is intended to be merely illustrative and not a definitive contribution to acoustics – in fact all the experiments described in the next page or two were conducted, including the construction of the resonators, during a single day, and could equally readily be carried out by the reader for his own instruction, given access to normal laboratory facilities.

The resonant pipe was a brass tube 1 metre long and of diameter 35 mm with a wall-thickness 1.7 mm. A small loudspeaker, excited by a variable-frequency oscillator, was placed 20 mm from one end and the excitation observed by listening at the other end. It is easy to hear the resonances and to tune the oscillator to the peak of response within about ± 2 Hz; with a few repetitions of each measurement a resonant frequency can be pinpointed within 1 Hz. The first few resonances were at the following frequencies, the figures in brackets being successive multiples of 169: 167 (169), 338 (338), 507 (507), 676 (676), 843 (845), 1016 (1014). One would need to study the matter more carefully to be convinced of any departure from a strict harmonic series. The end-correction is therefore sensibly independent of frequency in this range. That such a correction is needed is demonstrated by repeating the experiment with a tube exactly half the length, the resonant frequencies then being 328, 660, 992, etc., markedly lower than the even harmonics of the longer tube. If one adds 25 mm to the length of each tube the ratio of lengths, 1.95, matches the mean frequency ratio of corresponding harmonics. This is not an accurate determination of the end-correction but is nevertheless quite good. Better investigations suggest that 0.6 times the radius should be added at each end, i.e. 21 mm altogether in this particular case. The coefficient 0.6 is empirical and applies to an unflanged pipe, for which the theory is difficult on account of the awkward flow back outside the pipe. A pipe with a large plane flange on its end, while still troublesome, is not intractable to analysis, and was investigated by Rayleigh,[3] who gave 0.82 times the radius as the correction. If one makes the simplifying assumption that the flow pattern outside is exactly like that outside a hole in a thin plane sheet (see the discussion of Helmholtz resonators in chapter 2) the coefficient comes out at $\pi/4$, which is very close to Rayleigh's result. This is relevant to the next part of the experiment, to which we now turn.

23

The effect of small holes is shown in fig. 10; the holes were drilled one-quarter of the length from one end, at a pressure antinode of the second harmonic where the effect is strongest. It is immediately clear that a

Fig. 10. Effect of hole size on the resonant frequency of an open-ended tube.

very small hole can cause a considerable change in resonant frequency, as anyone knows who has played a recorder and had difficulty in closing every hole completely. The full curve is the theoretical result, with no adjustable constants, derived in the following way. A hole of radius r would behave like an ideal tube of length $\frac{1}{2}\pi r$ if the wall-thickness were negligible, and we have just seen that a long flanged tube has very nearly the same coefficient for its end-correction; we may confidently take $\frac{1}{2}\pi r + t$, t being the wall-thickness, as the effective length. If the characteristic admittance of the main tube, of radius r_0, is A_0, that of the hole is $A_0 r^2 / r_0^2$, and according to (10) the admittance it presents is given by

$$A_{\text{hole}} = -j(A_0 r^2 / r_0^2) \cot \left[k(\tfrac{1}{2}\pi r + t) \right] \doteq \frac{-2jA_0}{\pi k r_0^2} \frac{r^2}{r + 2t/\pi},$$

the approximation being justified since the effective length is very much less than a wavelength. The resonance condition (23) then takes the form

$$\cot kl_1 + \cot kl_2 = -\frac{2}{\pi k r_0^2} \frac{r^2}{r + 2t/\pi}. \tag{24}$$

When $r = 0$ (no hole), $k(l_1 + l_2) = 2\pi$; with the hole situated as it is, $kl_1 = \frac{3}{2}\pi$ and $kl_2 = \frac{1}{2}\pi$, so that $\cot kl_1 = \cot kl_2 = 0$. For small variations of k, then, (24) takes the form

$$(l_1 + l_2)\,\Delta k = \frac{2}{\pi k r_0^2} \frac{r^2}{r + 2t/\pi},$$

and, but for the wall thickness, Δk and hence $\Delta\omega$ would be proportional to the radius of the hole. With larger holes, this approximation fails, since the most effective hole cannot do more than shift the open end to the hole position, rendering the tube beyond ineffective and raising the frequency by a factor $(l_1 + l_2)/l_1$ which is $\frac{4}{3}$ here. It is at first sight somewhat surprising that a hole of diameter 7 mm in a tube 5 times as large should go halfway towards this limit. In woodwind instruments the holes are rather larger in proportion, but the wall thickness is also larger, so that they do not completely immobilize the tube beyond. This explains why finer tuning can be achieved by various combinations of open and closed holes, the open hole nearest the mouthpiece being always the principal factor in tuning.

To conclude the experiment, fig. 11 shows the changes in the resonant frequencies due to a side tube of variable length. The side tube, being closed by oil, is effectively open-circuited with admittance $jA_0(r^2/r_0^2)\tan kl_3$, l_3 being the actual length plus an end-correction at the junction with the main pipe. The broken lines on the graph show the unperturbed resonances of the pipe as horizontal lines, and the fundamental resonance of the side tube running through them. The coupling together of the resonators at the junction produces a characteristic effect which we shall meet several times – at each cross-over the lines are broken and rejoined with different connections. Thus the line ABC may be thought of as describing at the end A the resonance of the main tube in its third harmonic, slightly perturbed by the presence of the side tube, round B as the resonance of the side tube slightly perturbed by the main tube, and round C as the resonance of the main tube in its second harmonic. There is a continuous gradation as l_3 is changed. It will be observed that there are no experimental points over much of the section B; this is because the farther end of the main tube is so little excited that it is almost impossible to discern the resonance by listening there. This picture is confirmed, and can be made quantitative, by use of fig. 9(b). As l_3 is increased, the broken curve suffers a reduction in horizontal scale and the discontinuity moves steadily towards the origin, while the full curves stay fixed. Some resonances are strongly shifted and others less strongly. For those that have a pressure node at the point where the side tube enters, there is no coupling, and the lines in fig. 10 cross without influencing one another; the

(a)

(b)

Fig. 11. The effect of a side arm on the resonant frequency of an open-ended tube; (a) schematic diagram of experiment, (b) shift of certain resonances with length of side arm. The curve *ABC* is theoretical; the horizontal broken curves are resonances of the main tube when $l_3 = 0$, and the third broken curve is the resonance of the side arm alone.

fourth harmonic in the present experiment is very nearly such a resonance, characterized in fig. 9(b) by a branch that plunges almost vertically between two very closely spaced discontinuities. In fact no shift in frequency could be discerned as l_3 was changed. Finally we note that the line *ABC* in fig. 11 is the theoretical curve, again with no adjustable parameters; the theory is verified in a most satisfactory manner.

Radiation from an open end

The treatment of the end-correction to an open-ended acoustic resonator, following Rayleigh's analysis, is incomplete in so far as it assumes no radiation of power in the form of a wave. His solution of the field equation was quite correct, but the equation chosen for solution was itself an approximation. The pressure of the sound field obeys the wave equation

$$\nabla^2 p = \frac{1}{c^2} \partial^2 p / \partial t^2 = -k^2 p$$

for a harmonic wave, and provided the region in which the disturbance exists (e.g. for a distance of a few radii of the tube beyond the end) is much smaller than the wavelength, the right-hand side of this equation is so much less than the cartesian components of the operator on the left that it is reasonable to replace the wave equation by Laplace's equation, $\nabla^2 p = 0$. This eliminates the radiation field, and causes all external regions to vibrate in phase, with a flow pattern of exactly the same form as for irrotational flow of an inviscid, incompressible fluid which also obeys Laplace's equation. A two-dimensional example of such flow is exhibited in fig. 12 which shows how short a path is needed for the fluid to spread evenly into the available space. In three dimensions the same happens, with the flow velocity soon becoming radial and isotropic, and falling as $1/r^2$ at greater distances. It is only at distances comparable to the wavelength that the inadequacies of Laplace's equation become obvious, and if the orifice is small we may at that distance assume perfect isotropy of flow. We can therefore confine our discussion of the wave equation to spherically symmetrical solutions. It is quite possible at this point to proceed purely

Fig. 12. Two-dimensional irrotational flow emerging from parallel plates. The stream-lines define equal fluxes between each.[4]

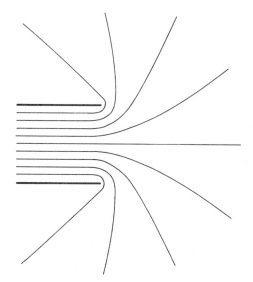

mathematically, by writing the wave equation in spherical coordinates, but we shall instead derive the required form from the basic physics, since we acquire in the process other useful relations.

Consider a spherically symmetrical radial displacement $\xi(r, t)$ applied to a volume element defined by the solid angle $d\Omega$ and the spheres at r and $r+dr$. Then the inner surface moves from r to $r+\xi$, and the outer surface to $r+dr+\xi+(\partial\xi/\partial r)\,dr$. The original volume $r^2\,d\Omega\,dr$ has become $(r+\xi)^2\,d\Omega(1+\partial\xi/\partial r)\,dr$, a relative dilation of $2\xi/r+\partial\xi/\partial r$, to first order in ξ. If the bulk modulus of the medium is K, the pressure at this point is given by

$$p=-K(2\xi/r+\partial\xi/\partial r)=-\frac{K}{r^2}\frac{\partial}{\partial r}(r^2\xi). \tag{25}$$

Now the force on unit volume of the medium is $-\mathrm{grad}\,p$, which in this case is a radial force of magnitude $-\partial p/\partial r$. Hence the equation of motion takes the form

$$\rho\,\partial^2\xi/\partial t^2=-\partial p/\partial r. \tag{26}$$

From (25) we have that

$$\frac{\partial^2 p}{\partial t^2}=-\frac{K}{r^2}\frac{\partial}{\partial r}\left(r^2\frac{\partial^2\xi}{\partial t^2}\right)=\frac{c_1^2}{r^2}\frac{\partial}{\partial r}\left(r^2\frac{\partial p}{\partial r}\right)$$

from (26), where $c_1^2 = K/\rho$. For oscillatory motion

$$\frac{\partial}{\partial r}\left(r^2\frac{\partial p}{\partial r}\right)=-k^2r^2p, \tag{27}$$

whose solution describing an outgoing spherical wave has the form,

$$p=\frac{a}{r}e^{ikr}. \tag{28}$$

Corresponding to this pressure variation the radial displacement ξ follows from (26), and the radial velocity $v=-i\omega\xi$ takes the form

$$v=\frac{ia}{\omega\rho r^2}e^{ikr}(1-ikr). \tag{29}$$

The difference between this and the solution of Laplace's equation lies in the phase variation with distance, described by e^{ikr}, and the extra term $-ikr$ in the coefficient. Much less than a wavelength from the centre, where $kr\ll 1$, v varies as $1/r^2$ (incompressible flow), but at great distances the second term is dominant and v propagates as a wave-motion, of the form e^{ikr}/r. We can now calculate the properties of the open-ended pipe as a radiator of sound, using the close-in behaviour to determine a in terms of the motion at the orifice, and the distant behaviour to determine the energy flux. If the radiation is weak we may assume the standing waves in the tube to be perfect and composed of travelling waves in both directions whose pressure amplitude is p_0. The associated velocity oscillations have amplitude $p_0/\rho c_c$ and at the velocity antinode just beyond the end of the pipe they add to give an oscillatory flux Φ over the whole orifice, of area S, where

$$\Phi = \int v \, dS = 2Sp_0/\rho c_1. \tag{30}$$

This must be matched to the flux in the close-in field described by (29), which we assume isotropic within a solid angle Ω (4π for all-round radiation, 2π for a flanged orifice). Hence

$$\Phi = 2Sp_0/\rho c_1 = ia\Omega/\omega\rho;$$

i.e. $$a = -2iSp_0k/\Omega. \tag{31}$$

At great distances the pressure, with amplitude $|a|/r$, and the velocity, with amplitude $k|a|/\omega\rho r$, vibrate in phase and are responsible for a mean flux of energy per unit area equal to $\frac{1}{2}k|a|^2/\omega\rho r^2$. Over the whole solid angle, then, the radiated power is P, where

$$P = \tfrac{1}{2}k|a|^2\Omega/\omega\rho = 2S^2k^2p_0^2/\rho c_1\Omega. \tag{32}$$

This result can be conveniently expressed in terms of the *radiation resistance* of the orifice. If a line of characteristic impedance Z_0 is terminated by a resistance R which is much less than Z_0, (7) shows that r^2, the fraction of power reflected, is approximately $1 - 4R/Z_0$, so that $4R/Z_0$ is absorbed in the resistor. We assign such a value to R as makes $4R/Z_0$ the fraction of the power arriving at the orifice which is radiated. Since the power carried by the travelling wave is $\frac{1}{2}p_0^2/Z_0$, we have that

$$\frac{R}{Z_0} = \frac{1}{4}\frac{2S^2k^2p_0^2}{\rho c_1\Omega}\frac{2Z_0}{p_0^2} = k^2S/\Omega, \tag{33}$$

since, according to (20), $Z_0 = \rho c_1/S$.

In the experiments described above, the cross-sectional area of the pipe was nearly $10^{-3}\,\text{m}^2$, and at the frequency of the second harmonic the wavelength was close to 1 m, so that $k \sim 6$ and the fraction radiated into the full solid angle was about 1.3%. If this radiation loss is the major cause of dissipation we may estimate Q for the resonance by noting that a wave reflected to and fro at this frequency will suffer 1.3% loss of energy for every wavelength traversed, and will be reduced by a factor e in something like 80 wavelengths, or 500 radians. Radiation therefore limits Q to 500. As it happens there is a greater source of dissipation in the viscous drag near the walls of the tube. Though straightforward enough, it would take us too far afield to develop the theory of this effect, and we shall merely note the result of the calculation – if the viscosity of the gas is η and the radius of the tube r_0 the value of Q due to this source of loss alone is $(\tfrac{3}{2}\omega\rho/\eta)^{\frac{1}{2}}r_0$, about 250 in the present case. The two sources of loss acting together limit Q to 170, which is still high enough to explain how, even by the rather crude device of listening, it was possible to determine resonant frequencies within about 1 Hz.

Similar problems of end-correction and radiation arise with electromagnetic systems, but in detail the differences are considerable. Unlike sound waves, electromagnetic waves cannot be propagated in metal tubes whose cross-section is small compared with the wavelength; the boundary conditions to be satisfied by the electric and magnetic fields do not allow a wave of constant amplitude over the whole cross-section. Typical transmission systems for long waves have at least two conductors, like the parallel wire and coaxial lines of fig. 1. Near the end the lines are distorted, bulging out from the planes normal to the propagation direction, as illustrated in fig. 13. When the wavelength is much longer than the lateral dimensions, the lines take up the same shape as in a charged capacitor of this form. In fact the determination of the end-correction is entirely analogous to the process already described for the acoustic resonator. The boundary conditions are troublesome, however, and there does not seem to exist any systematic compilation of solutions for electromagnetic end-corrections. In practice, of course, it is not hard to determine the natural frequency of a resonator by experiment and thus measure the end-correction directly – its calculation would be an exercise in applied mathematics hardly warranted by the usefulness of the results.

(a)   (b)

Fig. 13. Spread of field lines from the end of (a) a parallel wire transmission line and (b) the inner conductor of a coaxial line.

A long coaxial line carrying a standing wave alternates at quarter-cycle intervals between a pure electric field pattern and one that is purely magnetic; this is the analogue in electromagnetism of the exchange between potential and kinetic energy in a mechanical vibrator. The nodes of one field coincide with the antinodes of the other. If for one moment we ignore end-corrections, there can be no magnetic field at an open circuit, for such a field demands longitudinal currents in the conductors, but there is no objection to an electric field at this point. The field configurations are therefore as shown in fig. 14;

Fig. 14. Magnetic field and current, and electric field and surface charge density in one half-wavelength (marked by broken lines) of a standing wave on a long line.

the charge density on the surface of the conductors follows $\mathscr{E}$ and the current follows $\mathscr{B}$. The end-correction arises from the extra concentration of electric field lines, and therefore of surface charge, near the ends, as shown in fig. 15. In oscillation the extra charge must be alternately supplied and removed by the current, which does not fall to zero as a sinusoid but remains rather large until very near the end. The regions away from the end are thus fitted by a rather longer wavelength, and the resonant frequency is correspondingly reduced. For example, a resonator of this configuration, with outer and inner diameters of 14 and 1 mm, and an inner conductor 13 mm long, was found to have a fundamental resonance corresponding to a free space wavelength of 32 mm, rather than 26; the end-correction amounted to $1\frac{1}{2}$ mm at each end.

Fig. 15. Field lines, current and charge density on a short inner conductor of a coaxial line, to show origin of end-correction:

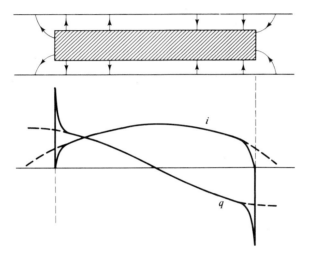

With the outer conductor extended in this way, no radiation is possible until the wavelength is reduced to the point where the outer conductor alone can carry propagating waves (waveguide modes).[1] In a tube much larger than the wavelength there are many different modes of waveguide propagation, each with its own critical free-space wavelength, λ_c. By λ_c we mean that the mode in question cannot be propagated at a frequency less than c/λ_c. The most readily propagated mode for a circular waveguide is the TE_{11} mode shown in fig. 16(a), for which λ_c is 3.41 times the radius of the tube. This is not one that can be excited by a perfectly centred coaxial line, for reasons of symmetry; in the coaxial mode $\mathscr{E}(-r) = -\mathscr{E}(r)$, while in the TE_{11} mode $\mathscr{E}(-r) = \mathscr{E}(r)$. The lowest mode that can be excited by the coaxial line is the TM_{01} of fig. 16(b), which has axial symmetry and a radial electric field to match that of the coaxial line. The critical wavelength for this mode is 2.61 times the radius; there was therefore no possibility of exciting it beyond the end of the inner conductor in the tube of 7 mm radius just mentioned, at a resonant frequency corresponding to a free

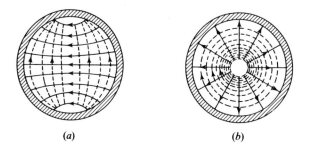

Fig. 16. Electric (full) and magnetic (broken) field lines in the transverse plane of a circular waveguide carrying (a) the TE_{11} mode and (b) the TM_{01} mode.

(a)   (b)

space wavelength of 32 mm; this resonator would be free of radiation losses until the free-space wavelength was reduced to 18 mm (or 25 mm if it was asymmetrical and could radiate a TE_{11} mode).

We need go no more deeply here into the theory and practice of waveguides, since it is enough to appreciate that they are one-dimensional wave propagators to which the analysis in the early pages of this chapter applies. They can be used as the basis of resonators by blocking off both ends with metal plates to act as short circuits; open-ended waveguides radiate strongly, having dimensions comparable with the wavelength, and are not suitable for high-Q resonators. The most popular waveguide is rectangular, operating in its lowest, TE_{10}, mode with the field pattern drawn in fig. 17. The critical wavelength is twice the width measured in a

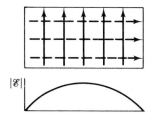

$|\mathscr{E}|$

Fig. 17. Electric (full) and magnetic (broken) field lines in the transverse plane of a rectangular waveguide. The electrical field strength varies sinusoidally across the longer dimension as indicated below.

direction normal to $\mathscr{E}$. The mechanical simplicity of waveguides at microwave frequencies ($\lambda < 20$ cm, say) makes them the obvious choice for illustrating various arrangements in which resonators interact with waves.

Resonators attached to transmission lines

Typical arrangements are shown schematically in fig. 18. In each case the input transmission line, on the left, is imagined to contain a matched attenuating section, so that the wave reflected from the resonator disappears without trace. Similarly, in the cases (c)–(f) where there is an output line on the right, this is imagined to be matched. The three arrangements (b), (d) and (f) parallel (a), (c) and (e), the former being waveguide devices in which the resonant section is defined by diaphragms pierced with coupling holes, while the latter have lumped resonant circuits. There is no essential difference, apart from the lumped circuits having only one resonant frequency. It is the external features that concern us, and the

Fig. 18. Resonant circuits
coupled to transmission
lines, and their waveguide
equivalents. A pierced
diaphragm in a waveguide
allows partial transmission
and partial reflection, and
serves to couple a
resonant cavity to the
waveguide; (a) and
(b) line terminated by
resonator; (c) and
(d) feed-through
resonator; (e) and (f) side
arm terminated by
resonator.

Fig. 18. Resonant circuits coupled to transmission lines, and their waveguide equivalents. A pierced diaphragm in a waveguide allows partial transmission and partial reflection, and serves to couple a resonant cavity to the waveguide; (a) and (b) line terminated by resonator; (c) and (d) feed-through resonator; (e) and (f) side arm terminated by resonator.

influence they have on the form of the resonance. The impulse response may be cited to indicate the differences due to the manner of excitation and detection. In (c), for instance, an incident impulse starts the resonator ringing, and what emerges on the output line is just a decaying oscillation; this arrangement has properties similar to those discussed in chapter 6. In (a), however, there is a reflected impulse followed by the oscillation, while in (e) there is a direct impulse as well as a later impulse and oscillation.

At this point we abandon the use of the impulse response in favour of the conventional approach by way of the response to a single-frequency input. The difficulty of handling the impulse response is particularly marked for the transmission line resonators, which have a large number of resonances at frequencies above the fundamental which we may suppose we are interested in. Instead of being excited to sinusoidal oscillation, the resonator may be thought of as reflecting the impulse to and fro, emitting a small fraction of its energy at each reflection. In general the impulse suffers progressive distortion on reflection, so that the emerging signal is a succession of pulses of gradually changing form. In principle, Fourier analysis of this output would tell one about all the resonances in one operation, but the game is not worth the candle.

Let us consider first the simplest case, (c) and (d), where the output signal is due entirely to the excitation of the resonator and is not mixed with any part of the input. We represent the resonator as a length of line connected by lossfree transducers to the input and output lines, and as in fig. 5 measure amplitudes at those points where the reflection coefficients are real and the transmission coefficients, as we now know, imaginary. We take these coefficients to be r_1 and it_1 for the left-hand junction, r_2 and it_2

129

Fig. 19. Notation used in analysis of feed-through resonator.

for the right; $r_1^2+t_1^2=r_2^2+t_2^2=1$. As fig. 19 shows, a wave of amplitude $A(1+\tfrac{1}{2}\varepsilon)\,\mathrm{e}^{-i\theta}$ starting from the inside of the left-hand junction has amplitude $r_2(1-\tfrac{1}{2}\varepsilon)A\,\mathrm{e}^{i\theta}$ after reflection at the other end, the factors $(1\pm\tfrac{1}{2}\varepsilon)$ taking account of attenuation due to imperfect conductors, dielectric etc, and $\mathrm{e}^{i\theta}$ being the phase change along the resonator; $\theta=kl$ if l is the length of the resonator. Now, by the same argument as was applied to fig. 6, we regard the wave leaving the left-hand junction inside the resonator as made up of a reflected part of the wave arriving and a transmitted part of the incident wave of unit amplitude:

$$A(1+\tfrac{1}{2}\varepsilon)\,\mathrm{e}^{-i\theta}=r_1r_2(1-\tfrac{1}{2}\varepsilon)A\,\mathrm{e}^{i\theta}+it_1.$$

Hence A is determined:

$$A=it_1/[(1+\tfrac{1}{2}\varepsilon)\,\mathrm{e}^{-i\theta}-r_1r_2(1-\tfrac{1}{2}\varepsilon)\,\mathrm{e}^{i\theta}], \qquad (34)$$

and
$$T=it_2A.$$

If we confine attention to high-Q resonators, $\varepsilon\ll1$ and $r_{1,2}\sim1-\tfrac{1}{2}t_{1,2}^2$, so that

$$T\sim-t_1t_2\,\mathrm{e}^{i\theta}/[1-(1-\alpha)\,\mathrm{e}^{2i\theta}],$$

where $\alpha=\varepsilon+\tfrac{1}{2}(t_1^2+t_2^2)$. The nth resonant peak occurs when $\theta=n\pi$ and the denominator takes the very small value α. Only small changes in θ are needed to increase this substantially, so that to a good approximation

$$T\sim\pm t_1t_2/(\alpha-2i\,\Delta\theta), \qquad (35)$$

which, as expected, is of the standard Lorentzian form. If a square-law detector is used to measure the power transmitted through the resonator,

76

$$TT^*=t_1^2t_2^2/[\alpha^2+4(\Delta\theta)^2], \qquad (36)$$

which falls to half its peak when $\Delta\theta=\pm\tfrac{1}{2}\alpha$. To traverse the peak at half power involves a relative change in θ, and hence in frequency, of $\alpha/n\pi$; therefore

$$Q=n\pi/\alpha. \qquad (37)$$

This is what one would expect from the natural decrement of a standing wave excited in the resonator. In going once through the resonator in both directions a travelling wave loses a fraction ε of its amplitude and 2ε of its energy by attenuation, as well as a fraction $t_1^2+t_2^2$ into the coupling lines, making 2α in all. Since during this journey the phase of oscillation advances by $2n\pi$ radians, the loss per radian, which is $1/Q$, is $\alpha/n\pi$.

There is a relationship between the transmission through the resonator and the contribution of coupling losses to the width. According to (35), the amplitude transmission coefficient, ζ, the value of $|T|$ at resonance, is $t_1 t_2/\alpha$. We define Q_0 as the unloaded Q of the resonator, $n\pi/\varepsilon$, when coupling to the external circuits is negligible. Then it follows immediately that

$$Q/Q_0 + \tfrac{1}{2}(t_1/t_2 + t_2/t_1)\zeta = 1. \tag{38}$$

If, therefore, the coupling diaphragms are systematically changed in such a way that the ratio t_2/t_1 stays constant (it may be easiest to achieve this by taking pains that the holes in the diaphragms are identical), Q will vary linearly with ζ, and by extrapolating the line to $\zeta = 0$ one can determine the value of Q due to the resonator losses alone. The resonators shown in fig. 20 are designed to allow t_1 and t_2 to be varied systematically while keeping their ratio constant. In both cases the resonant elements are wires; the

Fig. 20. Two examples of resonant wires coupled in feed-through fashion to coaxial lines. In both cases the total length of wire is a little less than half the free-space wavelength at the resonant frequency. Coupling is altered by raising or lowering the wire.

(a)   (b)

hairpin in (a) is a parallel wire transmission line open at the lower end and short-circuited at the upper, with magnetic coupling to the external coaxial lines which can be varied by moving the hairpin vertically. Similarly the coaxial resonator (b) is adjustable by vertical movement, but the coupling is electrical. An example of the verification of (38) is given in fig. 21. These resonators were used at very low temperatures to study the surface resistance of superconductors at microwave frequencies, and as the sample (the movable wire) was cooled below its superconducting transition temperature the resistance fell rapidly and Q rose to very high values.[5] By adjusting the coupling to keep the fraction of power transmitted at

Fig. 21. Variation of Q with transmission coefficient ζ, when coupling is altered. This curve, obtained with the resonator of fig. 20(a), shows a slight departure from linearity at weak coupling, due to direct coupling between the loops. With better design the fault is easily reduced to insignificance.

resonance constant, ζ and therefore Q/Q_0 remained constant – the variations of the measured values of Q therefore reflected, without the need for correction, relative values of Q_0 at different temperatures. It is perhaps worth remarking, in parenthesis, that these resonators, with their principal conductors in the form of thin wires, were designed to maximize the effect of resistive losses, as is highly desirable when the microwave resistance of superconductors is so small. Even so, values of Q in excess of 5×10^5 were recorded. In other experiments, aimed at achieving the largest Q by building resonant cavities with superconducting walls, values well above 10^{10} have been reached.[6]

One implication of (38) is that if there are no significant internal losses, so that Q is determined solely by the coupling, $\zeta = 2/(t_1/t_2 + t_2/t_1)$ and is unity if $t_1 = t_2$. But this is true only for lossless transducers. If the coupling to the external circuits is by way of lossy transducers the transmission is seriously reduced. Consider, for example, a line on which a resonant element is delineated by shunting two identical small resistors across it, as in fig. 22. From (7) we have that the reflected (or re-radiated) amplitude is $2\gamma - 1$, if $\gamma \ll 1$, and the transmitted amplitude must therefore be 2γ; r and t now are both real, instead of one being real and the other imaginary, as before. The calculation can be carried through in the same way as that leading to (35) and the answer is essentially the same, except that we can

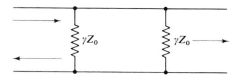

Fig. 22. Resonant length of line delineated by small shunt resistances, modelling the Fabry–Perot interferometer.

no longer take $r^2 + t^2 = 1$. The value of Q is still determined by the internal losses and by the loss on reflection, so that if we write $\alpha = \varepsilon + t'^2$ we must remember that t'^2 is the power not reflected, i.e. $1 - r^2$, but this includes both transducer loss and power into the output line. On the other hand, the factor $t_1^2 t_2^2$ in the numerator of (36) genuinely refers to the power entering the line and excludes what is lost in the resistors. For a given value of r, and therefore of α, the numerator of (36) is $(1 - r^2)^2$ for lossless transducers and $(1 - r)^4$ for resistors acting as transducers. The latter is smaller than the former by a factor $\frac{1}{4}(1 - r)^2$ if r is nearly unity, representing a very significant diminution in transmitted intensity, with the transducers absorbing almost all of what ideally they should transmit.

This presents a real problem in the design of reflecting coatings for optical interferometers. A Fabry–Perot interferometer, in which circular fringes are seen in transmission after multiple reflection between two optical flats, is an instrument governed by the same rules as we have been discussing.[7] A high value of Q is achieved in this case by using a spacing of many (perhaps 10^4–10^5) wavelengths between the optical flats (the transducers). With so large a value of n in (37), α need not be very small to obtain the resolution that a Q of, say, 10^6 confers. But if α is to be made $\frac{1}{10}$ by means of a resistive coating, such as a thin metal film may approximate to, the resulting transmission at resonance – the centres of the bright interference bands – is only $1/1600$ of the incident power. Good approximations to lossless reflecting coatings can be achieved by superimposing layers, about one-quarter wavelength thick, of transparent materials with alternately high and low refractive index. By such means 99% reflection is possible with almost all the remaining 1% of power transmitted without absorption.

Because the high Q-value in a Fabry–Perot interferometer is commonly realized through a large value of n rather than almost perfect reflectivity, the resonant modes lies close together and their responses overlap. To return to (34) rather than the approximation (35), putting $\varepsilon = 0$ and $r_1 = r_2 = r$, we have

$$A = \mathrm{i}t/(\mathrm{e}^{-\mathrm{i}\theta} - r^2 \, \mathrm{e}^{\mathrm{i}\theta}),$$

so that $$AA^* = t^2/(1 + r^4 - 2r^2 \cos 2\theta). \tag{39}$$

The denominator is represented geometrically by the square of the distance D between a point P sweeping round a circle of radius r^2 and a fixed point O at unit distance from the centre of the circle, as in fig. 23. The difference between this and the Lorentzian resonance is that in the latter the circle is replaced by its tangent at the nearest point. If r^2 is close to unity the difference is negligible except far from the peak where the response is relatively very weak. The form of (39) when $1 - r^2$ is not so small is shown in fig. 24. For low reflectivity the curve approximates to a sinusoid. Resolution of the resonances is never totally lost, but the contrast between maxima and minima is weakened.

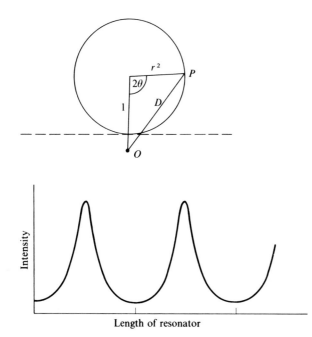

Fig. 23. Geometrical
construction to find
variation of transmitted
intensity with frequency in
the arrangement of fig. 22,
according to (39).

Fig. 24. Line-shape of
transmitted intensity
derived from (39) and fig.
23 with $r^2 = \frac{1}{2}$.

Intensity

Length of resonator

We have talked of resonances since this was the language with which the discussion started, but it will be realized that the conventional description of the variations of intensity is in terms of interference fringes. It is easy enough to make the logical connection between the two, by what may however appear an artificial device. In the optical interferometer θ is changed by altering not the frequency but the angle of incidence of a plane wave on the plates. The artificial device consists of regarding the wave incident at an angle to the axis as a wave travelling along the axis and having a variation of phase across its wavefront. The plane wave represented by $\xi = e^{i(\mathbf{k}\cdot\mathbf{r} - \omega t)}$ is to be thought of as moving along the z-axis:

$$\xi = \xi_0(x, y)\, e^{i(k_z z - \omega t)},$$

with its wavefront modulated according to the variations of $\xi_0(x, y)$,

where
$$\xi_0(x, y) = e^{i(k_x x + k_y y)}. \tag{40}$$

The utility of this device depends on the reflecting surfaces being plane and lying in the (x, y)-plane, causing no change to the form of $\xi_0(x, y)$ – it is only the motion in the z-direction that is changed by reflection, so that in this way oblique waves can be brought into the ambit of our treatment of one-dimensional systems. The argument is illustrated in fig. 25. The true wavelength is λ, but measured along z the 'longitudinal wavelength' is greater, $\lambda_\ell = \lambda \sec \phi$, while the phase variation in the (x, y)-plane is governed by a 'transverse wavelength' $\lambda_t = \lambda \csc \phi$. Changing ϕ for a wave of given frequency does not alter its real velocity, but its velocity along z is greater, being $c \csc \phi$. Because of this (put otherwise, because

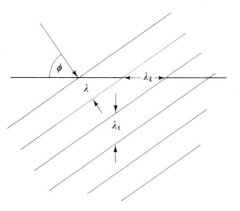

Fig. 25. Obliquely incident plane wave, to illustrate meaning of 'longitudinal wavelength', λ_ℓ, and 'transverse wavelength', λ_t.

the phase θ in (39) is $k_z l$, where l is the distance between the plates) θ can be changed by changing ϕ rather than ω. Hence fig. 24 is also a representation of the variation of intensity with angle when monochromatic light is allowed to fall on the interferometer from all directions.

To conclude this argument, let us note that one way of achieving omnidirectional illumination is to bring near the interferometer an extended monochromatic source (flame, discharge tube etc.). Since every point on the source is an independent emitter, the amplitude distribution in the (x, y)-plane is described by a function $\xi'(x, y)\, e^{-i\omega t}$ whose phase is random, however uniform the strength of the source may be. As a consequence of this randomness of phase, if we Fourier-analyse the spatial variation of amplitude we shall obtain a uniform spectrum. Now the functions $\xi_0 = e^{i(k_x x + k_x y)}$ are nothing else but the basis functions of this Fourier transformation, and we therefore expect all k_x and k_y to appear with equal intensity. The eye or camera, focussed at infinity, channels each Fourier coefficient to a different point on the retina or photographic plate, and thus shows in one single image the response of fig. 24, as a set of interference rings. Fig. 26 is just such a photograph, taken with an exceptionally well-coated pair of plates, so that the bright fringes are very sharp. Even if the appearance of the rings left the matter in doubt, conclusive evidence for the enhanced sharpness of the fringes comes from the fact that alternate rings are formed by two different spectral lines. Sinusoidal fringes, as produced by the Michelson interferometer, are incapable of this resolution; two sinusoids added together produce a single resultant sinusoid.

Let us turn to the next-simplest arrangement, fig. 18(a) or (b), in which the response of the resonator is revealed by variations in the reflected wave amplitude or power. We need not delay to understand the technicalities of separating the reflected from the incident wave – it is enough to know that there are commercially available devices for so doing. The analysis of case (b) can be taken up directly from that of (d), as expressed in (34). After making the same approximations as led to (35) and which are adequate for

Fig. 26. Photograph of circular Fabry–Perot fringes obtained with highly reflecting surfaces, to show very narrow transmission peaks.[8]

high-Q resonators, we have that

$$A \sim it/(\alpha - 2i\,\Delta\theta),$$

where $\alpha = \varepsilon + \frac{1}{2}t^2$, there being now only one external circuit to provide coupling loss. This value of A may also without extra error be taken as the amplitude of the wave falling on the transducer from inside, and hence

$$R = r + itA \sim 1 - t^2/(\alpha - 2i\,\Delta\theta). \tag{41}$$

Well away from resonance there is virtually total reflection from the transducer, and resonance reveals itself as a drop in the reflected power. At the very centre of the resonance, when $\Delta\theta = 0$,

$$R_{\min} = 1 - t^2/\alpha.$$

If $\frac{1}{2}t^2$ is adjusted to equal ε, so that the resonant width is equally due to internal dissipation and to coupling losses, $R_{\min} = 0$. For stronger or weaker coupling the reflected power does not fall to zero, but otherwise the shape is always the same. With a square-law detector the measured output is given by

$$RR^* = 1 - 2\varepsilon t^2/[\alpha^2 + 4(\Delta\theta)^2],$$

being a Lorentzian resonance subtracted from a constant, unity, as in fig. 27.

It is with the arrangement of fig. 18(e) and (f) that the form of the resonance becomes less attractive. There is no need to analyse the general case, in which a wave arriving at the junction from any of the three arms is partially reflected and partitioned between the two other according to the impedances they present. Instead, we shall suppose that the side arm, terminated by the resonator, has a characteristic impedance twice that of

Fig. 27. Absorption
resonance observed for
reflected wave in
arrangement of fig. 18(a)
or (b).

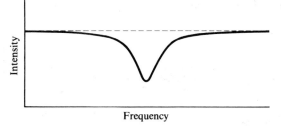

Fig. 27. Absorption resonance observed for reflected wave in arrangement of fig. 18(a) or (b).

the other two, which are equal. If the junction places all lines in series, in a ring, a wave incident from the side arm will see the sum of the two other characteristic impedances and will suffer no reflection, sharing its power equally between them. On the other hand a wave incident from the input line sees three times its characteristic impedance. According to (7), $r = \frac{1}{2}$, so that a quarter of the input power is reflected, while a further quarter carries straight on and the remaining half goes into the side arm, thus dividing the transmitted power in proportion to the impedances as is proper for resistances in series. If the incident wave has unit amplitude, that entering the side arm has amplitude $1/\sqrt{2}$, and is reflected as $R/\sqrt{2}$ according to (41). On dividing into two parts at the junction, an amplitude $\frac{1}{2}R\,e^{i\phi}$ enters the output line, the phase factor $e^{i\phi}$ being inserted to take account of the length of the side arm. The original wave that went straight past the junction has amplitude $\frac{1}{2}$, and combines with that from the side arm to give an output amplitude of the form:

$$T = \tfrac{1}{2}(1 + R\,e^{i\phi}). \qquad (42)$$

The behaviour of T is most readily followed diagrammatically. First we write R in the form, derived from (41),

$$R = 1 - C/(1 - 2iQ\,\Delta\omega/\omega_0), \qquad (43)$$

in which $C = t^2/\alpha$ and can take any value between 0 and 2. On an Argand diagram the second term of (43) is represented by a circle of diameter C, as in fig. 28 which shows both the strongly ($C > 1$) and the weakly ($C < 1$) coupled cases. As the point P passes quickly through B, at resonance, the characteristic dip in the reflected amplitude occurs. When the resonator terminates a side arm, (42) shows how R and unity are to be compounded with a phase difference depending on the length of the arm. Being concerned only with the amplitude, and not the phase, of T, we shall construct $T\,e^{-i\phi}$ which involves adding the unit vector $e^{-i\phi}$ to R as given by (43). If Q is a point on the unit circle in fig. 28, representing $-e^{-i\phi}$, the output $|T|$ is proportional to the length of the vector QP. As drawn in fig. 28(a), $|T|$, which is QA well below the resonant frequency, rises with frequency to a peak when P is on the opposite side of the circle to Q, then falls and continues to fall past the resonant point B, reaching a minimum at a frequency above resonance and then reverting to the initial value. This distorted resonance is reminiscent of the case we examined in chapter 6 (fig. 6.8), and for the same reason – the addition of a steady signal to a resonant response whose phase changes across the resonance produces a resultant that is highly dependent on the relative phases of the two at resonance.

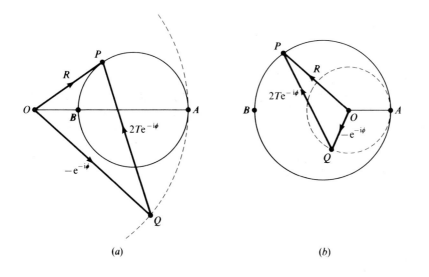

(a)   (b)

Fig. 28. Construction for transmitted amplitude in arrangement of fig. 18(e) or (f).

Symmetrical responses with this arrangement may be obtained by ensuring that Q lies on the line AB. If Q coincides with A, there is virtually no output except near resonance; this is because it has been arranged so that when the resonator is perfectly reflecting, the length of the side arm is $\lambda/4$ (or an odd number of quarter-wavelengths) and acts as an open circuit at the junction, causing total reflection of the input wave. As the resonator is excited, near resonance, the barrier becomes less than perfect, and what is transmitted is proportional to AP, a Lorentzian resonance just as would be obtained with the arrangement of fig. 18(a) and (b). By making the side arm of zero length or an integral number of half-wavelengths, Q is caused to fall on the opposite side of its circle from A. Off-resonance the side arm now behaves like a continuous metal sheet closing the hole and allowing the wave to carry on unreflected. Near resonance the response is essentially the same as R in (41), with a resonant dip in the signal. This is the configuration normally used for absorption wavemeters; a calibrated adjustable length of waveguide attached to the side of a transmission section of a microwave circuit provides an easy way of measuring frequency and, since Q-values of 10^4 are fairly easy to reach, a precision of at least 1 in 10^6 can be reached if needed. The mechanism for tuning the wavemeter, usually by moving a reflecting termination at one end of the cavity, presents design problems, about which much detailed information is available and need not be repeated here.[9]

Excitation of a resonator by a plane wave

To illustrate the most elementary form of what has been developed into a powerful general formalism, especially important for describing in wave mechanics the scattering of particles by atoms and nuclei, consider a Helmholtz resonator held in mid-air and excited by a plane sound wave.[10] The merit of this example lies in the sound wave being a scalar wave and much simpler to handle than the vector electromagnetic wave, and in the availability of real resonators much smaller than their natural wavelength; indeed the analysis closely follows the way in which such a resonator is (or was) used in practice.

The resonator interacts with the plane wave by virtue of the exaggerated motion of the air through its orifice near resonance. If the orifice is sealed the resonator is no more than a small passive obstacle, scattering a little of the sound wave but so little as to be negligible. The wave field can therefore be understood as composed of the undisturbed plane wave on which is superposed any wave radiated by the orifice – a picture of the process exactly parallel with our treatment earlier of an obstacle on a transmission line. In the present example we know that the oscillatory air-stream from the orifice will have spread evenly, to become an isotropic radial pulsation, before it has travelled anything like a wavelength away from the resonator; thus the only sort of wave that is radiated is spherically symmetrical. This would not be so if the dimensions of the resonator were comparable with a wavelength, and it is in this situation that the more general treatment mentioned above is brought into play. But this belongs more to the physics of wave-motion than of vibrators and we shall go no further than the simplest case where the re-radiated wave is isotropic. The response of the resonator is then defined by a single complex parameter, giving the amplitude and phase of the re-radiated wave relative to that of the incident plane wave. We shall take the problem in two steps. First, writing down the pressure variation with position and time when spherical and plane waves are superimposed, we shall see how the relative phase and amplitude affects the flow of energy. Then we shall relate the re-radiated and incident waves to the variations of pressure and velocity in the vicinity of the Helmholtz resonator itself.

Fig. 29(a) shows plane and spherical wavefronts, and it is immediately obvious how important is the phase relationship. As shown, they are in antiphase in the forward direction, so that the amplitude of vibration near the axis is reduced – fig. 29(b) is a graph of amplitude on the plane marked in the diagram, on the assumption that the spherical wave has one-tenth of the amplitude of the plane wave at the centre. If x denotes distance from the axis, and ϕ the energy flux per unit area, the total flux across the plane is $\int_0^\infty \phi \cdot 2\pi x \, dx$ or $\pi \int_0^\infty \phi \, d(x^2)$. It is easier to see the effect of the scattered wave by plotting the square of the amplitude against x^2, as in fig. 29(c). It now appears that with this phase relationship the oscillations of flux around the mean give alternately equal positive and negative contributions to the

(a)

(b)

(c)

Fig. 29. (*a*) A spherical wave superposed on a plane wave, in antiphase in the forward direction; (*b*) variation of intensity across the phase P when the spherical wave has an amplitude at the centre C one-tenth that of the plane wave; (*c*) the same plotted against x^2, x being measured from C.

integral, with an odd half thrown in at the beginning, so that the resultant effect of the oscillations is zero, provided their amplitude decays steadily with increasing x. On the other hand, if the crests of the spherical wave lie a quarter-wavelength behind those of the plane wave, the resultant will be negative and equal to half the first loop. This is the sort of phase relationship that results when the scattering object is lossy. Once the amplitude and phase of the scattered wave have been chosen, the loss of energy flux in the ongoing travelling wave, and the energy flux in the back-scattered spherical wave, can both be calculated; thus the total dissipation by the scattering object is in principle determined as what is needed to satisfy the energy balance. This means that for a given obstacle the amplitude and phase of the scattered wave are not arbitrarily disposable. In particular, if the obstacle is loss-free a unique relationship must exist between amplitude and phase. To make the argument quantitative it is convenient to start afresh and to calculate the energy flux outwards through the surface of a sphere of radius r, centred on the obstacle. For this purpose we need expressions for the pressure and velocity fields, particularly the radial velocity v_r since Φ, the flux per unit area of the sphere, is $\frac{1}{2}\,\mathrm{Re}\,[pv_r^*]$.

57

In terms of the polar coordinates defined by fig. 29(a), the two waves together produce pressure variations of the form

$$p(r, \theta, t) = \left(p_0\, e^{ikz} + \frac{a}{r}\, e^{ikr} \right) e^{-i\omega t}, \quad \text{where } z = r \cos \theta. \tag{44}$$

For the second term of (44), the spherical wave, integration of (26) gives ξ and hence v, which is $-i\omega\xi$:

$$v_{\mathrm{sph}} = \frac{a}{Z_0 r}\left(1 + \frac{i}{kr} \right) e^{ikr}, \quad \text{where } Z_0 = \rho c_1. \tag{45}$$

The corresponding expression for the plane wave takes the form:

$$v_{\mathrm{pl}} = (p_0/Z_0)\, e^{ikz}. \tag{46}$$

Note that the term $e^{-i\omega t}$ has been suppressed, according to convention, in (45) and (46). The radial component of velocity is $v_{\mathrm{sph}} + v_{\mathrm{pl}} \cos \theta$, so that

$$v_r = \frac{1}{Z_0}\left[\frac{a}{r}\left(1 + \frac{i}{kr} \right) e^{ikr} + p_0 \cos \theta\, e^{ikr \cos \theta} \right]. \tag{47}$$

For the total flux, P, through a sphere of radius r, we need to evaluate $\int_0^\pi 2\pi r^2 \Phi \sin \theta\, d\theta$:

$$P = \frac{\pi r^2}{Z_0} \int_{-1}^{1} \mathrm{Re}\left[p_0^2 \zeta + \frac{aa^*}{r^2}\left(1 - \frac{i}{kr} \right) + \frac{a^* p_0}{r}\left(1 - \frac{i}{kr} \right) e^{-ikr(1-\zeta)} \right.$$
$$\left. + \frac{ap_0}{r} \zeta\, e^{ikr(1-\zeta)} \right] d\zeta,$$

in which ζ has been written for $\cos \theta$. The first two terms are the separate fluxes of the plane and spherical waves, the former vanishing on integration; the last two terms result from the simultaneous presence of both. The integrations are straightforward, yielding the result

$$P = \frac{2\pi}{Z_0}(aa^* - p_0\, \mathrm{Im}\,[a]/k).$$

It is seen from this expression that if the re-radiated wave is weak, the flux of energy is determined by the second term and requires a component of the radial wave in phase quadrature with the incident wave. Since P is the mean flux outwards, $-P$ represents P_{abs} the energy absorbed in the scatterer:

$$P_{abs} = \frac{2\pi p_0^2}{k^2 Z_0}\left(\text{Im}\,[s] - ss^*\right), \tag{48}$$

in which s is written for ka/p_0, being a dimensionless measure of the amplitude of the scattered wave. Since the energy flux in the incident plane wave is $\frac{1}{2}p_0^2/Z_0$, P_{abs} is as much power as is conveyed through an area A_{abs} of wavefront, where

$$A_{abs} = 2P_{abs}Z_0/p_0^2 = \frac{4\pi}{k^2}(\text{Im}\,[s] - ss^*) = \frac{\pi}{k^2}(1 - |2s - \mathrm{i}|^2). \tag{49}$$

A_{abs} is the *absorption cross-section* of the scatterer, defined in a way that clearly allows it to be carried over to systems other than the particular acoustic resonator we have chosen as an example.

When the scatterer is lossless, but not necessarily weak, the vanishing of (49) forces a relationship between the phase and strength of scattering. For if s is written as $\alpha\,\mathrm{e}^{\mathrm{i}\phi}$, α being real, the condition $P_{abs} = 0$ implies that $\sin\phi = \alpha$. It is only necessary to know $\phi(\omega)$ to know all that matters of the scattering behaviour of a *small* loss-free scatterer. The word *small* must again be emphasized, since obstacles comparable with the wavelength can re-radiate waves that are not spherically symmetrical. In these circumstances each type of wave, described by a particular spherical harmonic, has its characteristic scattering phase $\phi(\omega)$, and the complete set of ϕs constitutes a description of the scattering process.

To confine attention to the Helmholtz resonator scattering only an isotropic wave, we need to fix s by reference to the properties of the actual scatterer. Let us idealize the resonator, replacing the single rather large aperture by a multitude of smaller apertures, spread evenly round the sphere, to give the same resonant frequency but with no awkward hiatus between R, the radius of the resonator, and a spherical shell at some rather larger R' where the flow can be considered to have evened itself out into a radial motion; with the idealized resonator R and R' coincide. The response of the resonator can be written immediately from (6.21) and the zero-frequency compliance. A pressure increment p immediately outside the sphere produces a volume change in the contained air, $\Delta V/V = -p/K$, where K is the bulk modulus, and a radial displacement $\xi/R = -\frac{1}{3}p/K$. Hence

140

$$\chi_0 \equiv \xi/p = -\tfrac{1}{3}R/\rho c_1^2, \quad \text{since } c_1^2 = K/\rho,$$

and the velocity response at frequency ω follows from (6.21):

$$\chi_v \equiv v/p = -\tfrac{1}{3}\mathrm{i}\omega\omega_0^2 R/\rho c_1^2\,(\omega^2 + \mathrm{i}\lambda\omega - \omega_0'^2), \tag{50}$$

where ω_0' is the resonant frequency found by this particular process of excitation (the reason for this remark will appear shortly). The damping term, $\mathrm{i}\lambda\omega$, in which $\lambda \equiv 2\omega''$, is to be understood as describing dissipative effects within the resonator, not radiative damping, since χ_v describes the velocity response to an oscillatory pressure measured immediately outside and acting only on what lies within.

It is χ_v that enables the re-radiated wave to be matched to the local conditions outside the resonator. From (44) and (47) we write the values of v_r and p at radius R, expanding the exponentials as power series in R and retaining only the leading terms ($\propto 1/R^2$) and the first correction ($\propto 1/R$). After averaging round the sphere we have

$$\bar{v}_r(R) \sim ia/\rho\omega R^2, \tag{51}$$

$$\bar{p}(R) \sim p_0 + a(1 + ikR)/R. \tag{52}$$

Then a takes such a value as makes $\bar{v}_r/\bar{p} = \chi_v$; after a little manipulation we find that

$$s \equiv ka/p_0 = -\lambda'\omega/[\omega^2 - \omega_0'^2(1 - \tfrac{1}{3}k^2R^2) + i(\lambda + \lambda')\omega], \tag{53}$$

where

$$\lambda'\omega = \tfrac{1}{3}\omega_0'^2 k^3 R^3.$$

The correction term to $\omega_0'^2$ is easily understood. As used here, ω_0' defines the resonant frequency when excitation is by a pressure oscillation immediately outside. But the free resonance is at a slightly lower frequency, since the local velocity field outside, falling as $1/r^2$, also contributes to the kinetic energy, which is therefore not entirely attributable to the air within the sphere at R. What appears in (53) as $\omega_0'^2(1 - \tfrac{1}{3}k^2R^2)$ is in fact the square of the true resonant frequency, ω_0. The difference between ω_0' and ω_0 is unimportant for the definition of λ'. What really matters is the extra term in the denominator, which depends on retaining the second term in (52); it is λ' which ensures that even a resonator without internal losses can never be excited beyond a certain limit, set by its ability to radiate. It is easily verified that if $\lambda = 0$, the denominator of (53) describes a resonator whose Q is governed by the same rate of dissipation as is expressed by (32), with Ω put equal to the total solid angle, 4π. Apart from the radiative loss, the resonator responds to the plane wave as it would to any other external pressure variation. It may be noted that λ' is not independent of frequency, like λ, but varies as ω^2. If we were to construct a model differential equation to simulate the behaviour, the radiation term would have to appear not in the first time-derivative of displacement but in the third.

It is interesting to look into this matter further, by setting up the differential equation for free vibrations of the resonator. For this we need to know the relation between p and ξ immediately outside, on the understanding that there are no external sources, so that all motion is to be attributed to spherical waves radiating outwards. Now an arbitrary time variation of $p(R)$ can be expressed as a Fourier series (or integral), each component of which describes the pressure at R due to a spherical wave of that frequency. We may therefore write $p(R, t)$ as the value at R of $p(r, t)$, where

$$p(r, t) = \sum_j \frac{a_j}{r} e^{i(k_j r - \omega_j t)}.$$

The point of this method of approach is that we can explicitly exclude converging waves of the type $(b/r)\,e^{-i(kr+\omega t)}$. Corresponding to $p(r, t)$ we have, from (26),

$$\rho\ddot{\xi}(r, t) = \sum_j \frac{a_j}{r^2}(1-ik_j r)\,e^{i(k_j r-\omega_j t)},$$

and in particular the displacement at R is given by

$$\rho\ddot{\xi}(R, t) = \sum_j \frac{a_j}{R^2}(1-ik_j R)\,e^{i(k_j R-\omega_j t)} = p(R, t)/R + \dot{p}(R, t)/c_1. \quad (54)$$

The pressure inside the resonator is $-3K\xi/R$, and the pressure differential, $p(R, t) + 3K\xi/R$, is responsible for accelerating the air in the vicinity of the holes. If the total mass of this air is written as $4\pi R^2 M$, M is in effect the mass per unit area of the surface, and the equation of motion (local dissipation being neglected) takes the form

$$M\ddot{\xi} + 3K\xi/R + p(R, t) = 0.$$

Eliminating p from this and (54), we have

$$R\dddot{\xi}/c_1(1+\gamma) + \ddot{\xi} + R\omega_0^2\dot{\xi}/c_1 + \omega_0^2\xi = 0, \quad (55)$$

where $\omega_0^2 = 3K/MR(1+\gamma)$, $\gamma = \rho R/M \doteqdot \frac{1}{3}k_0^2 R^2$ and $k_0 = \omega_0/c_1$.

The significance of γ has already been noted in connection with (53); it corrects the resonant frequency to allow for the inertia of the gas outside. As expected, the equation of motion contains a third-order term, but paradoxically its sign is such as to cause regeneration rather than dissipation. It is, however, more than compensated at ω_0 by the dissipative first-order term, and it is the difference that describes the back-reaction of the radiated wave on its source and thus governs the radiative loss. The resonant frequency derived from (55) is approximately $\omega_0(1-\frac{1}{6}k_0^3 R^3)$, which is the same as makes the denominator of (53) vanish.

A similar effect occurs when an accelerated charge radiates electromagnetic waves.[11] The field generated by the charge, when calculated to a sufficient order of approximation, includes an electric field acting on the charge itself with a force $\frac{2}{3}e^2\dddot{r}/c^3$. The equation of motion of a charged mass bound by an elastic force to a fixed centre thus has the form

$$-\tfrac{2}{3}e^2\dddot{x}/c^3 + m\ddot{x} + \mu x = 0, \quad (56)$$

with a natural frequency $\Omega \doteqdot \omega_0 - \frac{1}{3}i\,e^2\omega_0^2/mc^3$. The imaginary part describes the radiative power loss, which for an oscillatory dipole $p = p_0\,e^{-i\Omega t}$ amounts to $\frac{1}{3}\omega_0^4 p_0^2/c^3$, the classical result for the Hertzian dipole, normally obtained by evaluation of the energy flux in the distant radiation field. Unfortunately (56) also allows an exponentially growing solution and cannot therefore be a complete expression of the equation of motion. This is but one example of a ubiquitous problem – every naive attempt to extend

physical concepts from mechanical to electrodynamical processes ultimately founders on a divergence. Quantum electrodynamics provides a
new formalism for dealing with such difficulties.

After this digression we return to the acoustical resonator, for which (49)
and (53) yield the variation with frequency of the absorption cross-section:

$$A_{abs} = 4\pi c_i^2 \lambda \lambda' / [(\omega^2 - \omega_0^2)^2 + (\lambda + \lambda')^2 \omega^2]. \tag{57}$$

At resonance the re-radiated wave has its greatest amplitude:

$$s_{res} = i\lambda' / (\lambda + \lambda'),$$

and $$(A_{obs})_{res} = 4\pi\lambda\lambda' / k^2 (\lambda + \lambda')^2.$$

The highest possible absorption results when the holes in the resonator are
matched to the internal losses so that $\lambda = \lambda'$. Then

$$(A_{abs})_{max} = \pi / k^2. \tag{58}$$

This is the area of a circle of radius $1/k$, having a circumference equal to
one wavelength, and it is the largest that can be attained by any isotropic
radiator. The necessity for this maximum is inherent in (49), for $|2s - i|$ is
essentially positive. It is, of course, quite a large cross-section for absorption; the resonator described in chapter 2, with a radius of 9 cm and a
natural frequency of about 190 Hz (wavelength 1.77 m), is in principle
capable of extracting all the energy in a plane wave passing through a circle
of radius 28 cm.

The process just analysed has many analogues in different branches of
physics. The de Broglie wave governing particle behaviour in quantum
mechanics is a scalar wave whose frequency ω is $E/\hbar$ for a particle of total
energy E. A typical process corresponding to resonance is the absorption
of neutrons by nuclei.[12] The nucleus before the neutron enters is normally
in its ground state, with a definite energy, and the compound nucleus after
the neutron has entered has a wide spectrum of energy levels; if the
incident neutron has energy ΔE that allows the compound nucleus to fall
straight into a stationary state, as shown in fig. 30, the quantum mechanical
equivalent of resonance occurs; the internal vibration frequency is determined by the energy difference between the initial and final states, and if
this matches the neutron frequency the wave outside is depleted and the
'oscillation' within builds up; this reveals itself as a peak in the cross-
section of the nucleus for capturing neutrons. The Q of the peak is
determined by processes very closely analogous to those we have discussed. There is a natural breadth of resonance due to the possibility that
the neutron may be spontaneously re-emitted – this is the analogue of λ',
the radiation damping. It is also possible, however, that the compound
nucleus will fall into a lower state by, for example, emitting energy in the
form of a γ-ray – this is the analogue of λ, the internal dissipation, for the
neutron is then permanently captured. The working out of this problem,

23

Fig. 30. Formation of an excited compound nucleus by the absorption of a neutron having the correct energy ΔE.

along strictly parallel lines to those used here, yields the Breit–Wigner formula for the variation with energy of absorption cross-section:

$$\sigma_{n\gamma} = \frac{\pi \Gamma_{nr} \Gamma_{\gamma}}{kk_r[(E-E_r)^2 + \frac{1}{4}(\Gamma_{nr} + \Gamma_{\gamma})^2]}. \tag{59}$$

Here $\sigma_{n\gamma}$ parallels A_{abs} in (57); other parallels are as follows:

$$E \equiv \omega, \qquad E_r \equiv \omega_0, \qquad \Gamma_{nr} \equiv \lambda', \qquad \Gamma_{\gamma} \equiv \lambda.$$

The principal difference resides in the behaviour as ω or E goes to zero. In the acoustic case the cross-section tends to a constant, while with particles it increases as $1/k$, which for de Broglie waves is $1/E^{\frac{1}{2}}$. This difference results from the acoustic wave being non-dispersive ($\omega \propto k$) while the de Broglie wave is dispersive ($\omega \propto k^2$). It is profoundly important for the absorption of low energy (thermal) neutrons in reactors. These have a relatively large de Broglie wavelength, and there are a few nuclear species (e.g. Xe 135) with absorption cross-sections 10^6 times as great in area as the nucleus itself.[13] Only a small concentration of these is sufficient to reduce the efficiency of the reactor drastically, or even to prevent it from reaching criticality and becoming self-maintaining.

Electromagnetic scattering also behaves very similarly, if we overlook minor differences arising from the transverse character of the waves. A dipole feeding a resonant circuit (fig. 31) has an absorption cross-section

Fig. 31. Dipole feeding a resonant circuit.

which depends on the direction and polarization of the incident radiation, but the average of this cross-section over all directions is at most $\frac{1}{2}\pi/k^2$, one-half of the result expressed in (58). The factor $\frac{1}{2}$ may be considered a consequence of the polarization of transverse waves – an aerial cannot respond completely to all polarizations. If it has the maximum cross-section for one plane of polarization it is not responsive to the plane at right angles. The result (58) would apply if the radiation falling from every direction was optimally polarized, but for random polarization it must be halved. This average cross-section of $\frac{1}{2}\pi/k^2$ is achieved by every radio aerial that is perfectly matched to its feeder, as can be seen by a thermodynamic argument similar to that used by Kirchhoff in his theory of black-body radiation.[14] Imagine two cavities, each with walls maintained at the same temperature, and connected by a transmission line with an aerial at each end matched to the line. Any power picked up by aerial 1 from the black-body radiation in cavity 1 is transmitted to aerial 2 and radiated, without reflection, into cavity 2. Similarly cavity 1 is fed by whatever is picked up by aerial 2. Since there can be no resultant power transfer between cavities at the same temperature, both aerials must pick up the same amount; and the argument may be extended to any other aerial matched to its line. But the aerials are irradiated by isotropic radiation, and what they pick up is determined by the average absorption cross-section, which must therefore be the same for all matched aerials. Small aerials have a relatively small cross-section and pick up from a wide angular range; large aerials designed to receive only in a narrow pencil of directions have a correspondingly large cross-section for those directions from which anything is received.

Given that the aerial picks up as much power as would pass through a sphere of radius $1/\sqrt{2}k$ we may relate this to the energy density of radiation in the cavity. Consider a frequency range $d\omega$, and let $u_\omega\, d\omega$ be the energy per unit volume in this range. Draw a large spherical shell, lying between r and $r+dr$, and centred on the aerial. Of all the energy, $4\pi r^2 u_\omega\, dr\, d\omega$, in the shell at any instant, only a fraction $1/8k^2r^2$ is moving in such a direction as to pass through the little sphere at the centre. Hence the energy absorbed from this shell is $\frac{1}{2}\pi u_\omega\, dr\, d\omega/k^2$, and as the different contributions from the shell arrive during an interval of dr/c, the power absorption is $\frac{1}{2}\pi c u_\omega\, d\omega/k^2$. Now according to Planck's law,

$$u_\omega = \hbar\omega^3/\pi c^3(e^{\hbar\omega/k_B T}-1),$$

and hence the power absorbed is given by

$$P = \hbar\omega\, d\omega/2\pi(e^{\hbar\omega/k_B T}-1).$$

At low frequencies ($\hbar\omega \ll k_B T$) the Planck law may be replaced by that of Rayleigh–Jeans, to give

$$P = \tfrac{1}{2}k_B T\, d\omega/\pi = k_B T\beta, \tag{60}$$

where β is the bandwidth in Hz. This is reminiscent of the formula for

Johnson noise in a resistor (6.49), and not accidentally; for if we repeat **158** the notional thermodynamical experiment with one of the aerials replaced by a resistor having R equal to Z_0, to terminate the line, the other aerial will feed power into it at the rate (60). This must be accompanied by an equal amount of power generated by the resistor, and indeed a Johnson noise generator for which $\overline{V^2} = 4Rk_B T\beta$, feeding a line in series with itself, will generate just the right amount. This argument may be regarded as providing another derivation of Johnson noise or, alternatively, as confirming still further the idea of a maximum absorption cross-section, and throwing light on the close interrelation of different kinds of physical approach to a basic problem.

Anomalous dispersion

The interaction of an atom or molecule with visible light is an example of excitation by a wave whose wavelength greatly exceeds the size of the excited system. When the frequency is close to a characteristic absorption line (e.g. the Lyman series of ultraviolet lines in hydrogen or the two closely-spaced lines of the sodium doublet) the refractive index and absorption vary with frequency in a way that is fitted very closely by a resonance formula. Like neutron absorption, this is strictly a quantum phenomenon, yet one which has a precise parallel in classical physics. It is therefore appropriate at this stage, even though we make no attempt at developing the detailed argument, to state the quantum-mechanical result which validates the classical treatment, and then to examine briefly the consequences.

When a system occuping a quantum level with energy ε_k is irradiated with electromagnetic radiation of frequency ω, it is perturbed and acquires an oscillatory dipole moment. As is typical of quantum systems, the magnitude of this moment is not precisely definable for a single system; nevertheless the average taken over many identical systems is well-defined, and it is just this that matters when one makes an observation of the optical properties of a macroscopic sample containing many replicas of the atomic or molecular system under investigation. For such an assembly it is permissible to talk in classical deterministic terms when describing the evolution of the mean polarization. Moreover the solution of the quantum-mechanical problem shows that the polarization evolves exactly as if the system were a classical harmonic oscillator. Consider, then, a particle of mass m, carrying charge e, bound by an elastic spring to a fixed centre, so that under the influence of a varying electric field $\mathscr{E}_z(t)$ it obeys the equation of motion

$$m\ddot{z} + \mu z = e\mathscr{E}_z(t).$$

This is the same as the equation for a lossless pendulum, and the suscep- **129** tibility therefore takes the form (6.2) with ω'' equated to zero:

$$\chi(\omega) = 1/m(\omega_0^2 - \omega^2),$$

where $\omega_0^2 = \mu/m$. Now χ describes the amplitude of oscillatory displacement under the influence of an oscillatory force of unit strength, but for present purposes the *atomic polarizability* is more useful – the amplitude of the oscillatory dipole moment due to an oscillatory electric field of unit strength. Since the force is $e\mathcal{E}_z$ and the dipole moment is ez, the atomic polarizability, α, is $e^2\chi$:

$$\alpha(\omega) = e^2/m(\omega_0^2 - \omega^2). \tag{61}$$

This is the expression for a classical oscillator. The average behaviour of a single quantized system can be simulated by a collection of such oscillators; thus with a system initially in the energy state ε_k the possibility of a transition to another state ε_j introduces a term of the form (61) into the polarizability, with ω_0 taking such a value that the quantum $\hbar\omega_0$ is equal to the energy difference $|\varepsilon_j - \varepsilon_k|$; resonance, the vanishing of the denominator in (61), corresponds to the quantum energy exactly matching what is required to cause the system to change energy levels from ε_k to ε_j. When such a transition is possible the system is highly sensitive to the irradiating electric field, while at a slightly different frequency it is less susceptible but nevertheless capable of being disturbed.† One modification of (61) is, however, called for – different energy levels are differently coupled to the initial state ε_k, in the sense that a given strength of $\mathcal{E}_z$ may readily stimulate a transition to some levels and only very weakly, or not at all, to others. For each transition, therefore, we introduce an *oscillator strength*, f_{jk}, writing for the total polarizability of the kth state,

$$\alpha_k(\omega) = \frac{e^2}{m} \sum_j \frac{f_{jk}}{(\omega_{jk}^2 - \omega^2)}. \tag{62}$$

It is worth noting, in passing, that when all the transitions involve only one electron changing its energy level, $\Sigma_j f_{jk} = 1$; as the reader will verify easily enough, this means that the system responds to a sharp impulse P as if the electron were free, moving off with initial velocity P/m. The oscillator strengths are positive when the system is being excited at resonance to absorb a quantum and make a transition to a higher level, $\varepsilon_j > \varepsilon_k$; but negative when $\varepsilon_j < \varepsilon_k$ and the effect of resonant irradiation is to extract a quantum from the system. This reversal of sign is highly relevant to the operation of masers and lasers (see chapter 11 for a brief discussion).

346

---

† We have ignored the transient effects arising from switching on the force, quite improperly in view of the assumption of no intrinsic loss in the system. In practice, however, there are various processes that cause or simulate loss, and in normal spectroscopic observations, which involve protracted irradiation of the sample, it is proper to consider only the steady-state response. When, on the other hand, a short wave-group travels in a dispersive medium, the transient response of the molecules to the leading edge, and the residual oscillations left behind the trailing edge, are significant factors in determining the speed of propagation and the evolution of the shape of the group.

Having stated the general result in (62), let us now concentrate on the special case of an isolated spectral line, where only one term in (62) need be considered. If, as in (61), the resonant frequency is ω_0, each individual system reacts to irradiation at a nearby frequency ω in a way that is exactly analogous to the Helmholtz resonator or the tuned dipole; being set into vibration it re-radiates a spherical wavelet. In a simple gas each molecule behaves similarly, and the superposition of all the spherical wavelets and the incident plane wave leads to two separate effects. In the forward direction there is a constant phase relationship between the wavelets and the plane wave, so that at any point all the wavelets add together into a resultant that is uniquely phase-linked to the plane wave – put otherwise, their superposition yields a resultant plane wave which combines with and modifies the propagation characteristics of the incident wave. This is, in rough pictorial terms, how the gas contrives to modify the wave velocity, and is the origin of the refractive index. At any other angle than the forward direction, however, the wavelets arrive at a given point with a phase that depends on precisely where they originated. The randomly distributed gas molecules contribute randomly phased wavelets, and the intensity of the wave scattered out of the forward direction is the sum of the intensities due to each molecule separately. This scattering is normally very weak, though the scattering of about 10% of sunlight in its passage through the atmosphere is obvious as the blue of the sky.†[15] Solids and liquids, in which the molecules are more ordered, normally scatter even less by this mechanism, but much more because of particulate impurities or structural defects.

To derive an expression for the refractive index there is no need to add the forward-scattered wavelets, though this is a possible approach to the problem. It is easier, however, to consider simply the volume polarizability of the medium, $N\alpha$ if N is the number of molecules per unit volume. The dielectric constant is $1 + N\alpha/\varepsilon_0$ and the refractive index n is the square root of this. For a gas or any sparse distribution of scattering centres it is normally good enough to write

$$\delta n \equiv n - n_0 = \tfrac{1}{2}N\alpha/\varepsilon_0 = fNe^2/2m\varepsilon_0(\omega_0^2 - \omega^2) \tag{63}$$

in the vicinity of an absorption line of oscillator strength f. In writing this we have introduced n_0 to represent other non-resonant contributions to the refractivity, which can be taken as sensibly constant around ω_0, where

---

† The characteristic absorption frequencies of nitrogen and oxygen are in the ultraviolet so that, in (62), $\omega_{jk} \gg \omega$ and α is nearly independent of frequency (i.e. air has rather low optical dispersion). An oscillating dipole of given magnitude scatters as the fourth power of its frequency, and blue light is much more heavily scattered than red. Hence the colour of the sky and the reddening of the setting sun, seen through a thick layer of atmosphere. The scattering of light by clouds, on the other hand, involves obstacles (water drops) larger than the wavelength, and all colours are more-or-less equally scattered, so that clouds tend to be white, especially when viewed from above.

Fig. 32. Variation of
refractive index around
anomalous dispersion
peak.

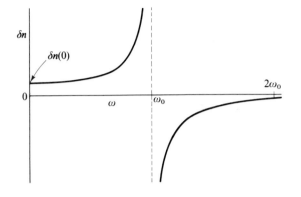

Fig. 32. Variation of refractive index around anomalous dispersion peak.

the absorption line results in a very strong dispersion. The characteristic anomalous dispersion curve, as given by (63), is shown in fig. 32.

It is will be observed that in traversing this curve δn drops from its low-frequency value $fNe^2/2m\varepsilon_0\omega_0^2$ – representing the static polarizability – to zero at frequencies well above ω_0. This can only happen if there is absorption somewhere in between, as follows immediately by applying the second of the Kramers–Kronig relations (5.9). The increment of refractive index, considered as a frequency-dependent complex number $\delta n' + \mathrm{i}\,\delta n''$, is a linear compliance to which the relation applies, and we see that

$$\delta n'(0) - \delta n'(\infty) = \frac{2}{\pi}\int_0^\infty \delta n''\,\mathrm{d}\omega/\omega \sim \frac{2}{\pi\omega_0}\int_0^\infty \delta n''\,\mathrm{d}\omega, \qquad (64)$$

the second, approximate, form being suitable for a narrow absorption line when $\delta n''$ can be taken as zero everywhere except very close to ω_0. It follows that the area under the absorption curve is $\frac{1}{2}\pi\omega_0$ times the refractive index loss in traversing the anomalous dispersion peak; and this result does not depend on any special assumption as to the origin of the loss, except that the system be linear, as is a very good assumption with light of ordinary intensity. If the intrinsic losses are very small, as assumed in setting up the model analysed here, the loss peak is narrow and correspondingly high. The anomalous dispersion curve (63) can therefore be followed very close to ω_0 and substantial variations of n observed. There are good accounts, with excellent photographs, in specialist optics texts of the ingenious experiments by which the phenomenon has been studied, and there is no need to say more about this aspect.[16]

It is, however, worth noting that the ideal behaviour expressed by (63) must break down in the frequency range where the loss is appreciable. To take a specific example, the motion of gas molecules imposes a Doppler shift on the resonant frequency, and the loss peak, however narrow intrinsically, is smeared into a Gaussian form, reflecting the Gaussian distribution of molecular velocities parallel to the line of sight. Simultaneously the rise of $\delta n'$ towards $\pm\infty$ on either side of the resonance is halted, $\delta n'$ becoming a continuous function passing through zero at resonance,

110

somewhat like χ' in fig. 5.4. This part of the variation is hard to observe experimentally because of absorption, as the following argument shows. If the sample traversed by the light has length X, and the complex wave-number is written as $(n'+in'')k_0$, k_0 being the wavenumber for the same frequency in free space, the phase change in the path X is $n'k_0X$, of which $\delta n' \cdot k_0X$ is attributable to the anomalous dispersion peak. At the same time the intensity is divided by $\exp(2\delta n'' \cdot k_0X)$, which we write as R. Thus the phase change ϕ which is to be measured, i.e. $\delta n' \cdot k_0X$, and the intensity loss R are related by the equation:

$$\phi = \frac{\delta n'}{2\delta n''} \ln R, \tag{65}$$

and it will be noted that the length X has disappeared. Restrictions on the value of ϕ are intrinsic and cannot be eliminated by redesign of the experiment.

Let us now define the width, $\Delta\omega$, of the absorption peak by replacing the real peak by a rectangle of the same area, with the same height $\delta n''_{\max}$ and width $\Delta\omega$ which must, according to (64), have the value

$$\Delta\omega = \frac{1}{2}\pi\omega_0[\delta n'(0) - \delta n'(\infty)]/\delta n''_{\max}. \tag{66}$$

It is at a frequency about $\frac{1}{2}\Delta\omega$ from the peak that (63) begins to fail, and at that point $\delta n'$ can hardly be greater than the value given by (63), which may be rewritten to yield the inequality

$$\delta n' \le \omega_0[\delta n'(0) - \delta n'(\infty)]/\Delta\omega = \frac{2}{\pi}\delta n''_{\max} \quad \text{from (66)}.$$

Hence, from (65),

$$\phi \lesssim \frac{1}{\pi}\frac{\delta n''_{\max}}{\delta n''} \ln R.$$

In the absorption region, then, where $\delta n'' \sim \delta n''_{\max}$, ϕ can only reach 1 radian at the cost of something like twenty-fold attenuation. The best experimental procedures for studying the anomalous dispersion curve rely on interferometry, and it is no easy matter to measure fringe shifts of $1/2\pi$ with any accuracy in the presence of heavy attenuation. It is more satis-factory to measure the shape and strength of the absorption peak by passing the light through a cell, and to use the Kramers–Kronig transform to evaluate n' from this.

110

8 Velocity-dependent forces

Coriolis and Lorentz forces

In this chapter are collected together a somewhat disparate assembly of oscillatory (better, periodic) systems which have one thing in common – non-reversibility; not simply the lack of reversibility because of dissipation, but that which results from velocity-dependent forces not in themselves capable of causing dissipation. To classify such systems together in a single category may not be significant, and the reader who finds it pointless has probably missed nothing of value. The systems themselves, however, have interest and importance in their own right, and the whole chapter may therefore be regarded (like much else in this book) as a catalogue of phenomena that fit only loosely into any tidy pattern but must nevertheless form part of the mental furniture of anyone who wishes to appreciate the variety that oscillatory phenomena can exhibit. To a physicist the section on nuclear magnetic resonance may seem intrinsically more important than the rest. The reader should therefore take heed that it is also considerably more difficult, perhaps the hardest part of the book; it may, however, be omitted without losing the thread of any subsequent argument.

The velocity-dependent forces considered here, we repeat, are not the dissipative forces of viscosity or resistance responsible for the middle term, $k\dot{\xi}$, in an equation of motion such as (2.23). They are, rather, the non-dissipative Coriolis force or Lorentz force – the former being involved when dynamical processes are observed from a rotating frame of reference, the latter being the force on a charged particle moving through a magnetic field.[1] Both forces act normal to the direction of motion and do no work, but nevertheless considerably affect the trajectory. They take similar forms:

Coriolis force – an observer in a frame rotating at constant angular velocity $\boldsymbol{\Omega}$ finds a particle of mass m, acted upon by a real force $\boldsymbol{F}$, to obey the equation of motion

$$m\ddot{\boldsymbol{r}} = \boldsymbol{F} - 2m\boldsymbol{\Omega} \wedge \dot{\boldsymbol{r}} - m\boldsymbol{\Omega} \wedge (\boldsymbol{\Omega} \wedge \boldsymbol{r}). \qquad (1)$$

The second term is the Coriolis force and the third the centrifugal force.

Lorentz force – a particle of mass m and charge e moving in a magnetic field $\boldsymbol{B}$ obeys the equation of motion

$$m\ddot{\boldsymbol{r}} = \boldsymbol{F} - e\boldsymbol{B} \wedge \dot{\boldsymbol{r}}. \qquad (2)$$

30

The force F includes the electrical force eE; the second term is the Lorentz force.†

The similarity in form of these forces means that under suitable conditions the effect of a uniform field B on the trajectory of a particle can be eliminated by observing its motion from a frame rotating at angular velocity $\Omega = -\frac{1}{2}eB/m$. We need not discuss what these suitable conditions are (essentially that the trajectory be localized so that r does not become too large, and that F be a central force field). This matter is fully dealt with in standard textbook discussions of *Larmor's theorem.*[2]

When Coriolis or Lorentz forces are present a lossless system ceases to be reversible. To elaborate this point, we note that if F is determined by position alone the equation of motion,

$$m\ddot{r} = F(r),$$

with solution $r = r(t)$, is equally satisfied by $r = r(-t)$, since only t^2 appears in the equation; if the particle can perform a certain trajectory it can equally well perform it backwards by being started at the end-point with its velocity reversed. This is not so when the equation of motion is (1) or (2), for reversing t reverses $\dot{r}$ and the sign of the Coriolis and Lorentz forces, but not of F; $\ddot{r}$ is no longer uniquely determed by r, and we expect a different trajectory in each direction. The simplest sort of vibrator, confined to a straight line or any other one-dimensional path, is of course unaffected by these transverse forces. To see their effect in the most elementary form we need to allow the vibrator two degrees of freedom. The Foucault pendulum provides an example. A conical pendulum can swing in a circle with the same angular velocity $\omega_0 = (l/g)^{\frac{1}{2}}$ clockwise or anticlockwise if it is viewed from a non-rotating frame. Set up at the North Pole, however, and viewed from above by an earthbound observer, it will seem to have angular velocity of magnitude $\pm\omega_0 - \Omega$ if Ω is the Earth's (negative) angular velocity, the positive sign applying to clockwise motion – the two frequencies are no longer equal and opposite. The trajectories in this simple case have the same shape, but this is exceptional.

When circular motion is to be described it is convenient to employ complex numbers, as discussed in chapter 3. The displacement of the bob of the pendulum is denoted by a complex number, r, the whole of which is significant; thus, $r = a\,e^{-i\omega t}$ represents clockwise motion (if $\omega > 0$) round a circle of radius a, and the two stable modes of the pendulum are of the form $a_1\,e^{-i(\Omega+\omega_0)t}$ and $a_2\,e^{-i(\Omega-\omega_0)t}$. If they are equally excited, with $a_1 = a_2 = a$, superposition of the two yields the characteristic precessional linear motion of the Foucault pendulum,

$$r = a\,e^{i(\Omega+\omega_0)t} + a\,e^{i(\Omega-\omega_0)t} = 2a\,e^{i\Omega t}\cos\omega_0 t.$$

13

62

---

† In this chapter and occasionally elsewhere, B and E refer to steady applied fields, which may be strong, b, $\mathscr{B}$ and $\mathscr{E}$ to weak fields superimposed on them.

The bob swings very nearly in a plane with its characteristic frequency ω_0, but the plane rotates with angular velocity $-\Omega$, i.e. once a day, at the Pole, once in cosec θ days at latitude θ where the Coriolis force is reduced by a factor sin θ. Because of this precession the linear motion is not a normal mode of the vibrator – it is the circular motions that are. It may be noted, though we shall not dwell on it at this stage, that when the Coriolis force is absent the two circular motions have as much claim as linear vibrations in perpendicular planes to be called the normal modes. The two modes are *degenerate*, i.e. have frequencies of the same magnitude, and any linear combination gives rise to a stable pattern of vibration at the same frequency. Rotation, through the Coriolis effect, removes this degeneracy and with it all ambiguity in the definition of normal modes.

When a system whose normal modes are circularly polarized is driven by an oscillatory force directed along a fixed line, to calculate the response one may first represent the force as the sum of two rotating forces of constant magnitude. If F_0 is a complex number, $F_0 \cos(\omega t + \varepsilon)$ is an oscillatory force acting along a line whose direction is the argument of F_0. It is dissected into two components

$$F_0 \cos(\omega t + \varepsilon) = A_L e^{i\omega t} + A_R e^{-i\omega t},$$

where $A_L = \frac{1}{2}F_0 e^{i\varepsilon}$ and $A_R = \frac{1}{2}F_0 e^{-i\varepsilon}$; the left- and right-handed components, A_L and A_R, have the same magnitude, $\frac{1}{2}|F_0|$, but opposite arguments. If the system responds to a rotating force by itself rotating in synchronism, a compliance χ may be defined to describe the magnitude and phase of the response. Thus $\chi_L A_L e^{i\omega t}$ and $\chi_R A_R e^{-i\omega t}$ are the two rotating components of displacement. Real χ means that force and displacement rotate together and no work is done, but an imaginary part to χ implies a component of displacement normal to the force; if the displacement lags behind the force energy is steadily transferred to the system.

When the circular modes are degenerate the magnitudes of χ_L and χ_R are equal, but since displacement lags behind force by the same angle for both senses of rotation their arguments are opposite: $\chi_L = \chi_R^*$. An applied force $F_0 \cos \omega t$, with F_0 real, then produces a response

$$r = \frac{1}{2}\chi_2 F_0 e^{-i\omega t} + \frac{1}{2}\chi_R^* F_0 e^{i\omega t}.$$

If χ_R is written as $\chi e^{i\phi}$,

$$r = \chi F_0 \cos(\omega t - \phi),$$

a linear response lagging in phase behind the force by the phase angle of χ. In this case χ_R and χ_L have the same significance as the compliance defined for linear motion.

With non-degenerate circular modes, χ_L and χ_R are not complex conjugates and the general form of the response to a linear force is elliptical motion, being the composition of two circular motions of different amplitude at equal but opposite speeds. In most of the examples in this chapter the two modes are so different in character that only one matters at any time. The result of linear excitation is then circular motion. In cases like this we shall take it for granted that, although we may talk in terms of response to rotating forces, in practice it is usually easier to excite the system with an oscillating force, constant in direction, but the resulting response is the same.

Whirling[3]

The Foucault pendulum is an exceedingly gentle illustration of the Coriolis effect. A more dramatic example, and one with great nuisance value, arises when an oscillatory system is spun at an angular velocity exceeding its natural vibrational frequency. It may then spontaneously vibrate and possibly fragment, with great destruction to itself and objects in its vicinity. Consider a very simple case, illustrated in fig. 1, a small mass suspended by a length of piano wire from the spindle of a d.c. motor. The dimensions

Motor

Fig. 1. Demonstration of whirling. A brass cylinder, $\frac{1}{2}$ in. diameter and $\frac{3}{4}$ in. long, soldered to a steel wire, 0.02 in. diameter and 4 in. long, is suitable. *Note warning in text.*

shown are suitable for private experimenting, though the observer should always interpose a cushion or screen between himself and the mass, which a moment's negligence may permit to fly off horizontally at great speed. This will happen if the motor is gradually accelerated through the critical angular velocity, equal to the natural vibration frequency of a few Hz. If, however, the wire is gently restrained with the fingers until this speed is exceeded, it will be found that the mass can continue to spin without significant horizontal deflection.

This arrangement is easily analysed, either in the laboratory frame of reference or in terms of a frame rotating with the motor spindle. The complex notation for horizontal displacement of the mass is convenient in both cases. To take the laboratory frame first, we suppose the wire to have a slight permanent set, so that the mass takes up a rest position r_0 from the axis when the spindle is not turning. When it is displaced to a neighbouring position r the linear restoring force due to the wire will have magnitude

$\mu(r_0-r)$, directed back towards the rest position; the equation of motion therefore has the form

$$m\ddot{r}+\mu(r-r_0)=0, \qquad (3)$$

with solution $r-r_0=A\,\mathrm{e}^{-i\omega_0 t}+B\,\mathrm{e}^{i\omega_0 t}$, the superposition of two circular motions in opposite senses at the frequency ω_0, equal to $(\mu/m)^{\frac{1}{2}}$.

Now let the spindle rotate at angular velocity Ω. Then the rest position varies with time as $r_0\,\mathrm{e}^{-i\Omega t}$, and (3) takes the form

$$m\ddot{r}+\mu r=\mu r_0\,\mathrm{e}^{-i\Omega t}, \qquad (4)$$

the equation of undamped harmonic motion about the origin, not about r_0, with a rotating driving force arising from the initial imperfection as measured by r_0. The steady-state solution of (4) represents the mass rotating about the origin at angular velocity Ω,

$$r=r_0\,\mathrm{e}^{-i\Omega t}/(1-\Omega^2/\omega_0^2). \qquad (5)$$

The resonant denominator is responsible for amplification of the initial permanent set. If Ω is sufficiently close to ω_0 any displacement r_0 can be amplified to the point where the linear approximation implicit in (5) fails. More probably, however, a new complication will set in when the wire is bent beyond its elastic limit, for the losses accompanying plastic flow may induce instability, r growing until the mass whirls around on a wire bent horizontal, or the wire breaks. We shall return shortly to the instability due to losses. For the present we consider only lossless systems, for which (5) shows that once the critical speed ω_0 is exceeded, the displacement falls again to manageable size. The process of acceleration through ω_0 is exactly analogous to the excitation of a resonant system by a gliding tone, treated in chapter 6; when Ω has reached a value much greater than ω_0, all that is left is a residual circular motion at ω_0, excited as Ω passed through the critical range.

148

The wire and mass in fig. 1 may be replaced by a length of wire (e.g. 30 cm of 24 SWG copper wire, straightened by pulling). In contrast to the former example, this has more than one resonant mode, as illustrated in fig. 2, each with its characteristic value of ω_0. These modes, though not

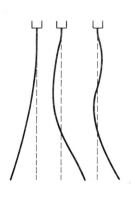

Fig. 2. The first three modes of a spinning wire, free at the lower end.

sinusoidal, constitute a suitable set of basic functions for a Fourier series describing the initial shape of the wire. A typical term in the series will have initial amplitude a_i, which will be changed to $a_i/(1 - \Omega^2/\omega_{0i}^2)$ by spinning. Whirling can now occur near any one of the characteristic frequencies. It is easy to demonstrate that in between resonances the wire will spin stably; one has only to restrain the lower end with the hand as the fundamental frequency is traversed, and to stop somewhere between this point and the next frequency, which is very well separated from it.

Even easier is to let the lowest few centimetres of the wire dip into a beaker of water, which supplies a damping term to the equation of motion and prevents the amplitude building up without limit close to resonance. It is then possible to follow the development of the pattern as Ω is increased slowly, and to see how each mode in turn is amplified and then sinks into insignificance as its resonant frequency is passed, to be replaced by the next higher mode. With a very fine wire completely immersed one can run through several resonances, though the wide spacing of the frequencies makes it difficult to follow a long sequence.

If the damping mechanism, instead of being, like a beaker of water, fixed in the laboratory frame, rotates with the spindle the behaviour is quite different; at all speeds above the lowest resonance the motion is unstable and whirling is inevitable. It is now convenient to examine the behaviour from the point of view of a rotating reference frame, in which the equation of motion takes the form

$$m\ddot{r} = \mu(r_0 - r) + 2im\Omega\dot{r} + m\Omega^2 r - \lambda\dot{r}. \tag{6}$$

The first term on the right, taken from (3), is the elastic restoring force, the second and third the Coriolis and centrifugal forces from (1) expressed in complex notation, and the fourth a viscous damping term proportional to velocity relative to the rotating frame. The steady-state solution is obtained by putting $\dot{r}$ and $\ddot{r}$ equal to zero, and is the same as (5) without the exponential. To determine whether this solution is stable we must examine the oscillatory modes, and it is now permissible to set r_0 equal to zero. Substituting the trial solution $e^{-i\omega t}$ we see at once that

$$\omega^2 + (2\Omega + ik)\omega + (\Omega^2 - \omega_0^2) = 0, \quad \text{where } k = \lambda/m. \tag{7}$$

The sum of the two roots, $-(2\Omega + ik)$, lies below the real axis, and the product of the roots is $(\Omega^2 - \omega_0^2)$, real and negative so long as $\Omega < \omega_0$; one of the roots must lie in the third quadrant and the other in the fourth. Both therefore have negative imaginary parts and cause the motion of the mass, displaced from equilibrium, to spiral in to the origin – the steady-state solution is stable when $\Omega < \omega_0$. When $\Omega > \omega_0$, however, the product of the roots is positive and real, so that one must lie above and one below the real axis. The former describes an outward spiralling motion, and the equilibrium is unstable.

This is readily demonstrated with the arrangement of fig. 1, by sliding a well-fitting PVC sleeve on to the wire. The plastic flow of the sleeve in bending is sufficient to induce instability; at speeds above ω_0 the mass, however well centred initially, gradually develops a displacement, spinning in an ever-increasing circle. The last stage is very sudden, probably aided by extra losses due to inelastic deformation of the wire. Since rotating machinery commonly carries its dissipative mechanisms around with it, this instability is a potential hazard that must always be borne in mind.

To return to the theoretical analysis, the transition from stable to unstable motion occurs at the angular velocity for which the elastic restoring force and the centrifugal force, both varying as r, balance. The two forces together define a parabolic potential energy function, which is bowl-like at low speeds, dome-like at high speeds. As the mass spins round, in the absence of losses, it is kept in its orbit, even on the dome, by the Coriolis force. When losses are introduced, the mass necessarily moves to orbits of lower energy, i.e. inwards when $\Omega < \omega_0$, outwards when $\Omega > \omega_0$. Looking at the phenomenon from the laboratory frame dispels any air of mystery which may attach to it. The effect is exhibited by a simple conical pendulum swinging in a vessel of fluid. If the fluid is caused to rotate more slowly than the pendulum bob in its circular orbit, the bob experiences a retarding force, and sinks towards the vertical. But if the fluid rotates faster than the bob, it urges the bob onwards and causes its orbit radius to increase. Exactly the same thing occurs with the whirling mass.

In the laboratory frame of reference, with the viscous damping fluid rotating, the equation of motion corresponding to (6) has the form, when $r_0 = 0$,

$$m\ddot{r} = -\mu r - \lambda(\dot{r} + i\Omega r),$$

since $\dot{r} + i\Omega r$ is the relative velocity of pendulum bob and fluid. Hence the natural frequencies are determined by the equation

$$\omega^2 + ik(\omega - \Omega) - \omega_0^2 = 0,$$

which may be written in the form

$$(\omega + \tfrac{1}{2}ik)^2 = (\omega_0^2 - \tfrac{1}{4}k^2) + i\Omega k. \tag{8}$$

As Ω is varied, the trajectory of ω is easily found. For if $\omega + \tfrac{1}{2}ik$ is written as $x + iy$, equating the real parts on both sides of (8) shows that

$$x^2 - y^2 = \omega_0^2 - \tfrac{1}{4}k^2, \tag{9}$$

the equation of a rectangular hyperbola. The trajectory of ω is shown in fig. 3, the asymptotes of the hyperbola intersecting at $-\tfrac{1}{2}k$, the origin of x and y, shown as Q in the diagram. From the imaginary part of (8), $xy = \tfrac{1}{2}\Omega k$; this, with (9), shows that if one solution of (8) is (x', y'), the other is $(-x', -y')$, so that the two values of ω_0 for any given Ω lie on a line through Q, for example A and A' are a possible pair. Only when $\Omega = 0$ do the two solutions mirror each other in the imaginary axis, which is the condition that must be satisfied if the circularly polarized normal modes are to combine into a linear normal mode. So long as $\Omega < \omega_0$, both solutions lie

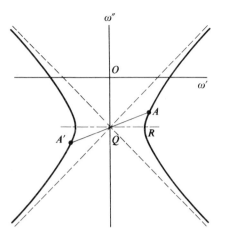

Fig. 3. Solution of (8).

below the real axis and represent decaying modes. But as Ω is increased above ω_0, A crosses over into the positive half-plane and the mode grows exponentially.

The gyro-pendulum

This example of a system with circular normal modes that are not time-reversed copies of each other is not important in itself, but it is easy to make and sufficiently amusing to be worth constructing; its discussion also serves as an introduction to the next section, dealing with a magnetized gyroscopic top, a matter of considerable physical significance. The gyro-pendulum illustrated in fig. 4 is essentially a rapidly spinning flywheel whose axis is mounted rigidly on a short pendulum which is free to swing in any direction. For small horizontal displacements described by the complex number r, the equation of motion is found by considering the horizontal components of angular momentum about the point of support, which changes because of the torque $imgr$, m being the total mass of flywheel and motor. The angular momentum is made up of two terms: (1) Lr/l, the horizontal component of the flywheel angular momentum, L, and (2) $-iml\dot{r}$, the angular momentum of the whole suspended mass as the pendulum swings. It is assumed that $r \ll l$ so that the normal linear approximation for pendulum motion is valid. It follows that, if losses are neglected,

$$\frac{\mathrm{d}}{\mathrm{d}t}(Lr/l - iml\dot{r}) = imgr,$$

i.e.
$$\ddot{r} + 2i\omega_g\dot{r} + \omega_0^2 r = 0, \tag{10}$$

where
$$\omega_0 = (g/l)^{\frac{1}{2}} \quad \text{and} \quad \omega_g = L/2ml^2.$$

The characteristic frequencies resulting from (10) have the form

$$\omega_{1,2} = \omega_g \pm (\omega_0^2 + \omega_g^2)^{\frac{1}{2}}. \tag{11}$$

Fig. 4. The gyro-pendulum. A brass flywheel, 1 cm thick and 12 cm in diameter, spinning at about 3000 r.p.m. gives good results. The rigid rod holding the motor hangs from a standard flexible joint used in radio sets; a steel wire through the middle serves to support the weight of motor and flywheel. The flywheel is about 25 cm below the flexible support. On the right the notation used in the text is illustrated. *Note: this is a potentially dangerous experiment.* The flywheel must be unflawed, well-balanced and securely pinned to the motor shaft, which must itself be strong (say $\frac{1}{4}$ in. diameter) to result the considerable gyroscopic torques.

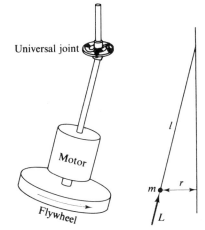

If L is positive (clockwise rotation of the flywheel when viewed from above), the faster mode, ω_1, given by the positive sign in (11), is clockwise and the slower, ω_2, anticlockwise.

When the pendulum is deflected and released it executes an angular looping motion, the result of both modes being excited together. Fig. 5 shows the calculated epicyclic trajectory when the parameters take such

Fig. 5. Typical trajectory of gyro-pendulum released from rest.

values that $\omega_0 = 2\omega_g$; the two frequencies in (11) are then in the ratio $(\sqrt{5}+1)/(\sqrt{5}-1)$, about 2.6. The pendulum is supposed to have started at rest in its position of maximum deflection; this involves the amplitudes of the two modes being in inverse ratio to their frequencies, so that their linear velocities cancel when they are in phase. The orbit is then sharply cusped, with n cusps per revolution, n being easily shown to be $1 - \omega_1/\omega_2$, 3.6 in this particular case. It may be observed that the pendulum has no orbital angular momentum at the cusps, but acquires angular momentum between cusps, in the reverse sense to the flywheel. This reflects the conservation of angular momentum about a vertical axis, the varying tilt of the flywheel demanding compensation in the orbit. By contrast every orbit of a simple pendulum is characterized by constancy of the orbital angular momentum. This is only possible when the circular modes are equal and opposite in frequency.

Another difference between the gyro-pendulum and a simple pendulum is that the potential and kinetic energies in the circular modes are no longer equal. Since $V = \frac{1}{2}mgr^2/l$ and $T_{1,2} = \frac{1}{2}m\omega_{1,2}^2 r^2$, $T_{1,2}/V = \omega_{1,2}^2/\omega_0^2$; in terms of n, the number of cusps, this takes the simple form

$$T/V = n - 1 \text{ in the fast mode}$$
$$1/(n-1) \text{ in the slow mode.}$$

With $n = 3.6$, as in the example just worked out, there is a considerable difference in the proportion of the energy in kinetic form. This shows itself as a difference in the damping rate for the two modes. If the pendulum is supported on a spongy base so that the point of support sways and is subject to viscous damping the fast mode is far more susceptible and decays much sooner than the slow. As a particular example, readily demonstrated, if the pendulum is set in an epicyclic orbit and then the whole apparatus picked up in the hands the cusps vanish almost at once, leaving the pendulum executing a slow circular orbit.

The gyromagnetic top[4]

We turn to a system which is more difficult to demonstrate in the laboratory, but which is of great theoretical importance since it models in terms of Newtonian dynamics the processes involved in *nuclear magnetic resonance*. We have already, in anomalous dispersion, met one characteristically quantum-mechanical process that can be replaced by an equivalent classical model, and here is another. At this stage the reader must be asked to accept that the equivalence can be strictly established. Once this is done, there are advantages in thinking about the phenomenon in classical terms, many of the essential results being intuitively acceptable as classical mechanisms when from a quantal standpoint they only emerge as mathematical consequences of a rather involved analysis. This is the justification for discussing in some detail the mechanics of a gyroscope which carries a permanent magnetic moment $\boldsymbol{\mu}$, parallel to its angular momentum $\boldsymbol{L}$. For this discussion we abandon complex notation in favour of vector notation.†

In a steady magnetic field $\boldsymbol{B}$ the torque, and therefore the rate of change of angular momentum, is $\boldsymbol{\mu} \wedge \boldsymbol{B}$; this causes the axis of the gyroscope to precess around the direction of $\boldsymbol{B}$ with angular velocity $\boldsymbol{\Omega}$ such that the resulting change in angular momentum, $\boldsymbol{\Omega} \wedge \boldsymbol{L}$, is equal to $\boldsymbol{\mu} \wedge \boldsymbol{B}$. If we

---

† Although the system is magnetic, no Lorentz force is involved explicitly, the only magnetic force being the velocity-independent torque exerted by $\boldsymbol{B}$ on $\boldsymbol{\mu}$. The velocity-dependent processes are still purely mechanical, being those characteristic of gyroscopic motion.

write γ for the (scalar) gyromagnetic ratio μ/L, it follows immediately that

$$\boldsymbol{\Omega} = -\gamma\boldsymbol{B}. \tag{12}$$

This result is important in showing that the precessional rate is the same for all orientations of the axis. For an atomic nucleus in a given energy state (normally its ground state, of minimum energy) γ is uniquely defined by the state, so that Ω/B is a characteristic property of the nucleus. The value of γ may be determined experimentally by perturbing the precessional motion by means of a weak additional magnetic field rotating at something like $\boldsymbol{\Omega}$; at resonance the orientation of the axis is systematically altered, and the resulting magnetic effects may be detected. The geometry of the arrangement is shown in fig. 6; $\boldsymbol{L}$ precesses around $\boldsymbol{B}$ with angular velocity $\boldsymbol{\Omega}$, and the extra field $\boldsymbol{b}$ rotates with angular velocity $\boldsymbol{\omega}$, also around $\boldsymbol{B}$.

Fig. 6. Illustrating the notation for the gyromagnetic top.

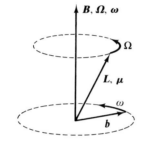

It is convenient to picture the behaviour from the point of view of an observer rotating with $\boldsymbol{b}$, who will see $\boldsymbol{b}$ at rest in his frame of reference. Any vector $\boldsymbol{A}$ which is constant in the laboratory frame will appear to him to be suffering change as, from his viewpoint, it rotates at an angular velocity $-\boldsymbol{\omega}$; the perceived rate of change will be $-\boldsymbol{\omega}\wedge\boldsymbol{A}$. This is for constant $\boldsymbol{A}$, and in general it must be supplemented by any actual rate at which $\boldsymbol{A}$ changes in the laboratory frame. Thus the vector $\boldsymbol{L}$, precessing momentarily with angular velocity $-\gamma(\boldsymbol{B}+\boldsymbol{b})$ about the vector resultant of $\boldsymbol{B}$ and $\boldsymbol{b}$, will seem to the observer to be changing at a rate given by the equation

$$\dot{\boldsymbol{L}} = -\gamma(\boldsymbol{B}+\boldsymbol{b})\wedge\boldsymbol{L} - \boldsymbol{\omega}\wedge\boldsymbol{L}$$
$$= -(\boldsymbol{\Delta}+\gamma\boldsymbol{b})\wedge\boldsymbol{L}, \tag{13}$$

where $\boldsymbol{\Delta} = \boldsymbol{\omega} - \boldsymbol{\Omega}$ and denotes the degree to which the rotation of $\boldsymbol{b}$ is mistuned to the precessional rate.

In the rotating frame $\boldsymbol{\Delta}+\gamma\boldsymbol{b}$ is fixed and the observer sees $\boldsymbol{L}$ precessing around the direction of the resultant, as shown in fig. 7. Since $\boldsymbol{\Delta}$ and $\boldsymbol{b}$ are perpendicular, the effective field has magnitude $(\Delta^2+\gamma^2b^2)^{\frac{1}{2}}/\gamma$, and the precessional angular velocity is $(\Delta^2+\gamma^2b^2)^{\frac{1}{2}}$, which we shall denote by χ. In the laboratory frame this pattern is seen as a slow perturbation of the principal precession. In case (a), where the cone does not include the

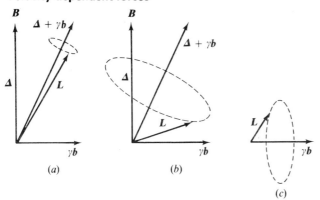

Fig. 7. Trajectories of L in the frame rotating with b.

direction of B, the vector L does not encircle the vertical axis, and in the laboratory frame the mean precession rate is ω; the moment does, however, alternately lag behind b and lead it, so that b alternately communicates energy to, and extracts it from the gyroscope. This is reflected in the periodic variations of the angle between μ and B, for while the kinetic energy is independent of orientation the potential energy has a magnetic term $-\mu \cdot B$. Case (a) is one in which b is strong enough, or well enough tuned in frequency, and L lies near enough to the optimum direction, parallel to $\Delta + \gamma b$, for the precession rate to be drawn into synchronism. In case (b), however, L lies far enough from $\Delta + \gamma b$ for the cone to encircle the vertical axis. The mean precessional rate is now nearer Ω than ω. In case (c) the rotation of b is precisely tuned to the unperturbed rate of precession, so that $\Delta = 0$; the tilting motion of the axis is symmetrically disposed about the horizontal plane, and occurs at a rate γb.

The response of the system to b may be recorded either as variations in the component of μ parallel to b or as variations in the transverse component. These measurements do not provide totally independent information because of energy conservation – the component of μ normal to both B and b determines the rate of energy flow to or from the gyroscope, and hence the rate of change of the tilt angle and the parallel component of μ. Let us concentrate on the transverse component of μ which, spinning at a mean rate of Ω or something near, will induce an oscillatory signal in any coil with its axis normal to B. It is of particular interest to take the special case when the moment μ starts parallel to B at time $t = 0$, so that the cone on which μ precesses in the rotating frame has the appearance of fig. 8(a). The projection of μ on the transverse plane follows round the ellipse of fig. 8(b) with frequency χ. The components parallel and perpendicular to b oscillate harmonically about a non-zero mean value represented by the centre of the ellipse. A little geometry shows that the components vary according to the expressions:

$$\mu_x = \mu \sin \theta \cos \theta (1 - \cos \chi t), \qquad \mu_y = \mu \cos \theta \sin \chi t, \qquad (14)$$

where θ is the tilt of the axis of the cone out of the transverse plane: $\tan \theta = \Delta / \gamma b$. This is in the rotating frame; in the laboratory frame (capital

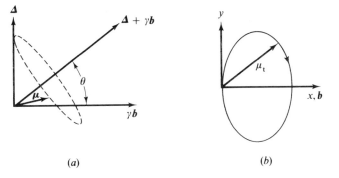

Fig. 8. Special case of fig. 7 when L and μ pass through the direction B. In (b) is shown the trajectory of μ_t the component of μ in the plane normal to B, as viewed in the rotating frame.

(a)　　　　　　　　(b)

subscripts) $\mu_X = \mu_x \cos \omega t + \mu_y \sin \omega t$ and $\mu_Y = \mu_y \cos \omega t - \mu_x \sin \omega t$. Therefore

$$\mu_X = \mu \cos \theta[\sin \theta \cos \omega t(1 - \cos \chi t) + \sin \omega t \sin \chi t]$$

$$\mu_Y = \mu \cos \theta[\cos \omega t \sin \chi t - \sin \theta \sin \omega t(1 - \cos \chi t)]. \tag{15}$$

It is easily seen from fig. 8(b) what these expressions mean. An observer in the laboratory frame sees μ_t spinning about the axis of B at the perturbing frequency ω, which is very nearly Ω, while its amplitude varies more slowly, with period $2\pi/\chi$, as μ_t traverses the elliptical locus. When $\gamma b \ll \Delta$, so that θ is nearly $\pi/2$, the ellipse shrinks to a small circle, and the transverse component μ_t oscillates harmonically:

$$\mu_t = 2\gamma b\mu \sin (\tfrac{1}{2}\chi t)/(\Delta^2 + \gamma^2 b^2)^{\frac{1}{2}} \sim \frac{2\gamma b\mu}{\Delta} \sin (\tfrac{1}{2}\chi t).$$

As this spins, the components in the laboratory frame vary according to the expressions

$$\mu_X = \frac{2\gamma b\mu}{\Delta} \sin (\omega + \tfrac{1}{2}\chi)t,$$

and

$$\mu_Y = \frac{2\gamma b\mu}{\Delta} \cos (\omega + \tfrac{1}{2}\chi)t. \tag{16}$$

146　　These should be compared with (6.31) and fig. 6.12(b). The behaviour is almost exactly the same as a lossless harmonic vibrator excited near resonance, and exhibiting a beat pattern due to the superposition of the steady-state response and the transient excitation.

In (16) the beat frequency is determined by the departure of the perturbing signal b from resonance, and depends little on the amplitude of b. This is true only when $\gamma b \ll \Delta$, but as the perturbation approaches resonance and this inequality fails, the beat pattern changes its character. The circular locus of μ_t becomes markedly elliptical and ultimately, just on resonance, a straight line, when $\theta = 0$ and the circular orbit in fig. 8(a) lies in a vertical plane. The axis of the gyroscope, viewed in the laboratory frame, if its starts parallel to B begins to precess in a steadily widening cone, passes through the transverse plane and converges onto the opposite direction to B; the process then repeats itself in reverse, and so on as long

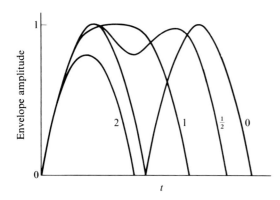

Fig. 9. Envelope curves showing the development of the component of $\boldsymbol{\mu}$ in the plane normal to $\boldsymbol{B}$ when the moment starts, at time $t = 0$, parallel to $\boldsymbol{B}$. The amplitude of b is the same for all, but Δ is different; the numbers attached to the curves are values of $\Delta/\gamma b$. An envelope amplitude of 1 means that $\boldsymbol{\mu}$ lies in the transverse plane. In the absence of dissipative effects etc. the curves repeat themselves indefinitely.

as the resonant perturbation is maintained. A series of cases, with b constant and Δ varying through resonance, is shown in fig. 9, which gives only the envelope of the oscillations at frequency ω, i.e. the amplitude of μ_T. The saturation of the oscillations is clearly exhibited, and the reason for saturation is obvious: μ_T can never exceed μ. As saturation is approached the non-linearity of the system also reveals itself in the way the beat frequency does not fall to zero at resonance but reaches a minimum value γb. In contrast to fig. 6.12, the envelope becomes markedly non-sinusoidal as resonance is approached.

This is a quite different type of non-linear vibrator from anything we have met so far, in that it is isochronous but saturable. In this respect the behaviour is nothing like that of a particle in a non-parabolic potential, which may possibly be isochronous, but which is not saturable., The difference may be illustrated by the response to a gliding tone. In chapter 6 we showed how a lossfree harmonic vibrator, excited by a constant-amplitude signal whose frequency varied linearly with time ($\omega = \omega_0 + \beta t$), ended up after passage through the resonance with an amplitude of vibration proportional to $\beta^{-\frac{1}{2}}$. With the gyromagnetic top the corresponding behaviour is a change in tilt of the axis relative to $\boldsymbol{B}$. When the frequency sweep is rapid the change is small and proportional to $\beta^{-\frac{1}{2}}$ – this is the linear approximation to the gyroscope equation which causes it to follow harmonic vibrator theory exactly. But if β is small enough for the predicted change to be well outside the linear approximation (for example, apparently conferring kinetic energy far in excess of the maximum permitted to the top) the matter obviously needs reconsideration.

149

In fact, in this limit the behaviour is very simply understood. Rather than sweeping the frequency, keeping $\boldsymbol{B}$ constant, let us keep ω constant and lower $\boldsymbol{B}$ slowly from a value well above resonance to one well below; this is an entirely equivalent process, and one that is carried out in certain procedures (*Adiabatic passage*) used in the study of nuclear magnetic resonance.† In the rotating frame the situation at some instant will resemble that illustrated in fig. 8(a). Let B now be suddenly changed by a small amount δB so as to depress the orientation of $\Delta + \gamma b$, without causing any immediate variation of the direction of L. If, when this change is made, L is somewhere on the upper half of the cone whose axis is $\Delta + \gamma b$, the depression of the axis means that it must now sweep out a slightly wider cone; conversely if it happens to be on the lower half the cone is narrowed. When B is changed by a large number of small increments over a time that allows many cycles of precession, there is on the average no first-order change in the cone angle – its axis simply follows the direction of $\Delta + \gamma b$. Second-order change in the cone angle can be made as small as desired by changing B sufficiently slowly. This process of adiabatic passage results in the component of M parallel to B simply reversing in sign as B goes through the resonant value and $\Delta + \gamma b$ swings from vertically upwards to vertically downwards.

Nuclear magnetic resonance

No attempt will be made to give a systematic account of the vast field of study opened up by the observation of magnetic resonance in solids, liquids and gases. We shall instead point out a few features that exemplify in practice the characteristics of this rather special oscillatory system. Adhering to the classical model we imagine a sample of water as a rather dense array (7×10^{28} to the cubic metre) of protons, the nuclei of the hydrogen atoms, and forget for the moment everything else. Each proton behaves like a magnetic top, with angular momentum L equal to $\frac{1}{2}\hbar$, or 0.527×10^{-34} Js, and magnetic moment μ equal to 1.41×10^{-26} J/T; the value of γ is 2.75×10^{8} T^{-1} s^{-1}, which means that in a field of 1 T the precession rate is $\frac{1}{2}\gamma/\pi$ or 42.6 MHz. If each precessed independently in a perfectly uniform field, all would have the same angular velocity Ω, so that in a frame rotating at this speed the whole distribution of moments could be represented by a stationary array of points on the surface of a sphere of unit radius, each point marking the end of a vector drawn from the centre. When this idealized system is perturbed by a field, b, rotating with angular velocity ω, an observer rotating with b will see each point moving in a circle with the same angular velocity $\Delta + \gamma b$ about the same tilted axis, as in fig. 7. The whole distribution of points therefore turns bodily about this axis, and any resultant moment M of the whole assembly turns with it. This makes it a

---

† The term *adiabatic* is used in thermodynamics and in quantum mechanics with subtly different meanings. This particular usage is of the latter kind, and implies that the parameters of a physical system (in this case B) are changed so gradually that the system evolves smoothly during the process. Thus far there is no conflict between thermodynamic and quantum mechanics. But it is further implied in quantum mechanics that the system is isolated from random processes such as collisions; thermodynamics does not require this restriction – indeed it would often make the process non-adiabatic.

very simple matter to calculate how the system responds to the perturbation.

Let us suppose that $\boldsymbol{B}$ alone is applied, and the system then allowed enough time (which may prove quite long) to settle into thermal equilibrium. Because the energy of a moment pointing along $\boldsymbol{B}$ is less, by $2\mu\boldsymbol{B}$, than that of one pointing against $\boldsymbol{B}$, there will be a slight excess of the former in thermal equilibrium. Making here our one concession to quantum mechanics (for the sake of simplicity rather than necessity) we recognize that only parallel and antiparallel states are permitted to the proton, and that these, according to the Boltzmann distribution, will be occupied so that out of N moments,

$$N\,e^X/(e^X + e^{-X}) \quad \text{will point along } \boldsymbol{B}$$

$$N\,e^{-X}/(e^X + e^{-X}) \quad \text{will point against } \boldsymbol{B}, \text{ where } X = \mu B/k_B T.$$

The resultant total moment M is due to the difference in these two occupations:

$$M = N\mu \tanh X \sim N\mu^2 B/k_B T \quad \text{if } X \ll 1. \tag{17}$$

In a field of 1 T at room temperature, X is about 3.5×10^{-6}, so that the resultant moment is as if only this tiny fraction were aligned with the field, the rest being randomly oriented. Nevertheless it is quite enough to give a good signal when appropriately perturbed.

One method of detecting $\boldsymbol{M}$ is to apply the field $\boldsymbol{b}$ at the resonant frequency for just as long $(\pi/2\gamma b)$ as is needed to cause each moment, and therefore $\boldsymbol{M}$, to turn through $\pi/2$ about $\boldsymbol{b}$ in the rotating frame. Immediately afterwards $\boldsymbol{M}$, instead of pointing along $\boldsymbol{B}$, lies in the transverse plane and stays in it while it precesses (in the laboratory frame) with angular velocity $\boldsymbol{\Omega}$. Any pick-up coil surrounding the sample, with its axis in the transverse plane, will now have induced in it an oscillating e.m.f. at frequency Ω. If the applied field were perfectly uniform and the nuclei free from any perturbing interactions, the oscillation would persist indefinitely. In practice, however, it decays, either through inhomogeneity of the field which causes differently located nuclei to precess at different rates or, if this effect has been reduced to insignificance, through interactions of the nuclei among themselves and with their surroundings. This method is analogous to determining the impulse response function of a vibrator. Alternatively the spin assembly may be excited by a coil producing a continuous oscillatory magnetic field transverse to $\boldsymbol{B}$, and the response detected by means of a similar coil normal to the first. This arrangement, due to Bloch, avoids swamping the signal induced by the spinning moment, since the input and output coils are not magnetically coupled. The resonant response which can be traced out by varying the input frequency is related to the impulse response by the usual formula, provided the amplitude of excitation is low enough for linearity to be maintained.

The interactions between nuclei deserve some discussion here, since they serve to exemplify a number of points of general physical interest. For a fuller treatment the reader should seek specialized texts – the matter is complex and abounds with subleties. At first sight one might think that, isolated from everything else, the assembly of nuclei could not alter their total moment M by mutual interactions, for one would imagine the total angular momentum associated with their spin to be conserved. This is a misconception, however – the interaction between two magnetic moments is non-central and transfers angular momentum to the translational degrees of freedom. A typical example is shown in fig. 10; both moments are subjected to a clockwise torque, the sum of the two being compensated by an equal anticlockwise torque arising from translational forces experienced by each magnetic moment in the inhomogeneous field of the other. It is therefore possible for a molecule like water, containing two protons, to exchange spin angular momentum for rotation of the molecule as a whole. Since translational motions are readily exchanged during collisions, this process provides a mechanism by which the precessing moment of the assembly can be dissipated by conversion into bulk rotation.

Fig. 10. Two protons with moments oriented as shown by the arrows in the circles; each produces a magnetic field at the other's location shown by the broken lines, leading to forces F and torques G.

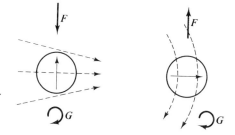

Interaction between the protons in a molecule is the dominant dissipative process in water, but interaction between neighbouring molecules is also effective. Whatever the detailed mechanism, it is clear that the points representing individual moments can wander round the unit sphere in a diffusive manner, so that a lopsided distribution, established in the manner described, is gradually restored to the equilibrium configuration. In certain cases, of which we shall soon examine one in more detail, there is reason to expect that the decay of the transverse moment will be exponential, but this is not a general rule.

The process just outlined, whereby an ordered assembly of spins relaxes to thermal equilibrium, is slow compared to the precessional rate, for the magnetic field at one proton due to the other in a water molecule is only about 1 mT, an extremely weak perturbation of the main field B. The response function is normally, therefore, a long train of oscillations and the resonance peak is correspondingly narrow; under the best conditions, indeed, Q-values of more than 10^9 are observed (see fig. 11). We shall now look into the matter in some detail, and it is important to recognize at the

outset a distinction between relaxation processes that conserve the component of M parallel to B and those that do not. If the assembly has been polarized with M lying away from the direction of B, it is possible for the individual moments to wander along lines of latitude (with respect to B as polar axis) and for the transverse component to decay while the longitudinal component stays constant; this involves no change in the polarization energy $M.B$, while relaxation of the longitudinal component requires energy exchange between the spins and the liquid as a whole. Let us estimate the energy exchange possible when spin angular momentum is converted into molecular angular momentum, a process strongly affected by the rate at which the molecule is turning at the time. We picture the molecule as a rigid body to which the protons are attached, without any constraint on their rotation except for their mutual interaction. As the molecule tumbles about under the impacts of its neighbours the proton spins stay pointing in the same direction, or precess in whatever magnetic field they experience, quite unaffected by the gross molecular motion.†
Now suppose the interaction between the spins to cause δL of angular momentum to be transferred from them to the molecule at a time when its angular velocity is ω_m. The accompanying energy exchange, $\omega_m \cdot \delta L$, occurs at the expense of the polarization energy, the only available source since the interaction energy is so small, and we therefore write $\omega_m \cdot \delta L = B \cdot \delta\mu$. But $\delta\mu = \gamma\delta L$ and $\Omega = \gamma B$, so that $\delta\mu \cdot (\Omega - \omega_m) = 0$; the change in spin moment can only take place in the plane normal to $\Omega - \omega_m$. With slowly rotating molecules, having $\omega_m \ll \Omega$, the change is almost entirely confined to the plane transverse to B; transverse relaxation is fast relative to longitudinal. This is what happens in viscous liquids, but in mobile liquids ω_m may considerably exceed Ω, and there is then no serious restriction to the direction of $\delta\mu$. More detailed justification of this elementary analysis will follow, in which it will be seen that increasing the tumbling speed does not make longitudinal relaxation easier – on the contrary both types of relaxation are inhibited, but the transverse more so than the longitudinal; it is in this way that they become comparable, as the Q of the resonance rises towards the enormous value quoted above.

Early in the history of nuclear magnetic resonance Bloch appreciated the possibility of transverse and longitudinal relaxation occurring at different rates, and introduced in his formulation of the equations of motion for the total moment M two characteristic relaxation times: T_1, the longitudinal relaxation time describing thermal equilibration in which the component of M parallel to B is changed, and T_2, the transverse relaxation time describing the diffusive decay of the transverse component. His introduction of these relaxation times was heuristic, expressing the essential idea in the simplest way, without any pretence that it described any real situation

---

† This is not so if the nucleus (not a proton) has an electric quadrupole moment, for this is acted on by the intramolecular electric field gradients which try to lock the spin direction relative to the molecule. We shall not consider the effects of this complicating process.

exactly; all the same, it does very well in a large number of applications.

We write Bloch's equations in complex form, using M_1 (real) to denote the parallel component of $\boldsymbol{M}$ and M_2 (complex) to denote the magnitude and direction of the transverse component. In addition a complex number b represents the rotating field in the transverse plane. The translation of the vector product $\boldsymbol{M} \wedge \boldsymbol{B}$ into this notation looks rather awkward but is quite straightforward. We quote the resulting equations and then expound them:

$$\dot{M_1} + \gamma \operatorname{Im}[M_2 b^*] + (M_1 - M_0)/T_1 = 0, \tag{18}$$

and
$$\dot{M_2} + i\gamma(M_2 B - M_1 b) + M_2/T_2 = 0. \tag{19}$$

If $b = 0$, the equations express the exponential relaxation of the magnetization to its equilibrium value M_0 in the steady field B. According to (18), M_1 returns to M_0 with relaxation time T_1; and according to (19), M_2 precesses with angular velocity Ω ($= \gamma B$), while decaying to zero with relaxation time T_2. The perturbing terms containing b are the components of the vector product $\boldsymbol{M} \wedge \boldsymbol{B}$, and show how M_1 is changed by the interaction of M_2 with b, M_2 by the interaction of M_1 with b.

When b rotates at angular velocity ω we may write it as $b' \, e^{-i\omega t}$, b' being taken as real, and look for a steady-state solution in which M_1 is constant and M_2 varies as $(M_2' + iM_2') \, e^{-i\omega t}$. Substituting this trial solution we find that

$$M_1 = M_0 - \gamma b' T_1 M_2'' \quad \text{from (18)}, \tag{20}$$

and
$$(1 - i\,\Delta T_2)(M_2' + iM_2'') = i\gamma M_1 b' T_2 \quad \text{from (19)}, \tag{21}$$

where, as before, Δ is written for $\omega - \Omega$. When M_1 from (20) is substituted in (21) and the real and imaginary parts of (21) separated, the solution emerges in the form

$$-M_2'/\Delta T_2 = M_2'' = \gamma b' T_2 M_0/(1 + \Delta^2 T_2^2 + \gamma^2 b'^2 T_1 T_2). \tag{22}$$

The amplitude of the moment in the transverse plane is $(M_2'^2 + M_2''^2)^{\frac{1}{2}}$:

$$|M_2| = \gamma b' T_2 M_0 (1 + \Delta^2 T_2^2)^{\frac{1}{2}}/(1 + \Delta^2 T_2^2 + \gamma^2 b'^2 T_1 T_2). \tag{23}$$

There is no need to describe the various techniques used to test and record the character of the response to a steady oscillatory field b, since these are fully dealt with in specialized texts. We shall consider only the salient physical processes that determine this character. So long as $\gamma b' \ll (T_1 T_2)^{-\frac{1}{2}}$ the last term in the denominator is negligible and the behaviour is a linear resonant response with a Q-value of $\frac{1}{2}\Omega T_2$; M_2 lags behind b by $\cot^{-1}(\Delta T_2)$ and causes the assembly of spins to receive energy continuously, to be dissipated as heat by the relaxation mechanism. At resonance the dissipation is greatest, with a phase lag of $\pi/2$ so that the power input W_{res} is $\Omega M_2'' b'$. From (22), without restricting b' to a small value,

$$W_{\text{res}} = \gamma^2 b'^2 T_2 M_0 B/(1 + \gamma^2 b'^2 T_1 T_2), \tag{24}$$

and in this state, according to (20), M_1 is reduced from its equilibrium value of M_0 to $M_0 - \delta M_1$, where

$$\delta M_1/M_0 = \gamma^2 b'^2 T_1 T_2/(1 + \gamma^2 b'^2 T_1 T_2). \tag{25}$$

This change in M_1 implies that the energy of the assembly is increased by $B\,\delta M_1$. If this is constantly being dissipated to an external energy sink (e.g. molecular motions) at a rate $B\,\delta M_1/T_1$, comparison of (24) and (25) shows that, as must of course be true in the steady state, the assembly of nuclei is losing as much energy as it gains.

There is a limit to the power dissipation set by the saturation of the system, which shows itself in the term in b'^2 in the denominator of (22). However large b' may be, W_{res} cannot exceed $M_0 B/T_1$, when M_1 has fallen to zero. The amplitude of response, as measured by $|M_2|$ in (23), reaches a maximum at resonance when $\gamma^2 b'^2 T_1 T_2 = 1$; this maximum value is $\frac{1}{2} M_0 (T_2/T_1)^{\frac{1}{2}}$, which is quite sizeable when T_1 and T_2 are comparable, as in many liquids. In solids, however, T_2 is commonly much shorter than T_1 and the signal is correspondingly weak.

As already remarked, T_1 and T_2 may be long in mobile liquids, making possible resonances of very high Q indeed. The experimental curve in fig. 11 has peaks as little as 0.04 Hz wide at a frequency of 200 MHz, so that $Q \sim 5 \times 10^9$. With a decay time of seconds the systematic traversing of the spectrum would be a very slow process, but the acquisition of data is greatly accelerated by use of the impulse response, which is the super-position of decaying oscillations at each of the frequencies present. The information needed to obtain the spectrum shown here was provided by a single shot, with the impulse response recorded for 109 seconds only; from then on the spectrometer was available for other samples, while the response function was Fourier-analysed by computer, in accordance with the principle expressed by (5.4). The fine structure in the spectrum arises from the fact that the various protons (hydrogen nuclei) in the molecule of *ortho*-dichlorobenzene are coupled by their magnetic moments (it is essentially a quantum-mechanical feature that such coupling results in discrete lines) but the whole range of the spectral distribution, of which only a part is shown here, occupies less than 100 Hz; the frequencies differ by one part in 2×10^6 at most. The impulse response is therefore a virtually pure oscillation at a frequency of 200 MHz, with an envelope showing a slow but complicated beating pattern having a typical time-scale of 10 ms, and decaying with a time-constant of about 8 s. It is only necessary to record the variations of amplitude and phase of the basic oscillation at such intervals as will reveal the modulations of the envelope, and something like 10^4 data points suffice. To determine the phase a comparison oscillator is used (details of the electronic circuitry need not concern us) whose frequency stability must be such that no significant phase drift occurs in 109 s. This is no trivial matter, but well within the capability of modern techniques. It is perhaps more impressive to find that a superconducting solenoid can be designed and shimmed so that over the 5 mm extent of the

106

Fig. 11. Example of extremely sharp proton resonances at 200 MHz in a magnetic field of 4.7 T. The whole spectrum shown occupies a band of 10 Hz and the lines are typically 0.04 Hz wide at half-peak ($Q \sim 5 \times 10^9$). The spectrum was obtained with a WP 200 spectrometer (Bruker, Spectrospin Ltd.), the sample being a dilute solution of *ortho*-dichlorobenzene, $C_6H_4Cl_2$, in deuterated acetone, $(CD_3)_2CO$; having no hydrogen, the solvent contributes no resonances in this frequency range.

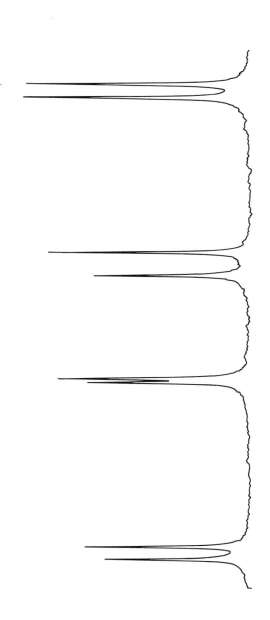

sample its field of 4.7 T is uniform to two parts in 10^{10}, as is necessary if the lines are not to be broadened as a result of protons in different regions of the sample precessing at different frequencies. In fact this degree of perfection is probably not quite achieved, the last refinement resulting from spinning the sample about an axis parallel to the field at about 30 Hz. Then, to first order of approximation, each proton experiences on the average the field strength on the axis of rotation. The small periodic fluctuations as it rotates impose a frequency modulation on the response, but so long as the modulation is slight enough for the phase of the signal to be shifted by no more than, say, one radian, the sidebands introduced by the modulation are relatively weak. Although these sidebands introduce 'ghost' lines separated from the main resonance by the frequency of rotation, they will remain insignificant provided this criterion is satisfied. What it amounts to is that if the inhomogeneity of the magnetic field is barely enough to broaden the line by an amount equal to the rotation frequency, rotation can eliminate the effects of the field gradient without introducing excessive sidebands; but it cannot work wonders – only improve what is already very good.

The physical mechanism of relaxation

We turn now to the estimation of T_1 and T_2 for the comparatively straightforward case of a mobile liquid such as water. Since even here the argument involves a number of steps we shall first give a synopsis and then proceed to fill in some of the details.

(i) We consider only the interaction between the two protons in the same water molecule. The molecule is in continuous random rotation as part of its thermal motion, the characteristic angular velocity being perhaps $10^{12}\,\text{s}^{-1}$, much greater than the resonant frequency. In this tumbling, the spin axes are not locked to the molecule but stay almost fixed in the laboratory frame, the slow changes of orientation that each spin undergoes being caused by the perturbing field of the other proton in the molecule. As the direction of the line joining the protons varies rapidly and irregularly, this perturbing field suffers equally rapid random fluctuations. Fourier analysis yields a continuous spectrum, of which only the components of lower frequency have time to cause a significant precession of the spin.

(ii) A point on the surface of the unit sphere, representing the spin axis of one proton, wanders randomly as a consequence of this perturbation, and the points representing all protons in the sample behave statistically as though they were governed by a diffusion constant D; if each point wanders independently the flux of points is described by a vector equal to $-D$ grad n, where n is the number of points per unit area, varying with position.

(iii) A polarized distribution of points on the sphere, such as describes a resultant magnetic moment M, evens itself out by this diffusive process, and we shall show that, left to itself, M decays as $e^{-t/T}$, where $T = 1/2D$.

(iv) The calculation of D proceeds in three stages and makes use of ideas already presented in earlier chapters: (a) a particle moving on a line with nearly random velocity, $v(t)$, will be shown to be governed by a diffusion constant equal to $\frac{1}{2}\pi I_\omega(0)$, where $I_\omega(0)$ is the low-frequency limit of the intensity spectrum of v; (b) the intensity spectrum itself will be derived from the autocorrelation function of the tumbling motion of the molecules, supplemented by (c) an explicit calculation of the strength of the magnetic interaction between the two protons.

(v) The outcome of this analysis is that when the tumbling is much more rapid than the precessional motion

$$T_1 = T_2 = 160\pi^2 a^6/9\mu_0^2\gamma^4\hbar^2\tau_c, \tag{26}$$

where a is the separation of the protons in the molecule and τ_c is the correlation time for the tumbling motion, the time it takes for memory of an earlier orientation of the molecule to be lost. For water, $a \doteqdot 0.15\,\text{nm}$ and $\tau_c \doteqdot 2 \times 10^{-12}\,\text{s}$, while γ for the proton is $2.675 \times 10^8\,\text{T}^{-1}\,\text{s}^{-1}$.† This yields a value of about 10 s for T_1 and T_2. The experimental value at room temperature is 3.6 s, and the difference is easily accounted for by the approximate value for τ_c and the neglect of interactions between protons on different molecules.

To put flesh on these bones, we start with (iii). Let the position of a point on the sphere be defined by polar angles (θ, ϕ), and let us consider M_1, the component of magnetization parallel to the polar axis. The ϕ-variable is irrelevant to M_1, all that matters being the mean concentration $n(\theta)$ around the line of latitude defined by θ. The diffusive flux across this line, whose length is $2\pi \sin\theta$, is $-2\pi \sin\theta \cdot D\, \partial n/\partial\theta$. Since the number of points is conserved, the variation of this quantity with θ determines the rate at which n changes:

$$\sin\theta\, \partial n/\partial t = D\frac{\partial}{\partial\theta}(\sin\theta\, \partial n/\partial\theta). \tag{27}$$

† The correlation time can be determined quite independently from the frequency-variation of the dielectric constant. In a slowly oscillating electric field, such that $\omega\tau_c \ll 1$, the electrical dipoles of the water molecule have time to follow the field changes, and ε is large; but when $\omega\tau_c \gg 1$ there is too little time available, and ε is much smaller. The similarity in magnitude of $1/\tau_c$ and the mean angular velocity of tumbling is coincidental, and means only that a water molecule can rotate through a substantial angle before collisions randomize its motion. In a viscous liquid this coincidence would not occur.

If sin θ is crossed out on both sides, this turns into the normal equation for diffusion on a line; the presence of sin θ makes allowance for the geometry of a spherical surface. The magnetization is given by the equation

$$M_1 = 2\pi\mu \int_0^\pi n \sin\theta \cos\theta \, d\theta, \tag{28}$$

so that $\quad dM_1/dt = 2\pi\mu \int_0^\pi \frac{\partial n}{\partial t} \sin\theta \cos\theta \, d\theta$

$$= 2\pi D\mu \int_0^\pi \frac{\partial}{\partial\theta}\left(\sin\theta \frac{\partial n}{\partial\theta}\right) \cos\theta \, d\theta, \quad \text{from (27),}$$

$$= 2\pi D\mu \int_0^\pi \sin^2\theta \frac{\partial n}{\partial\theta} \, d\theta, \quad \text{on integrating by parts.}$$

A further integration by parts reduces the integral to $-\int_0^\pi 2n \sin\theta \cos\theta \, d\theta$, so that from (28) one sees that

$$dM_1/dt = -2DM_1.$$

Hence $\qquad M_1 = M_0 e^{-t/T}, \quad \text{where } T = 1/2D. \tag{29}$

This is the result quoted in (iii).

To relate D to the velocity fluctuations ((iv)a of the synopsis), consider a distribution of points whose density varies in one direction (which we label s) only. Then the diffusive motion obeys the equation:

$$\partial n/\partial t = D \, \partial^2 n/\partial s^2.$$

The spread of an initially compact group of points follows the solution

$$n \propto e^{-s^2/4Dt}/t^{\frac{1}{2}}, \tag{83}$$

and the mean square displacement of a point in time t,

$$\overline{s^2}(t) = \int_0^\infty s^2 n(s,t) \, ds \Big/ \int_0^\infty n(s,t) \, ds = 2Dt. \tag{30}$$

This result is equally applicable to motion along a line drawn on a sphere; in any case we shall be dealing with small displacements only in calculating D, and the curvature is irrelevant.

Let us now calculate $\overline{s^2}(t)$ in terms of the velocity spectrum. If $\dot{s}(t)$ is a random real function with intensity spectrum I_ω, the components in a narrow range $\delta\omega$ combine to give a nearly periodic oscillation of $\dot{s}$ having the form $(2I_\omega\delta\omega)^{\frac{1}{2}} \cos(\omega t + \phi)$, ϕ being a random phase. By integration the oscillatory displacement due to the components in the narrow band is given by the expression

$$\delta s(t) = (2I_\omega\delta\omega)^{\frac{1}{2}}[\sin(\omega t + \phi) - \sin\phi]/\omega,$$

the integration constant being supplied to make $\delta s(0)$ vanish. This can be written in the form

$$\delta s(t) = 2(2I_\omega \delta \omega)^{\frac{1}{2}} \sin(\tfrac{1}{2}\omega t) \, \mathrm{Re} \, [e^{i(\frac{1}{2}\omega t + \phi)}]/\omega,$$

and interpreted as the real part of a randomly phased vector of magnitude $2(2I_\omega \delta \omega)^{\frac{1}{2}} \sin(\tfrac{1}{2}\omega t)/\omega$. Summation of all similar contributions from different ranges $\delta \omega$ to give an average resultant is achieved by adding the squares, so that the mean resultant vector has length R^2, where

$$R^2(t) = \int_0^\infty 8I_\omega \sin^2(\tfrac{1}{2}\omega t) \, d\omega/\omega^2.$$

Since R is the length of a vector on a vibration diagram, of which only the real part is physically significant, we have that

$$\overline{s^2}(t) = \tfrac{1}{2} R^2(t) = 4 \int_0^\infty I_\omega \sin^2(\tfrac{1}{2}\omega t) \, d\omega/\omega^2. \tag{31}$$

The presence of $\sin^2(\tfrac{1}{2}\omega t)/\omega^2$ shows that the integral is dominated by values of $\omega \lesssim 2\pi/t$. If I_ω is flat at low frequencies, falling off at higher frequencies, the evaluation of (31) may be troublesome when t is short, but for longer times it is easy since I_ω may be given its low-frequency limit $I_\omega(0)$. Physically this means that memory of previous motion persists for a time (the reciprocal of the frequency at which I_ω begins to fall off) and during this interval the motion of the point on the sphere is not random; but if one is concerned with motion over a much longer interval, the short-term memory becomes insignificant compared with the random walk. In a viscous liquid the orientation of a molecule may stay sensibly constant for quite a while, during which the mutual interactions cause the spins to drift systematically through large angles; this invalidates our model of the process as essentially diffusive. In water, on the other hand, the combination of rapid tumbling and frequent collisions means that the systematic drifting is constantly disrupted and can be safely neglected. We are therefore justified in writing, from (31),

$$\overline{s^2}(t) = \pi I_\omega(0) t.$$

The variation of $\overline{s^2}$ as t, having the same form as (30), enables us to identify D immediately:

$$D = \tfrac{1}{2} \pi I_\omega(0), \quad \text{as quoted in (iv)}a. \tag{32}$$

We now have a picture of the points executing diffusive motion in the form of a nearly random velocity whose component in one direction, $\dot{s}$, is described by the intensity spectrum I_ω, which we now estimate (part (iv)b of the program). It is clear that during an interval of time shorter than is needed for the molecule to rotate very far the interaction between two protons is constant and v is also constant. There is therefore persistence of memory during a correlation time τ_c, which measures the time during which the molecule tumbles far enough and randomly enough to forget where it was before. We expect the autocorrelation function of $\dot{s}$ to decay (probably exponentially but the precise manner is unimportant) with some characteristic time τ_c. This is a case already discussed in chapter 4, where it is shown that $I_\omega = I_\omega(0)/(1+\omega^2\tau_c^2)$. By integrating over all positive frequencies we have that

96

$$\overline{\dot{s}^2} = \int_0^\infty I_\omega \, d\omega = \tfrac{1}{2}\pi I_\omega(0)/\tau_c.$$

The value of $\overline{\dot{s}^2}$ depends only on the strength of the interaction between the dipoles. The effect of shortening τ_c is to spread I_ω over a wider frequency range and thus to lower its value at those low frequencies that are effective in the diffusive process. This is the reason for the especially sharp resonances in mobile fluids. As follows from (29) and (32)

$$T = 1/2\overline{\dot{s}^2}\tau_c. \tag{33}$$

This completes the program except for (iv)c, the evaluation of $\overline{\dot{s}^2}$ in terms of the magnetic field exerted by one proton on another. We shall carry out the calculation in cartesian coordinates, concentrating on v_y, the y-component of velocity on the unit sphere. Let one proton define the z-axis, as does A in fig. 12, and let the other, at B, have an arbitrary orientation.

Z

a

Y

.B

A

X

Fig. 12. Notation for calculating mutual torques of two magnetic dipoles.

By letting B occupy all positions on a sphere of radius a (the fixed separation of the protons in the molecule), and also allowing the spin at B to assume all orientations, we span the range of torques that one proton can exert on another, the latter having a definite orientation. The field at A due to the proton at B has an x-component (which is all that matters in calculating v_y) given by

$$B_x = \frac{\mu_0}{4\pi}\left[\frac{3x}{a^5}(\mu_x x + \mu_y y + \mu_z z) - \frac{\mu_x}{a^3}\right],$$

$\mu_{x,y,z}$ being the components of the dipole moment of the proton at B. The y-component of the torque on the dipole μ at A is then μB_x, and it is this that causes the point representing the dipole at A to move on the unit sphere with velocity $v_y = \mu B_x/L$:

$$v_{yA} = \frac{\mu_0 \mu}{4\pi L}\left[\frac{3x}{a^5}(\mu_x x + \mu_y y + \mu_z z) - \frac{\mu_x}{a^3}\right]. \tag{34}$$

At the same time the dipole at A exerts a torque on that at B causing it to move with velocity v_{yB}:

$$v_{yB} = \frac{\mu_0 \mu}{4\pi L}\left[\frac{3z}{a^5}(\mu_z x - \mu_x z) + \frac{\mu_x}{a^3}\right]. \tag{35}$$

Because the spins are coupled in this way their random motions are not independent, and the preceding analysis which assumed independence would be incorrect without allowance for this feature. To take an extreme example, if it were the case that the total spin angular momentum was conserved the two points would undergo such correlated random motion that no relaxation of the total moment occurred. As it happens, the correlation here has the reverse effect of enhancing the change in total moment over what would take place if each point wandered independently. The way to take account of correlation is to ascribe the whole movement to the proton at A, writing $v_y = v_{yA} + v_{yB}$; similarly when treating the motion of the proton at B we should add in the consequent motion of A. This looks like double counting but is not. What we are doing is to imagine in the first place that there is no correlation, in which case the two spins wander independently, each with a velocity spectrum of the form that would result from (34) alone, averaged over all arrangements of the second spin. We then recognize that the positive correlation represented by (35) adds to the magnitude of the wandering of each. By subsuming (35) in (34) we ensure that the variations of the total moment are correctly described, without having to abandon the model of independent motion.

This is the justification for writing v_y as $v_{yA} + v_{yB}$:

$$v_y = \frac{3\mu_0 \mu}{4\pi a^5 L}\left[\mu_x(x^2 - z^2) + \mu_y xy + 2\mu_z xz\right]. \tag{36}$$

Each dipole in the assembly oriented, like that at A, along the z-axis experiences a rapidly fluctuating motion v_y as the tumbling of the molecule takes B round the sphere of radius a centred on A. The mean value of v_y vanishes, since $\bar{x} = \bar{y} = \bar{z} = 0$ and $\overline{x^2} = \overline{y^2} = \overline{z^2} = \frac{1}{3}a^2$. To describe the magnitude of the fluctuating velocity we use $\overline{v_y^2}$, averaging not only over x, y and z but over all orientations of the dipole at B, by putting $\overline{\mu_x^2} = \overline{\mu_y^2} = \overline{\mu_z^2} = \frac{1}{3}\mu^2$. In this way we derive the mean square value of v_y for z-pointing dipoles:

$$\overline{v_y^2} = \frac{9\mu_0^2 \mu^2}{16\pi^2 a^{10} L^2}[\overline{\mu_x^2(x^2 - z^2)^2} + \overline{\mu_y^2 x^2 y^2} + 4\overline{\mu_z^2 x^2 z^2}],$$

all cross-terms vanishing when the averages are taken; this reduces to

$$\overline{v_y^2} = \frac{9\mu_0^2\mu^4}{80\pi^2 a^6 L^2},\tag{37}$$

when use is made of the averages $\overline{x^4} = \overline{z^4} = \frac{1}{5}a^4$ and $\overline{x^2y^2} = \overline{x^2z^2} = \frac{1}{15}a^4$. Hence, finally, by use of (33) we have

$$T = 40\pi^2 a^6 L^2 / 9\mu_0^2\mu^4\tau_c$$

$$= 160\pi^2 a^6 / 9\mu_0^2\hbar^2\gamma^4\tau_c, \quad \text{as quoted in (26),}$$

since $\mu = \gamma L$ and $L = \frac{1}{2}\hbar$ for a proton.

There is no distinction in this calculation between T_1 and T_2. If τ_c is much smaller than the precession period a laboratory observer and one in a frame rotating at the precessional angular velocity could hardly detect any difference in the rapidly fluctuating interaction between protons, and would therefore see no distinction between diffusive motion parallel to, and transverse to, $\boldsymbol{B}$. If, on the other hand, the tumbling is slow, the rotating observer sees both protons maintaining the same spin orientation in his frame, while the line joining them rotates round $\boldsymbol{B}$ at $-\boldsymbol{\Omega}$, the angle to $\boldsymbol{B}$ only changing slowly. It is then appropriate to average the torque over one period of precession before considering the tumbling motion. When this is done (we omit the details, which are quite straightforward) and the changes in both spin directions added, it is found that there is no change in the component of spin parallel to $\boldsymbol{B}$ – as expected, the diffusive motion is confined to the transverse plane.† This means that as τ_c is increased, by an increase in viscosity, for example, T_1 and T_2 decrease together, according to (26); but when $\Omega\tau_c$ rises to unity and above, T_2 continues to decrease while T_1 increases again. In highly viscous liquids the resonance line is not very sharp, being determined by a short T_2, while the rate of equilibrium is slow, being determined by a long T_1. These two effects combine to make detection of the resonance line difficult, since only very weak fields can be used without saturating the system.

With this we leave our discussion of nuclear magnetic resonance, which has been slanted towards the description of the phenomenon as a thing in itself, a fascinating example of a physical resonance process that is not time-reversible nor strictly linear, yet is still amenable to exact analysis. We

---

† This is one of these cases where the quantum-mechanical viewpoint leads to an easier intuitive understanding of the process. The energy levels of the proton are separated by $\hbar\Omega$, and transitions between them can only be excited if the perturbing field contains components with frequency around Ω. When $\Omega\tau_c \gg 1$, the spectrum of the perturbing interaction falls off at a frequency around $1/\tau_c$, well below Ω, and consequently the transitions, which are the quantal description of changes in the component of spin parallel to $\boldsymbol{B}$, cease to take place. On the other hand, motion round the transverse plane involves no energy changes, only disturbance to the phases of wavefunctions, and is controlled by the zero-frequency strength of the perturbation, i.e. $I_\omega(0)$, as assumed in the derivation of (26).

return to our catalogue of velocity-dependent forces and their effects as revealed by resonant systems.

Cyclotron resonance

The simplest example of the Lorentz force is provided by a free charged particle moving in a uniform magnetic field. Motion along the direction of B is unaffected, but in the plane transverse to B the particle describes a circle. The complete motion is therefore a circular helix. We shall now disregard the motion along B and concentrate on the transverse plane. If the particle has a transverse component of velocity v and moves in a circle of radius r, the centripetal acceleration mv^2/r is caused by the Lorentz force eBv, so that

$$r = mv/eB = p/eB, \tag{38}$$

where p is the momentum of the particle, Written in terms of p, (38) holds even when v is large enough to produce relativistic changes of mass. So long as $v \ll c$, however, and m is constant, (38) shows that the angular velocity, v/r, is independent of v, taking the value.

$$\omega_c = eB/m. \tag{39}$$

This equation defines the *cyclotron frequency*, ω_c.

This is a system which, although two-dimensional, has only one natural frequency, for the equation of motion has the form, in complex notation,

$$m\ddot{r} + ieB\dot{r} = 0, \tag{40}$$

with the general solution

$$r = r_0 + A\,e^{-i\omega_c t}, \; r_0 \text{ and } A \text{ being constants.}$$

The arbitrary positioning of the centre of the orbit, r_0, might be considered to account for the other degree of freedom that normally results in a second normal node; here the second mode has zero frequency.

In a cyclotron the particle being accelerated is not subjected to a uniform accelerating force, but instead the force is confined to one part of the large orbit.[5] This introduces extra complications which we shall not consider at this moment. Instead we investigate the response of electrons in a semi-conductor, where the orbits in a moderate field, say 1 T, are very small – less than 10^{-7} m in radius – so that an oscillatory electric field applied to the sample is virtually certain to be uniform over the trajectory of any one particle. If the semiconductor is pure, cold and shielded from light, the electron density can be made low enough for the charge displacements due to the electric field to have negligible disturbing influence on that field. Moreover, the electrons can traverse their orbits several times between collisions with impurities or other lattice irregularities, and this allows the resonance to manifest itself. In the absence of collisions, the equation of motion takes the form of (40), supplemented by a rotating field (or one rotating component of a linear oscillatory field)

$$m\ddot{r} + ieB\dot{r} = eE\,e^{-i\omega t}, \tag{41}$$

with the steady-state solution

$$\dot{r} = V\,e^{-i\omega t}, \quad \text{where } V = ieE/m(\omega - \omega_c). \tag{42}$$

The velocity is always normal to E, as it must be in a lossless system as assumed here. The resonant denominator vanishes when $\omega = \omega_c$, the amplitude of response growing without limit.

To take account of collisions we make the usual assumption of randomness – an electron, chosen at any instant, has a probability $e^{-t/\tau}$ of continuing to move without collision for more than a time t. This is the consequence of assuming that an electron always has a chance dt/τ of colliding in the next interval dt, irrespective of its past history. We assume further that collisions are catastrophic, in the sense that afterwards the electron is equally likely to move off in any direction. Let us now apply the impulse response method to a whole assembly of electrons, calculating their response to the rotating field $E\,e^{-i\omega t}$ from how they would respond if it was switched on only during the interval from t to $t+dt$. If a typical electron has velocity $v_0\,e^{i\theta}$ before, it will have $v_0\,e^{i\theta} + (eE\,e^{-i\omega t}/m)\,dt$ immediately after. It now travels in its cyclotron orbit so that at a later moment t_0, if it has not collided meanwhile, its velocity will be $v(t_0)$, where

$$v(t_0) = [v_0\,e^{i\theta} + (eE\,e^{-i\omega t}/m)\,dt]\,e^{-i\omega_c(t_0-t)}.$$

When we apply this result to the whole assembly of electrons, moving randomly before the impulse, we must average evenly over all values of θ and multiply by $e^{(t-t_0)/\tau}$ so as to count only those that have survived until t_0 without collision. The others, being randomly scattered, have zero mean velocity afterwards, and contribute nothing to the mean velocity of an electron at time t_0, which is now seen to be given by $\Delta\bar{v}$, where

$$\Delta\bar{v}(t_0) = (eE/m)\,e^{-i\omega t + (i\omega_c + 1/\tau)(t-t_0)}\,dt.$$

To proceed from here to the mean velocity at t_0, when the rotating field has been applied at all times up to t_0, simply involves integrating $\Delta\bar{v}$ from $-\infty$ to t_0:

$$\bar{v}(t_0) = \frac{eE\tau}{m}\,e^{-i\omega t}/[1 + i(\omega - \omega_c)\tau]. \tag{43}$$

The collisions give rise to a standard Lorentzian resonant response, with $Q = \frac{1}{2}\omega_c\tau$. The phase difference between E and $\bar{v}$ shows that the collisions allow energy to be absorbed by the electrons. The resonance does not merely alter the dielectric constant of the semiconductor – this is the effect of the lossless part of the response – but also causes dielectric loss. This can be picked up experimentally by placing the sample in a microwave cavity excited at its own resonant frequency, and varying the applied field until ω_c

passes through the cavity frequency.[6] The additional loss lowers the response level of the cavity, and in this way the cyclotron resonance can be plotted out, not by keeping ω_c fixed and varying the exciting frequency ω but, as is common also in NMR, keeping ω fixed and varying ω_c. It is clear from (43) that this is only a technical difference. An example of what may be observed is shown in fig. 13. The appearance of several resonances indicates that the electrons in a semiconductor do not behave as free particles. Some

Fig. 13. Cyclotron resonance in germanium at 4 K and a frequency of 24 GHz (Dresselhaus, Kip and Kittel[7]).

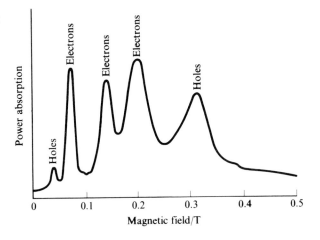

indeed (*holes*) execute orbits in the reverse sense as if they were positively charged. Electron orbits and hole orbits can of course be distinguished by using circularly polarized rather than linearly polarized microwaves to probe them. Each resonance shows up only with one sense of polarization. This is technically more difficult, but valuable as a diagnostic tool, for each group of electrons or holes has a different cyclotron frequency, as if each had its own characteristic mass, different from that of a free electron. Knowledge of the effective mass to be associated with each group, and the way this mass changes with orientation of the magnetic field, provides numerical information that can be compared with quantum-mechanical calculations of the dynamical behaviour of electrons in the semiconductor, so that the approximations which are unavoidable in the latter may be tested and the theoretical tools thereby improved. Cyclotron resonance has indeed been a valuable aid to the development of semiconductor physics.

Helicons

The last exhibit in our menagerie of curious oscillatory systems also owes its character to the Lorentz force, revealed in this case by the Hall effect.[8] A free charged particle, acted upon by uniform perpendicular magnetic and electric fields, executes a helical orbit whose axis is parallel to $\boldsymbol{B}$ but drifts at constant speed E/B in a direction normal to both fields. Consequently the assembly of free electrons in a metal, n per unit volume, if they suffered no collisions would drift bodily at this speed, setting up a current density neE/B. Using a complex coordinate system to describe motion in the plane transverse to B, we can write that if E is real, the current density J is purely imaginary, $J = ineE/B$, and the conductivity in the transverse plane is also purely imaginary:

$$\sigma_\perp \equiv J/E = ine/B. \tag{44}$$

This is the Hall conductivity, the component normal to E in the transverse plane. In a real metal there are collisions which jolt the electrons from one orbit to another and allow a steady drift of orbit axes in the direction of E. In general, then, the conductivity in the transverse plane is complex, but in pure monovalent and trivalent metals (Cu, Al) at low temperatures and in strong magnetic fields the imaginary (Hall) part may overwhelmingly dominate the real part, with Hall angle (being the angle between J and E) less than 1° away from the perfection of 90°. We may now forget the microscopic interpretation and discuss wave propagation in a metal in terms of the macroscopic parameter, the complex conductivity.

The phenomenon that concerns us here is propagation of a travelling electromagnetic wave along the direction of B. When $B = 0$ electromagnetic waves cannot enter a metal, but are reflected and penetrate only a very shallow surface layer (*skin effect*);[9] we must therefore first understand how a magnetic field so changes the situation as to allow penetration. The essential condition for wave propagation to occur is that the Hall angle shall be large, so let us begin with the ideal case of no electron scattering and a Hall angle of 90°, $\sigma_\perp$ being purely imaginary, is, say. In view of the earlier examples in this chapter it should come as no surprise that the wave is circularly polarized; $\mathscr{E}$ and $\mathscr{B}$, the field vectors of the wave-motion, are constant in amplitude but spin around the z-axis, as in fig. 14, which shows travelling waves in free space moving away from and towards the observer. In both cases the sense of circular polarization has been chosen so that a fixed observer facing along the arrow of B would see $\mathscr{E}$ and $\mathscr{B}$ rotating clockwise. Then for both waves the displacement current, $J_d = \varepsilon_0 \partial\mathscr{E}/\partial t = -i\omega\varepsilon_0\mathscr{E}$ in complex notation, and points at $\pi/2$ to $\mathscr{E}$ in a clockwise sense. Suppose now that the wave travels through a metal with perfect Hall conductivity; $\mathscr{E}$ will excite a current $\sigma_\perp\mathscr{E}$, i.e. $is\mathscr{E}$, parallel to J_d, and the total current will be $-i(\omega\varepsilon_0 - s)\mathscr{E}$ or $-i\omega\varepsilon\varepsilon_0\mathscr{E}$, the effective dielectric constant of the medium being given by

$$\varepsilon = 1 - s/\omega\varepsilon_0.$$

The imaginary character of $\sigma_\perp$ when the Hall effect is complete allows the electrons to contribute to the real dielectric constant. This is in marked contrast to the behaviour when no magnetic field is present and $\sigma_\perp$ is real; then s is imaginary and the effective dielectric constant is dominated by its imaginary component. The wave velocity, proportional to $\varepsilon^{-\frac{1}{2}}$, is complex

Fig. 14. Fields, $\mathscr{E}$ and $\mathscr{B}$ and displacement current, $\mathbf{J_d}$, in circularly polarized travelling waves and, on the right, their resultant when combined to form a standing wave.

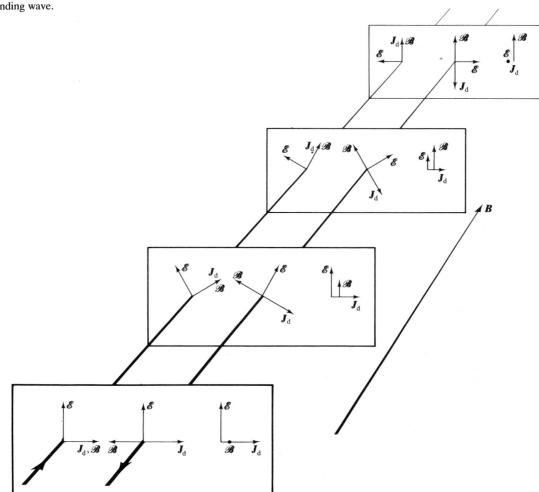

and describes a heavily damped wave – this is the skin effect. When s is real and negative, however, as for electrons when the Hall effect is complete, the additional term in ε raises the dielectric constant of the medium to a high positive real value, decreasing the velocity of propagation by a factor $(1 - s/\omega\varepsilon_0)^{\frac{1}{2}}$, without attenuation. The magnitude of the dielectric contribution of the electrons is so huge for metals under normal conditions of experiment that we are justified in dropping the free space displacement current, writing

$$\varepsilon = -s/\omega\varepsilon_0 = -ne/\omega\varepsilon_0 B \qquad (45)$$

and
$$\omega^2/k^2 = c^2/\varepsilon = -\omega/\mu_0 s.$$

Hence
$$k^2 = -\mu_0 \omega s, \qquad (46)$$

which is the dispersion law for a *helicon* (or *magneto-ionic*) wave.[10] Only one sense of circular polarization can be propagated in either direction, that which makes k^2 positive. The other sense, with positive s, has purely imaginary wave-number, which describes the surface excitation accompanying total reflection when an attempt is made to inject the wrong wave into the medium.

    79

    The magnitude of the wave velocity, which varies as $\omega^{\frac{1}{2}}$, is worth estimating for a typical case, aluminium in a field of 3 T. Aluminium behaves in a strong field as if n were equal to one hole per atom, i.e. s is positive and $n = 6 \times 10^{28}$ m^{-1}. Hence $s = 3.2 \times 10^9$ Ω^{-1} m^{-1}. At a frequency of 1 Hz (polarized in the opposite sense so that ω is negative) ε, according to (45), is 0.5×10^{20} – a truly enormous dielectric constant that reduces the velocity of wave propagation to 4.2 cm/s! A second consequence is that the ratio $\mathcal{E}/\mathcal{B}$, i.e. the wave impedance, is smaller by $\varepsilon^{\frac{1}{2}}$ than for a wave in empty space.

    Because of the very low wave velocity, low-frequency standing waves can be set up in samples of quite modest size; a plane aluminium slab 2.1 cm thick in a field of 3 T normal to the faces would be one-half wavelength thick at 1 Hz and would have a fundamental resonance at this frequency. The field configuration for the standing wave is shown in fig. 14; because of the low wave impedance relative to that of the free space outside, the slab behaves like an open-circuited transmission line, and its surfaces must coincide with nodes of $\mathcal{B}$. When the two travelling waves are superposed at some instant, the resultant $\mathcal{E}$ has the same direction everywhere and its amplitude varies sinusoidally with position, just as in a plane polarized electromagnetic wave in free space; similarly $\mathcal{B}$ varies sinusoidally with its nodes and antinodes interchanged relative to those of $\mathcal{E}$. However, $\mathcal{B}$ and $\mathcal{E}$ are parallel rather than perpendicular, and as time proceeds they do not oscillate in amplitude like a plane standing wave, but spin around $\mathbf{B}$ keeping constant amplitude. The configuration of $\mathcal{B}$ confers on the sample as a whole a magnetic moment, transverse to $\mathbf{B}$, and spinning at the resonant frequency. By the inductive effect of this moment the helicon standing wave is easily detected. The experimental arrangement in

Fig. 15. Crossed coils and rectangular metal crystal, for generating and detecting helicons.

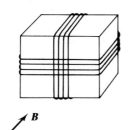

B

fig. 15, similar in principle to Bloch's nuclear induction arrangement, provides for one coil to excite helicons and another, normal to it, to detect them. By sweeping the frequency over a wide range a set of resonances can be detected at frequencies proportional to the squares of the integers, in accordance with the dispersion law (46). The odd harmonics are strong and the even weak (or absent altogether) with this arrangement, for the total magnetic moment is zero if there are an even number of loops, with alternating direction of magnetization, in the standing wave pattern. The resonances are not, of course, ideally sharp since electron scattering causes $\sigma_\perp$ to have a real component. The dispersion relation is now complex and, as discussed in chapter 3, in a freely oscillating standing wave k must be taken as real; (46) now shows that the phase angle of ω is equal and opposite to that of $\sigma_\perp$. Hence the natural decrement of the free helicon oscillations or, what is equivalent, the width of the resonance, provides a measure of ω''/ω' and hence of $\sigma_\perp''/\sigma_\perp'$, the tangent of the Hall angle. This makes the helicon a useful tool in studying the conductivity of metals in strong magnetic fields, with the advantage over conventional methods that no electrical contacts need to be made to the specimen.[11]

60

It is easy enough to solve the field equations for a standing wave in a parallel-sided slab of infinite lateral extent, but for a real-sample, a finite slab or a body of any other shape, there are formidable complexities of applied mathematics. Even for the cylinder and the sphere the boundary conditions that must be satisfied are very troublesome indeed, but we need not attempt to discuss why, or how the difficulties have been overcome. It is enough for our purposes to note the general field pattern in the fundamental resonance of a sphere, as shown schematically in fig. 16. The transverse

Fig. 16. Current lines and resulting magnetic moment in the lowest helicon mode of a sphere. The whole pattern spins about **B** at the helicon resonant frequency.

B

M

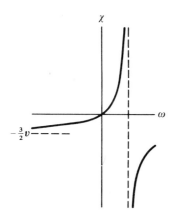

Fig. 17. Schematic diagram of susceptibility of a sphere, showing first helicon resonance.

magnetic moment is generated by circulating currents, and as with the slab the whole pattern of currents and moment spins around $\boldsymbol{B}$. If the sphere is subjected to a rotating field $\boldsymbol{b}$, lying in the plane transverse to $\boldsymbol{B}$, the resulting magnetic moment rotates with $\boldsymbol{b}$ but with a phase lag (so that there is a continuous energy flow into the sphere, dissipated as heat by electron scattering). The susceptibility χ, defined as $\mu_0 M/b$, is in general complex; its frequency variation shows a typical resonant peak at the fundamental mode frequency and, of course, at all resonances of higher order, of which the sphere exhibits a considerable and complicated variety.

The variation of χ to just beyond the first resonance is shown schematically in fig. 17; electron scattering has been assumed negligible, so that χ is wholly real,. It will be noted that when ω is in the opposite sense to the helicon rotation, χ tends to the limit $-\frac{3}{2}V$, V being the volume of the sphere. This is the value for a sphere that totally excludes magnetic field. At very low speeds, however, whether positive or negative, the induced currents are so weak that the applied field penetrates almost completely, and rotation of this uniform field within the sphere induces currents which are very simple in form, circulating around the equatorial axis parallel to $\boldsymbol{b}$ and with a strength proportional to distance from this axis. To quote the final result, if a is the radius of the sphere,

$$\chi \approx \tfrac{1}{5}\mu_0 V a^2 \omega \sigma_\perp = \tfrac{1}{5}V k^2 a^2, \tag{47}$$

in which k is taken from (46) and is the wavenumber of the helicon of frequency ω.

The calculation of χ in this low-frequency limit involves none of the complexities that are demanded by the resonance, and (47) becomes of interest and practical value when one finds that by coupling the helicon oscillation to mechanical motion the resonant frequency can be made as low as one wishes. Imagine the sphere mounted on a flexible cantilever, as in fig. 18, so that the interaction of the helicon moment $\boldsymbol{M}$ with the strong field $\boldsymbol{B}$ bends the cantilever and causes the sphere to turn through a small angle about an axis normal to $\boldsymbol{M}$ and $\boldsymbol{B}$ (the lateral displacement of the

Fig. 18. Sphere supported on flexible cantilever to give 'soft helicon' resonance.

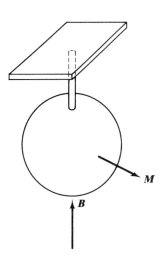

sphere is irrelevant – no currents are induced by the unrotating motion of a conductor through a uniform field). As **M** rotates about **B** with angular velocity ω, the axis of bending rotates with it, and the orientation of the sphere continually alters so that its originally vertical diameter describes a narrow-angled cone with vertical axis. If C is the torsion constant of the cantilever, the half-angle of the cone, θ, is equal to MB/C. In fig. 19(a) the diameter is shown as a broken line, displaced from the vertical by θ. From

Fig. 19. (a) The interaction of **B** and **M** bends the cantilever of fig. 18 and tilts the originally vertical diameter of the sphere. This is equivalent, from the sphere's point of view, to (b) where **B** is shown tilted. As **M** spins round **B**, **B** in (b) spins round the vertical axis.

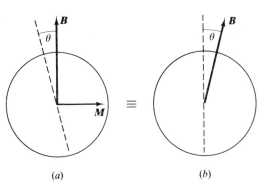

the point of view of the sphere, it is as if **B** had been tilted through $-\theta$, and as **M** rotates the sphere sees **B** describing a cone, as if the steady vertical **B** had been supplemented by a rotating transverse **b** equal to $B\theta$. Such a field establishes a moment $M = \chi B\theta$, and if the system is displaced and then allowed to oscillate freely, it must do so in such a way that the moment induced by its motion and the moment that is responsible for the bending of the cantilever are one and the same:

$$M = \chi B\theta \quad \text{and} \quad \theta = MB/C;$$

hence

$$\chi = C/B^2. \tag{48}$$

The natural frequency is that at which χ takes this required value. If the cantilever is weak enough for χ to lie in the range for which (47) applies,

$$\omega = 5C/\mu_0 B^2 V a^2 \sigma_\perp. \tag{49}$$

When B is large, ω varies as $1/B^2$ and with sufficiently small C the rotation period can be made as long as one likes, a value of more than 3 minutes being easily achieved with an aluminium sphere of 1 cm radius on a fairly flexible wire cantilever. The decrement per cycle is independent of the strength of the cantilever, if electron scattering is the only significant dissipative effect in the system, for (48) shows that χ and hence, from (47), $\omega \sigma_\perp$ must be real, just as for the rigidly mounted sphere.

The frequency only varies as $1/B^2$ when B is strong enough; in very weak fields the cantilever has enough rigidity to resist the magnetic torque and, according to (46) and (44), $\omega \propto B$. This is the *hard helicon* regime, to be contrasted with the *soft helicon* regime to which (49) applies. In between, ω has a maximum but the exact calculation of χ has not been carried to the point where the full variation with B can be computed and compared with experiment.

It is perhaps tempting to visualize the conical rocking of the sphere and the circulating current lines as analogous to a precessing gyroscope, but the analogy is unsound. There is no significant transverse angular momentum associated with the helicon. The precessing gyroscope is subjected to a resultant torque and precesses so that the rate of change of angular momentum equals the torque; in the soft helicon experiment the sphere is subjected to no torque, that due to MB being balanced by what the bent cantilever exerts. The similarity is more pictorial than physical.

9 The driven anharmonic vibrator; subharmonics; stability

Linear systems whose parameters are independent of time possess, as has been abundantly illustrated already, well-defined normal modes from which their motion can be synthesized by superposition; and the response to an applied force, varying with time, can be written in terms of the response to each separate Fourier component of the force. The same is not true of non-linear systems, since superposition is no longer a valid procedure for synthesizing the response. Every anharmonic system responds differently to a given form of time-dependent force, and even when the response has been found in any special case it will not scale up unchanged in response to an amplification of the force. Thus the response to a sinusoidal force is in general non-sinusoidal, the waveform changing with the amplitude of the force. There are very few general statements that can be made about the character of the response. One cannot even assert that the oscillations of the vibrator will have the same fundamental frequency as the applied force – it may respond at a subharmonic frequency, i.e. an integral submultiple of that of the force, or the response may be asynchronous to the point of randomness. Even when order prevails, with regular vibration at the fundamental or subharmonic frequency, changing the amplitude or frequency of the applied force to an infinitesimal degree may have the effect of throwing the response into an entirely new pattern.

The problems presented by asynchronous and unstable response have attracted much attention from mathematicians, and there is a very large body of analysis aimed at the systematic diagnosis and classification of pathological states.[1] It is difficult to reach a just evaluation of this work since the aims of a mathematician may be very different from those of a physicist or engineer. It is no denigration of pure mathematics to deny its usefulness, but when non-linear vibrations are under discussion one may feel that because it looks as if it deals with physical problems it must stand or fall on its practical utility. This is unfair to the mathematician, whose interest may be aroused primarily by the structural features underlying a great variety of systems, but so deeply that the topological and other techniques appropriate to their elucidation are inappropriate to working out the details in any particular case. The contrast with harmonic systems is striking; here the elementary mathematical tools that are useful in special cases also serve (if nothing more sophisticated comes to hand) to develop the general theory – consequently harmonic systems have long ceased to interest pure mathematicians. Anharmonic systems, on the other hand, are still full of mathematical interest, and this is enough to ensure that the average practising physicist or engineer will have the greatest difficulty in quarrying useful material from the professional literature. In time, it is to be hoped, the systematic assimilation of instabilities and related critical phenomena into a coherent pattern will lead to the sort of qualitative appreciation of their role that can be incorporated as a basic element in the scientist's world-picture.[2] The recent appearance of semi-popular articles expounding *catastrophe theory* points to an earnest desire on the part of mathematicians to break away from the traditional reductionist method that has given physical science its power. By contrast, these visionaries in the world of instabilities and critical phenomena appear to be seeking a new constructive approach to putting the pieces together again – the quest is for common patterns running through enormously diverse phenomena which will enable these phenomena to be described as complete entities and not as an ensemble of individually analysable fragments. This is an ambitious program which, even if only partially successful, will undoubtedly have far-reaching consequences. No revolution, however, is successful until its achievements have been consolidated, and it is too early to express confidence in the outcome of this particular high venture.

This preamble must be our excuse for shirking anything like a systematic account of what has been achieved by all the work to date. Instead we shall exhibit a few characteristic features of anharmonic systems by means of special examples which can be analysed in sufficient detail, and whose stability can be investigated, without excessive labour. The intent behind this procedure is to alert the reader's mind to the complexities that a real system may exhibit, at the same time warning him that he will usually find a complete solution a very taxing matter, involving long and tedious calculations. In practice it often suffices to be aware of the pitfalls, so that one ceases to rely on mathematics alone and develops a proper respect for experiment and observation.

Slightly anharmonic vibrators

The pendulum is a typical slightly anharmonic vibrator, with a restoring force proportional to $\sin \theta$ rather than directly proportional to the displacement θ. The resonant frequency decreases with amplitude and the response to a driving force of constant amplitude and variable frequency is correspondingly distorted from the Lorentzian form of resonance. Simple though it is, the pendulum is not a very convenient candidate for demonstrating anharmonicity in the laboratory, if only because its period is so long as to slow down the observations to an exasperating degree. A much more satisfactory example is a stretched wire executing transverse vibrations. It is almost inevitable that the tension will increase with deflection, and the frequency with amplitude as a result; by anchoring the ends of the

19

string in different ways the degree of anharmonicity can be adjusted. A suitable arrangement is shown in fig. 1; it is more elaborate than necessary, being designed for a dual purpose. In this chapter we shall consider only its response to a transverse exciting force at the end, but in the next chapter we shall discuss the response to a longitudinal force; the arrangement shown allows both forms of excitation to be applied without otherwise altering the system. The electromagnetic impeller acts on a comparatively massive bar, so that the amplitude of its vibrations is not seriously affected by the motion of the string, even at resonance. To a good approximation the amplitude of motion at the end is independent of frequency, and the argument illustrated in fig. 7.8 shows that the response would be of normal Lorentzian form were it not for the variation of tension and, as already noted, the non-linear damping at higher velocities.

299

173
31, 140

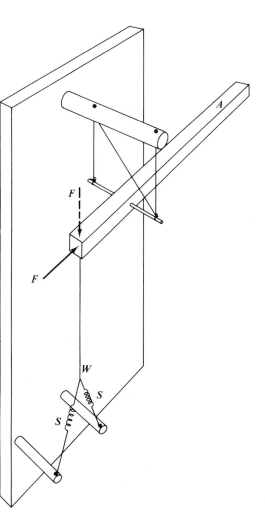

Fig. 1. Arrangement for studying vibrations of a stretched wire. The massive horizontal beam A is suspended by strings from a pivot one-third of the length from an end, and serves to tension the wire W, which at the lower end is attached to two springs S (this separates the two transverse modes and avoids parasitic excitation). An electromagnetic impeller applies an oscillatory force F either transversely (full arrow) or longitudinally (broken arrow; see chapter 10 for a discussion of this). Suspending A as shown makes the effective inertia of the beam the same for both directions of F. The diagram is not to scale – in the actual experiment A was about 25 cm long, while W was 150 cm.

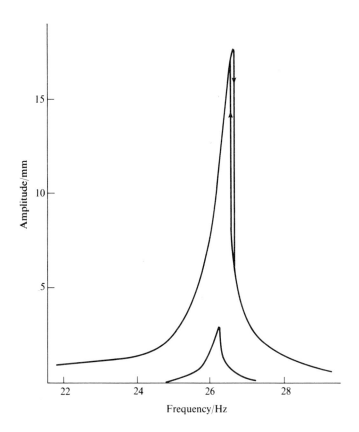

Fig. 2. Amplitude of centre of wire as a function of frequency. The upper curve was obtained with the force F ten times as strong as for the lower curve.

Fig. 2 shows typical response curves for the fundamental mode (a single half-wave loop) at two different levels of excitation. The variation of natural frequency with amplitude is clearly revealed by the canting of the top of the curves towards higher frequencies; so marked is the effect at the higher excitation that the amplitude of vibration has become 3-valued in the narrow frequency range 26.6–26.7 Hz. Setting aside for a moment this new effect, which is the most interesting aspect of the problem, let us first enquire whether the asymmetry has the expected magnitude. A complete solution of the problem of the vibrating wire when non-linearity is taken into account is very difficult, for a sinusoidal shape may no longer be assumed and one must recognize that the shape actually changes in the course of the cycle. A rough treatment, however, which should be nearly correct starts from the approximation that the wire oscillates like a simple vibrator with a non-linear force law; for small displacements we write the potential energy in the form

$$V = \tfrac{1}{2}\mu\xi^2(1+\beta\xi^2), \tag{1}$$

the correction term $\beta\xi^2$ being symmetrical with respect to positive and negative displacements. Corresponding to (1) the force depends on displacement according to the expression:

$$F = -\mu(\xi + 2\beta\xi^3).$$ (2)

14   The general result for the oscillation period, T, derived from (2.11), shows that when the amplitude of vibration is ξ_0 and the total energy E is $\frac{1}{2}\mu\xi_0^2(1+\beta\xi_0^2)$,

$$T = 2T_0/\pi \int_0^{\xi_0} d\xi/[\xi_0^2(1+\beta\xi_0^2) - \xi^2(1+\beta\xi^2)]^{\frac{1}{2}}.$$

When $\beta = 0$, $T = T_0$ for all ξ_0, and we may find the first correction to T_0 by Taylor expansion, writing

$$T(\xi_0) \approx T_0 + \beta(dT/d\beta)_{\beta=0}.$$

Differentiating under the integral sign, keeping ξ_0 constant, and then putting β equal to zero, we find that

$$(dT/d\beta)_{\beta=0} = -T_0/\pi \int_0^{\xi_0} (\xi_0^4 - \xi^4) \, d\xi/(\xi_0^2 - \xi^2)^{\frac{3}{2}} = -\tfrac{3}{4}\xi_0^2 T_0.$$

Hence $$T(\xi_0) = T_0(1 - \tfrac{3}{4}\beta\xi_0^2 + \cdots).$$ (3)

Since $\beta\xi_0^2$ is the only dimensionless parameter in the problem, there is no other term in ξ_0^2 that could have been overlooked by treating β rather than ξ_0^2 as the expansion parameter. According to (2) and (3), when the force constant at maximum amplitude ξ_0 has increased by a fractional amount f, the period has decreased, and the frequency increased, by $\tfrac{3}{8}f$.

In the experiment that produced the results in fig. 2 the wire was virtually inextensible and the mass at the top large enough to be little influenced by the wire; when the wire was displaced, therefore, the springs at the bottom were stretched. Direct measurement showed that when the vibration amplitude was 17.5 mm the spring tension was increased by 5%. We might therefore expect the frequency to increase by $\tfrac{3}{8}$ of this, i.e. 1.9%; perhaps a little more than is observed but near enough, in view of the approximate theory and rather inaccurate measurements. The effect can be magnified by removing the springs and giving the wire firmer anchorage at the bottom. The result is shown in fig. 3, where the instability and hysteresis are very clearly exhibited.

Even so, the frequency shift at the peak is still only 4%, and the form of the vibration must remain closely sinusoidal. We are therefore reasonably justified in supposing that if we were to sweep the frequency of the exciting force through resonance while adjusting its amplitude at every stage to maintain the vibration amplitude constant, the variation of force with frequency would be the same as for a linear resonant system (see fig.

138   6.10(d)). Translated into the terms of the experiment, if the force is constant the width of the resonant peak at any level of response is

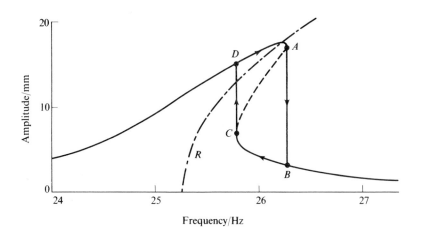

unaffected by the non-linearity. A Lorentzian peak, for example, still keeps the form (6.18),

$$P(\omega) = P_0/[(\omega - \omega')^2 + \omega''^2],$$

but now ω' is a function of P, for example, $\omega' = \omega'_0(1 + \frac{3}{4}\beta P)$ if $P = \xi^2$ and the period varies as in (3). In fact, the peaks in fig. 2 are probably not Lorentzian since the lower peak is only three-quarters as wide as the higher **140** peak at half amplitude. The loss due to air resistance in the form of eddies increases with velocity at a considerably faster rate then viscous drag alone would predict, and the effective Q becomes correspondingly lower at large amplitudes. The extra losses are very noticeable in practice, the wire humming loudly when strongly excited.

We come at last to the many-valued character of the response. It is readily seen that the portion AC of the curve, shown broken in fig. 3, is unrealizable. At a point on this part of the line a small displacement upwards, i.e. an increase in amplitude at constant frequency, brings the point inside the resonance curve – the force needed to maintain the amplitude is less than that required for a point on the line. Consequently the amplitude increases further, and goes on doing so until the upper branch of the resonance curve is reached. By the same argument, if the point is displaced downwards it proceeds down until it reaches the lower branch. On the other hand, any point not on the broken line represents a stable response, small departures being followed by a natural reversion to the line. There is no difficulty, then, in understanding why the response curve takes a different form over part of its range according as the frequency is increased or decreased, or why there are jumps at the ends of the stable ranges.

If the curve of fig. 3 is traced out by increasing the frequency, the collapse of the amplitude from A to B takes a few seconds, being the natural decay time of any transient oscillation of the wire, and is accompanied by a marked beat pattern. By contrast on the return path the transition of the amplitude from C to D takes about the same time but is more nearly monotonic. Since the system is non-linear it is not quite permissible to discuss the behaviour in terms of the superposition of free transient excitations and the driven vibration at the excitation frequency; nevertheless the analogy will serve to explain the difference qualitatively. At most amplitudes along AB the driving frequency differs from the natural frequency (shown as the line R) by something comparable to, or greater than, the resonance width; this is the condition, illustrated in fig. 6.12, for beats to be clearly exhibited when transients are excited. On the other hand, CD cuts the line R and lies so close to it over most of its course that the long period of the beats prevents their showing themselves.

145

An example of subharmonic excitation

If a small loudspeaker is laid on its back, with a lightly loaded pin gently pressed against it by means of a spring, as in fig. 4, excitation at (say) 400 Hz with a signal generator can impart sufficient acceleration to cause the pin to lose contact over part of the cycle; as the critical amplitude is exceeded the emitted note suddenly becomes louder and harsher. With increased excitation there comes a point (possibly after a noisily unsettled interim) when a strong note sounds out an octave, a twelfth, a fifteenth or even more below the exciting frequency. The pin is now bouncing high enough to return only after 2, 3, 4 or more cycles. Each subharmonic is stable over a limited range of excitation, and it normally gives way to the next in the sequence, so that the note sounded drops by discrete steps, spelling out the harmonic series in inversion.

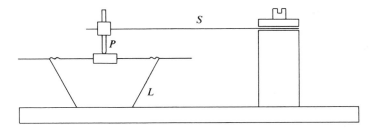

Fig. 4. A loudspeaker, L, lying on its back has a light pin, P, held by a spring, S, against the centre of its cone.

A simple, indeed oversimplified, picture of the process is shown in fig. 5, where the loudspeaker cone is assumed heavy enough to execute simple harmonic motion unperturbed by the bouncing pin. The pin is shown making contact with the cone every other cycle of the latter, and therefore responding in the octave subharmonic mode. The exact phase at which impact occurs is such that the upward movement of the cone compensates for the inelastic character of the bounce, the pin arriving and leaving at the same speed. The trajectory shown by a broken line is obviously unstable, for if the impact is slightly delayed it occurs when the cone is rising faster; the pin therefore rises next time to a greater height and arrives for the next

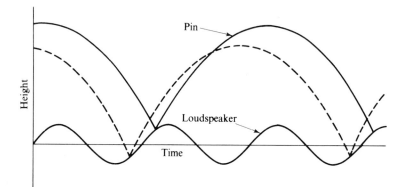

Fig. 5. Bouncing of the pin in the arrangement of fig. 4, generating the octave subharmonic.

bounce still further delayed. By contrast, the full curve represents a trajectory which, if delayed at one bounce, suffers correction and has the possibility of being restored to the correct phase. But one must not be rash in assuming stability, for overcorrection may lead to the phase of impact oscillating between advance and delay with ever-increasing amplitude. It is necessary therefore to examine the stability with greater care. Since the problem arises frequently we shall break off at this stage to develop rather general tests for stability, and return to this example later to illustrate their application.

Behaviour of simple systems near equilibrium

The simple systems we have encountered so far in this book have mostly been described by second-order differential equations, not necessarily linear. The general solution of such an equation contains two arbitrary constants, and any particular solution may therefore be specified by the values of those constants or, what is equivalent, by the values, at some point, of two physical variables. Thus the behaviour of a free harmonic vibrator is completely determined once the displacement and velocity have been fixed at some instant. We shall concentrate attention on such systems, without at first making explicit the two physical variables by which the state is defined. Let them be labelled x and y; then when we say that specifying x and y at some instant determines the subsequent behaviour, we mean that the motion can be represented by a well-defined trajectory on the (x, y)-plane, passing through the specified point (x, y). Normally different trajectories do not intersect, though there may be one or more singular points onto which trajectories converge; such a point is likely to be one at which the system is in stable equilibrium, while a point from which trajectories diverge is likely to be one of unstable equilibrium. Other types of singularities are possible, and our first task is to catalogue the permitted varieties, of which there are five only for a two-parameter system.

If (x_0, y_0) is an equilibrium point (which need not be stable), the system established exactly in this state can remain there indefinitely if undisturbed, but if shifted to a neighbouring point will in general find itself not in equilibrium, so that as time proceeds x and y will change. Since $\dot{x} = \dot{y} = 0$ at (x_0, y_0), we expect a Taylor expansion of $\dot{x}$ and $\dot{y}$ in the vicinity of (x_0, y_0) to start with linear terms:

$$\dot{\xi} = \alpha_1 \xi + \alpha_2 \eta, \qquad \dot{\eta} = \alpha_3 \xi + \alpha_4 \eta, \tag{4}$$

where $\xi \equiv x - x_0$ and $\eta \equiv y - y_0$. Provided ξ and η are not too large it will usually be adequate to terminate the series at this point – the behaviour has been *linearized* for the purpose of describing small displacements around (x_0, y_0).[†] The possible trajectories are defined by (4), which can be manipulated into separate and identical differential equations for the time variation of ξ and η:

$$\ddot{\xi} - (\alpha_1 + \alpha_4)\dot{\xi} + (\alpha_1 \alpha_4 - \alpha_2 \alpha_3)\xi = 0,$$

$$\ddot{\eta} - (\alpha_1 + \alpha_4)\dot{\eta} + (\alpha_1 \alpha_4 - \alpha_2 \alpha_3)\eta = 0.$$

30   These are nothing but the harmonic oscillator equation (2.23), with k equal to $-(\alpha_1 + \alpha_4)$ and ω_0^2 equal to $(\alpha_1 \alpha_4 - \alpha_2 \alpha_3)$. There is here, however, no physical requirement for either k or ω_0^2 to be positive. We shall write $\alpha_1 + \alpha_4$ as T, to indicate that it is the trace of the α-matrix relating $(\dot{\xi}, \dot{\eta})$ to (ξ, η), and also write $\alpha_1 \alpha_4 - \alpha_2 \alpha_3$ as D to indicate that it is the determinant of this matrix:

$$\ddot{\xi} - T\dot{\xi} + D\xi = 0 \quad \text{and similarly for } \eta. \tag{5}$$

The five regimes for the solutions of (5) are shown in fig. 6. Two stable regimes occupy the quadrant in which D is positive and T negative; two correspondingly unstable regimes occupy the quadrant in which D and T are both positive; and the last which is unstable, occupies the half-space in which D is negative. The criteria for the different regimes follow from the exponential solutions of (5). If $\xi = \xi_0 e^{pt}$,

$$p^2 - Tp + D = 0,$$

so that
$$p = \tfrac{1}{2}[T \pm (t^2 - 4D)^{\frac{1}{2}}]. \tag{6}$$

Focal stability. $D > 0$, $T < 0$ and $T^2 < 4D$; the solutions are complex, causing ξ and η to execute decaying oscillations. Since the relative amplitude and phase of ξ and η are not determined by T and D alone, but

---

[†] In later applications we shall often compress (4) into the form:

$$d\eta/d\xi = (\alpha_3 \xi + \alpha_4 \eta)/(\alpha_1 \xi + \alpha_2 \eta),$$

from which t has been eliminated. This enables trajectories to be drawn on the (ξ, η)-plane to describe the solutions of (4), but there is no indication of the sense in which a given trajectory is followed. Accordingly, a spiral trajectory may imply either the convergence of the system onto a point of stable equilibrium or the divergence from a point of unstable equilibrium. To avoid this ambiguity we shall always ensure that the coefficients in the compressed form keep the signs they have in (4).

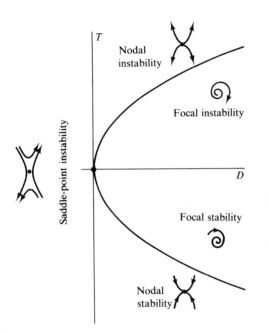

Fig. 6. Domains in which the solutions of (4) show different types of stability or instability.

depend on the magnitudes of the αs, the general form of the trajectory is an elliptical spiral converging onto the origin. The lightly damped harmonic vibrator is of this type.

Focal instability. $D>0$, $T>0$ and $T^2<4D$; as above, but the oscillations grow and the elliptical spiral diverges from the origin. The harmonic vibrator with positive feedback, or some other gain mechanism overriding viscous losses, is of this type.

Nodal stability. $D>0$, $T<0$ and $T^2>4D$; the solutions are real so that the oscillations are replaced by exponential decay of ξ and η. The trajectories converge on the origin without spiralling round it. We shall examine the form of these trajectories in more detail soon. The overdamped harmonic vibrator is of this type.

Nodal instability. $D>0$, $T>0$ and $T^2>4D$; as above but with trajectories diverging from the origin. A vibrator with very strong positive feedback is of this type.

Saddle-point instability. $D<0$; the solutions are real, one value of p being positive and the other negative. The positive solution guarantees instability, and the trajectories are of roughly hyperbolic form, avoiding the origin. A 'vibrator' with negative force constant, for example a ball sitting on top of a hill rather than in a valley, is of this type. No amount of damping or gain can stop it rolling off.

The general character of the trajectories just catalogued – spiral, non-spiral or hyperbolic; converging or diverging – is not altered by rotating or scaling the axes, and indeed the two parameters T and D are invariant under such coordinate transformations. It is helpful to use this fact to derive the form of the trajectories in more detail, transforming the coordinates so as to give (4) the simplest form. When the coordinates are rotated through an angle ϕ, followed by scaling up along one axis by a factor γ and down along the other by the same factor, the new variables X and Y are linearly related:

$$\dot{X} = \beta_1 X + \beta_2 Y, \qquad \dot{Y} = \beta_3 X + \beta_4 Y, \tag{7}$$

where

$$\left. \begin{array}{l} \beta_1 = \alpha_1 \cos^2 \phi - (\alpha_2 + \alpha_3) \cos \phi \sin \phi + \alpha_4 \sin^2 \phi, \\[6pt] \beta_2 = \dfrac{1}{\gamma}[a_2 \cos^2 \phi + (\alpha_1 - \alpha_4) \cos \phi \sin \phi - \alpha_3 \sin^2 \phi], \\[6pt] \beta_3 = \gamma[\alpha_3 \cos^2 \phi + (\alpha_1 - \alpha_4) \cos \phi \sin \phi - \alpha_2 \sin^2 \phi], \\[6pt] \beta_4 = \alpha_4 \cos^2 \phi + (\alpha_2 + \alpha_3) \cos \phi \sin \phi + \alpha_1 \sin^2 \phi. \end{array} \right\} \tag{8}$$

It is clear that ϕ can always be chosen to that $\beta_1 = \beta_4$; the required choice is such as to make $\tan 2\phi$ equal to $(\alpha_1 - \alpha_4)/(\alpha_2 + \alpha_3)$. Further, unless either β_2 or β_3 then vanishes, γ can be chosen so that $\beta_3 = \pm\beta_2$ (since γ may have either sign the absolute signs of β_2 and β_3 are irrelevant). When these choices have been made, (7) takes either of two forms:

Case (a): $\quad \beta_3 = -\beta_2;\qquad T = 2\beta_1 \quad$ and $\quad D = \beta_1^2 + \beta_2^2.$

$$\left. \begin{array}{l} \dot{X} = \beta_1 X + \beta_2 Y, \\ \dot{Y} = -\beta_2 X + \beta_1 Y. \end{array} \right\} \tag{9}$$

Case (b): $\quad \beta_3 = \beta_2;\qquad T = 2\beta_1 \quad$ and $\quad D = \beta_1^2 - \beta_2^2.$

$$\left. \begin{array}{l} \dot{X} = \beta_1 X + \beta_2 Y, \\ \dot{Y} = \beta_2 X + \beta_1 Y. \end{array} \right\} \tag{10}$$

We discuss the two cases separately.

Case (a): T^2 is always less than $4D$, and the system exhibits either focal stability or focal instability. If X and Y are components of the radius vector, R, (9) shows that (since $R^2 = X^2 + Y^2$),

$$\frac{\mathrm{d}}{\mathrm{d}t}(R^2) = 2X\dot{X} + 2Y\dot{Y} = TR^2. \tag{11}$$

Further, if we write $\tan \theta = Y/X$,

$$\sec^2 \theta \, \mathrm{d}\theta/\mathrm{d}t = (X\dot{Y} - Y\dot{X})/X^2 = -\beta_2 R^2/X^2,$$

therefore $\qquad\qquad \mathrm{d}\theta/\mathrm{d}t = -\beta_2.$

The representative point moves on an equiangular spiral, the radius vector increasing or decreasing exponentially, according to (11), while rotating at a constant speed, $-\beta_2$. For the spiral to converge, $T<0$. All this agrees with what we know already of decaying harmonic oscillations. The general form of trajectory in this case is an equiangular spiral distorted into elliptical form by scaling differently along any two orthogonal directions.

Case (b): it is convenient to carry out a further rotation of axes, through $\pi/4$, writing $X=(u+v)/\sqrt{2}$ and $Y=(u-v)/\sqrt{2}$. Then (10) takes the form

$$\dot{u}=(\beta_1-\beta_2)u, \qquad \dot{v}=(\beta_1+\beta_2)v. \tag{12}$$

These equations, showing that u and v develop exponentially and independently, define a non-spiralling trajectory which is stable only if $\beta_1-\beta_2$ and $\beta_1+\beta_2$ are both negative. Then $\beta_1^2-\beta_2^2$, i.e. D, must be positive and β_1, and hence T, must be negative. This is the case of nodal stability. In the coordinate system chosen, solution of (12) shows that

$$v=Au^{(\beta_1+\beta_2)/(\beta_1-\beta_2)}=Au^\gamma, \tag{13}$$

where A is an arbitrary constant, and $\gamma>0$. In the original coordinate system each value of γ defines a family of similar algebraic curves which may have any orientation, and describe the relaxation to equilibrium of an overdamped system. When $\beta_1-\beta_2$ and $\beta_1+\beta_2$ are both positive the trajectories have the same form, but the direction of motion is reversed; this is the case of nodal instability.

Finally we must consider the behaviour when $\beta_1-\beta_2$ and $\beta_1+\beta_2$ have opposite signs, i.e. when β_1 lies between $\pm\beta_2$, and $D<0$. Then γ in (13) is negative, and the trajectories avoid the origin. If $\beta_1=0$ they are rectangular hyperbolae; for non-vanishing β_1 they are not hyperbolae but have orthogonal asymptotes. Transforming back into the original coordinate system does not necessarily preserve this orthogonality.

The above analysis has assumed displacements from equilibrium small enough to be described by the linear equations (4), and this is permissible under all conditions save one – when the representative point lies exactly on the line dividing focal stability and instability, the horizontal axis in fig. 6. The example of an oscillating circuit with positive feedback will serve to illustrate the point. When the feedback is adjusted so that the total resistive loss is exactly cancelled ($Q=\infty$) the circuit may oscillate at any amplitude, poised between stability and instability, but only so long as everything is strictly linear. Commonly enough, however, the feedback is not so strong at higher amplitudes because the amplifier begins to saturate, and a strong oscillation experiences more resistive damping than the feedback can compensate. Consequently the system, which appears to be in neutral equilibrium if only linear terms are taken note of, is really in stable equilibrium; provided the higher order damping can be expressed by a power series expansion, so that a little damping is present at all amplitudes of oscillation, the current oscillations will decay to zero, though not exponentially, the effective time-constant steadily increasing as equilibrium is approached.

Now let us suppose the feedback to be increased slightly so that according to the linear theory the system is unstable. Oscillations will grow, but eventually the more rapidly increasing non-linear loss will compensate the linear gain, and the amplitude will saturate and thereafter remain constant. The representative point on the (x, y)-plane will execute a *limit cycle* about the point of unstable equilibrium. Further increase of positive feedback causes the limit cycle to expand and perhaps change its shape, so that the oscillations become non-sinusoidal. Eventually the limit cycle itself may become unstable so that the mode of oscillation switches to something entirely different. But this is well beyond the scope of what we are considering at this moment, which is confined to the almost-linear regime where the limit cycle will approximate to an ellipse and the waveform to a sinusoid.

An alternative form of behaviour along the line of infinite Q is for the total loss to become negative rather than positive as the amplitude grows. Then the system is unstable on this line, though the growth is not exponential, being slow at first and faster later. With slightly less feedback the linear lossy terms guarantee stability, but only while the amplitude is small enough for them to dominate the non-linear gain. In this case equilibrium is *metastable*, being stable for small displacements but unstable for larger.

Examples of instability

Since the problem, as formulated in the linearized first-order equations (4), has turned out to be reducible to the second-order differential equation of a simple vibrator, it is proper to indicate how the behaviour of the vibrator may be expressed as first-order equations. For this purpose we introduce displacement x and velocity v $(= \dot{x})$ as coordinates; then if μ is the force-constant and λ the viscous damping constant for a particle of mass m, as in (2.31),

$$\dot{x} = v \quad \text{and} \quad \dot{v} = -\mu x/m - \lambda v.$$

Interpreting ξ in (4) as x, η as v, we have that

$$\alpha_1 = 0, \qquad \alpha_2 = 1, \qquad \alpha_3 = -\mu/m, \qquad \alpha_4 = -\lambda/m.$$

Hence $$T = -\lambda/m, \qquad D = \mu/m.$$

In the notation of (2.23), $T = -k$ and $D = \omega_0^2$; the identity of (2.23) and (5) is immediately apparent. There is no need to discuss focal and nodal stability further – they are the underdamped and overdamped cases of harmonic vibrator behaviour. The vibration can be represented as a trajectory on the (x, v)-plane (*phase plane*), which turns out, as we shall discover, to be a very useful form of representation for non-linear vibrators. The general character of the trajectories is of course the same as shown in fig. 6. To realize the unstable trajectories a source of energy is needed. Very commonly this is provided by some external agency, the power pack of an amplifier in an oscillatory circuit, for example; but the source may be internal, and the growth of instability accompanied by the conversion of stored energy into kinetic energy of vibration or heat.† If the energy source acts so as to make k negative, spontaneous vibration (focal instability) or nodal instability is the consequence. These will be discussed more fully in chapter 11. Saddle-point instability is a very common occurrence in the natural world, without the intervention of external energy sources. The collapse of a structure, involving the conversion of potential or elastic energy into, first, kinetic energy, is normally of this type, and it is no great extension of the concept to allow phase transitions such as spontaneous changes of crystal structure to be classified as saddle-point instabilities. These matters are not strictly within the compass of a book on vibration, but in view of their importance we shall devote a little space to some straightforward examples.

The Euler strut, shown in fig. 7, is a classic example of a system in which the force-constant μ may be varied at will from positive to negative, so that a critical point can be found at which μ vanishes and stability fails. The elastic strip, initially straight, remains vertical so long as the load on it is not too great. Above a certain critical load, however, it becomes unstable and, if displaced, continues to move away from the vertical, the lowering of potential energy as the mass sinks being more than enough to supply the stored elastic energy in the bent strut; the excess then provides kinetic energy for the weight to move further from equilibrium. Finding the equation of motion of the mass is a matter of determining the horizontal restoring force when it is displaced. With the notation of fig. 7(c), the

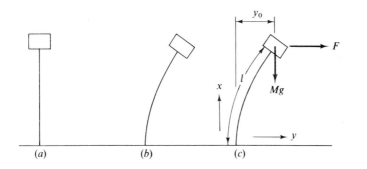

Fig. 7. The Euler strut: (a) subcritical load; (b) supercritical load; (c) notation used in analysis.

curvature at a point on the strut is, for small displacements, d^2y/dx^2, and to create this curvature a bending moment $R\,d^2y/dx^2$ is needed, R being the flexural rigidity of the strut, constant for a strut of uniform cross-section.

---

† The 'one-shot' pulsed laser is an example. Irradiation of the active material stores energy by exciting the molecules to higher states, and this energy is then emitted in the form of a short burst of monochromatic light as the molecules undergo spontaneous and phase-coherent oscillation.

Hence the shape of the strut is fixed by the equation

$$R\, d^2y/dx^2 = Mg(y_0 - y) + F(l - x), \qquad (14)$$

with the general solution:

$$y = y_0 + F(l - x)/Mg + A \cos qx + B \sin qx,$$

where $q^2 = Mg/R$ and A and B are constants. When $x = 0$ both y and dy/dx vanish. Hence $A = -Fl/Mg - y_0$ and $B = F/Mgq$. By putting $x = l$, $y = y_0$ we find y_0 and hence the force constant:

$$\mu \equiv F/y_0 = M\omega_p^2/(\tan ql/ql - 1), \qquad (15)$$

where ω_p is $(g/l)^{\frac{1}{2}}$, the frequency of a simple pendulum of length l.

The critical mass M_c is that which reduces μ to zero, by making ql equal to $\pi/2$. Hence

$$M_c = \tfrac{1}{4}\pi^2 R/gl^2.$$

The behaviour is easy to demonstrate with a straight length of steel strip, having a suitable mass attached to one end, and clamped in a vice. As the length of free strip is increased the natural frequency is seen to fall steadily, though the equilibrium state of the strip remains vertical. But after the point has been reached when $\mu = 0$ the vertical strip is unstable, and two new equilibrium positions now exist, one on each side. If the strip is perfectly elastic the bent equilibrium form may be worked out by rewriting the differential equation (14) with $F = 0$ and with the correct form for the curvature replacing the linear approximation d^2y/dx^2. There is something slightly academic about devoting great attention to this problem in applied mathematics, since most materials show brittleness or plasticity at small deformations; unless the strut is very thin in relation to its length the final configuration after the critical load is exceeded is liable to be some scattered fragments or a heap of tangled wreckage.[3] This is no disparagement of the elastic theory which, used sensibly, i.e. within the limitations of Hooke's law, has great value in predicting the critical load, but one must not suppose that what happens afterwards is necessarily contained in the linearized model. This is, however, a digression; our immediate concern is that the point of instability is a singular point in the physical behaviour, in the sense that continuity fails here, whether or not there is a Hookean regime beyond.† It is not possible to calculate what happens by

† The academic scientist may be satisfied to determine the stability of a system by means of linearized equations that correctly describe small perturbations from the stationary state. A practising engineer concerned with, for example, the flow of cooling gas in the core of a nuclear reactor, must worry about the possibility that the stable regime may not be proof against rather larger perturbations such as might occur in unusual circumstances. Unfortunately the systematic search for possible catastrophes over the whole range of variation of a large number of parameters is a formidable, often impracticable, undertaking. One of the characteristics of a great engineer is his capacity to foresee and guard against changes which routine analysis would overlook.

extrapolation of the properties through the critical point, even by very small steps. Thus a graph of the equilibrium position as a function of load shows a bifurcation at right angles (fig. 8). Symmetry about the vertical is maintained, but only in the sense that the positions available for equilibrium form a symmetrical pattern. Between what is allowed, and what actually occurs, symmetry is lost, for when two symmetrically disposed alternatives offer themselves a choice must be made between right and left. The outcome may be arbitrary or conditioned by external influences, but whichever choice is made the symmetry is inevitably broken. Obviously this critical point precludes extrapolation, since even though y_0 is a continuous function of load, its derivative with load is not.

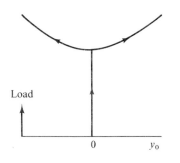

Fig. 8. Bifurcation of the solutions of the Euler strut problem at the critical load.

With other critical phenomena even y_0 need not be continuous; the hypothetical curve of fig. 9 is a perfectly possible form which would result in a sudden jump of y_0 at the critical load. The vibrating wire discussed at the beginning of this chapter shows a similar instability (see fig. 3), with hysteresis just like that indicated in fig. 9. This is hardly a structural instability, and perhaps a clearer illustration is provided by the sort of steel measuring tape that rolls into a small pocket container. The tape unrolls into a rather rigid straight strip by virtue of its slightly curved cross-section.

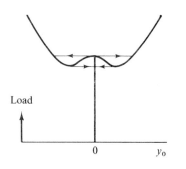

Fig. 9. An alternative form of bifurcation, with discontinuities and hysteresis.

Let the tape be placed flat on a table with its curvature concave upwards, and then the end gently pushed further and further beyond the edge. Quite suddenly, with a loud crack, the tape will kink and collapse to the floor. On pulling it back, it will remain kinked and hanging limp until much less projects over the edge, when sudenly it will jump back to its nearly horizontal, unkinked form. The transitions are sharp and the hysteresis very marked. This is not an easy problem to analyse in detail, but the short rope ladder shown in fig. 10 is quite simple; the diagram shows what torque is required to maintain a given angular displacement, and how it depends on past history. It is left to the reader to satisfy himself of the correctness of the graph. The inflected isotherms, below the critical temperature, in van der Waals' theory of imperfect gases, provide another example of an instability leading to a finite discontinuity. The fact that this is a simple model of a first-order phase transition may be taken as an indication of how the study of instability need not be confined to elementary dynamical systems but can be extended to include phenomena from all fields of science (and beyond).

Fig. 10. Two parallel strings separated by equally spaced rods are held in tension, and the bottom rod is twisted through θ about a vertical axis. The graph shows how the torque G on the bottom rod varies with θ.

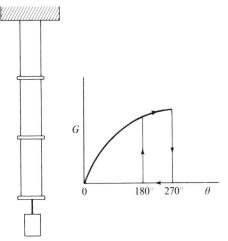

Returning to fig. 10, one observes that as G is increased the graph of G versus θ becomes horizontal at the point of instability – for small changes in θ near this point there is virtually no change in G, so that the differential torsion constant for small changes in displacement is passing through zero at the critical point. This is analogous to what happens with the strut, where μ passes through zero at the critical load. The singular character of the phenomenon results, of course, not from the mathematically well-behaved variation of μ itself, but from the fact that the physically interesting consequences are described by the frequency, $\omega_0 \propto \mu^{\frac{1}{2}}$. The variation of ω_0 with the load on the strut, as determined from (15), is shown in fig. 11. Since, near the critical load, μ is proportional to $M_c - M$, $\mu^{\frac{1}{2}}$ falls to zero with vertical tangent, another illustration of the impossibility of extrapolating further. The linear fall to zero of an elastic modulus or equivalent parameter is a clear diagnostic sign of approaching instability. Fig. 12 shows how the frequency of one of the lattice vibrational modes in strontium titanate (as determined by inelastic neutron scattering) falls with

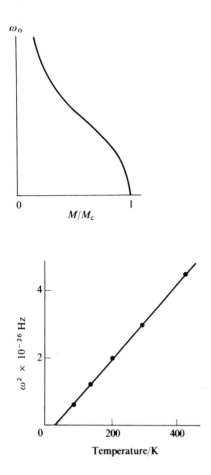

ω_0

0

1

M/M_c

Fig. 11. Vibration frequency of the mass on an Euler strut.

Fig. 12. Characteristic frequency of one of the vibrational modes in strontium titanate, showing how ω^2 falls linearly to zero as the temperature falls towards the structural transformation point at 35 K (R. A. Cowley[4]).

$\omega^2 \times 10^{-26}$ Hz

4

2

0   200   400

Temperature/K

decrease of temperature, and fig. 13 shows the decrease to zero of the reciprocal of the magnetic susceptibility in nickel. Extrapolation to zero in each case yields a value, fairly close to what is determined by direct observation, of the temperature at which a phase transformation occurs, with changes in crystal symmetry and other, non-structural, properties; strontium titanate changes its crystal structure, barium titanate becomes ferroelectric, nickel ferromagnetic. The properties in the new phases, below the critical temperature, are not deducible from those above by extrapolation. It would take us too far afield to discuss these applications any further, but before leaving the topic a word of warning is in order, to avoid giving too strong an impression of a close analogy between phase transformations and static instabilities. In the vicinity of the critical temperature, for these and a host of other materials exhibiting phase instability, the linear approach of the modulus to zero breaks down, as shows up clearly in fig. 13. The *order parameter*, used to characterize a structural change in a solid, is a variable that might be expected to behave like the frequency in our simple examples, its square falling linearly to zero

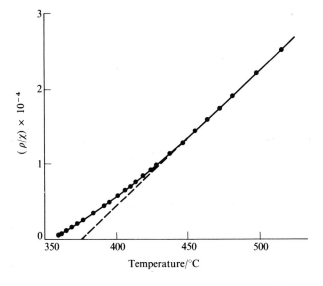

Fig. 13. Reciprocal of the magnetic susceptibility χ/ρ of nickel above its ferromagnetic Curie point, 358 °C (P. Weiss and R. Forrer[5]).

at the critical point. But, and fig. 14 shows this for two different materials, it is more usual for the cube to fall linearly in the very last stages before the transition. It is not surprising that a real material, built of a vast number of molecular units, should not be exactly modelled by a dynamical system of very few units. The study of real critical points is one of great difficulty and fascination, but must be ruled out of court here.

To conclude this presentation of examples illustrating the theory of stability in simple systems, let us note that dynamical equilibrium may also

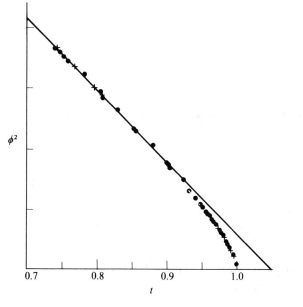

Fig. 14. The square of the order parameter, ϕ, in LaAlO$_3$ (crosses) and SrTiO$_3$ (points) as a function of reduced temperature, t, defined as T/T_c where T_c is the transition temperature. Just below T_c, $\phi \propto (1-t)^{\frac{1}{3}}$ (K. A. Müller and W. Berlinger[6]).

be brought into the same framework. A rigid body, for instance, spinning freely, obeys Euler's equations:[7]

$$I_x \dot{\omega}_x = (I_y - I_z)\omega_y\omega_z, \tag{16}$$

$$I_y \dot{\omega}_y = (I_z - I_x)\omega_z\omega_x, \tag{17}$$

$$I_z \dot{\omega}_z = (I_x - I_y)\omega_x\omega_y, \tag{18}$$

where I_x, I_y, I_z are the principal moments of inertia, in descending order of magnitude, and the components of $\boldsymbol{\omega}$ are the instantaneous angular velocities in the same coordinate system, fixed in the body. There are stationary solutions, the dynamical analogue of static equilibrium, when $\boldsymbol{\omega}$ lies along any of the axes, and $\dot{\boldsymbol{\omega}} = 0$, but examination of the equations shows that rotation about the y-axis, the intermediate axis of inertia, is unstable. For with $\boldsymbol{\omega}$ close to this axis, ω_x and ω_z are much smaller than ω_y, and (17) shows that to first order in these small components ω_y does not change. Hence (16) and (18) form a pair of equations of the same form as (4), with $\omega_x \equiv \xi$ and $\omega_z \equiv \eta$. Here α_1 and α_4 vanish, while $\alpha_2 = (I_y - I_z)\omega_y/I_x$ and $\alpha_3 = (I_x - I_y)\omega_y/I_z$. If I_y is intermediate between I_x and I_z, $\alpha_2\alpha_3$ is positive, D is negative and the motion shows saddle-point instability. On the other hand, the same analysis applied to motion about the axes where I is largest or smallest gives a positive value to D, while $T = 0$. The motion is oscillatory without decrement – in fact the axis of rotation wanders in an elliptical cone about the x- or z-axis, so that ω_y and ω_z (or ω_x) oscillate sinusoidally in phase quadrature.

The slightly anharmonic vibrator revisited

Let us use our stability criteria to verify the physically intuitive picture of mode-switching in the vibrating wire. We take as a model a vibrating mass whose motion is harmonic except for a weak variation of frequency with amplitude, $\omega(x_0) = \omega_0 + \gamma z_0^2$, as if the force-constant, averaged over a cycle, were $m\omega_0^2(1 + 2\gamma x_0^2/\omega_0)$. When a force F is applied at frequency ω_1, lagging in phase at some instant by ϕ behind the displacement, the in-phase component $F \cos \phi$ diminishes the restoring force so that the instantaneous vibration frequency is $\omega_0 + \gamma x_0^2 - F \cos \phi/2m\omega_0x_0$. The quadrature component $F \sin \phi$ enhances the damping term, which we write as having an amplitude $\lambda\omega_0x_0$. The equations for the development of ϕ and x_0 therefore take the form:

$$\dot{\phi} = \omega_0 + \gamma x_0^2 - F \cos \phi/2m\omega_0x_0 - \omega_1 = -\Delta + \gamma x_0^2 - B \cos \phi/x_0; \tag{19}$$

and $$\dot{x}_0 = -\tfrac{1}{2}\lambda x_0(1 + F \sin \phi/\lambda\omega_0x_0)/m = -Ax_0 - B \sin \phi, \tag{20}$$

where $\Delta = \omega_1 - \omega_0$, $A = \tfrac{1}{2}\lambda/m$ and $B = \tfrac{1}{2}F/m\omega_0$. The steady state of vibration is obtained by putting $\dot{\phi}$ and $\dot{x}_0$ equal to zero. In such a state let $\phi = \Phi$ and $x_0 = X$; then

$$\sin \Phi = -AX/B \quad \text{and} \quad \cos \Phi = (-\Delta X + \gamma X^3)/B. \tag{21}$$

249

By squaring and adding we derive a cubic equation in X^2, giving the possibility of three distinct amplitudes, as expected from fig. 3:

$$\gamma^2 X^6 - 2\Delta\gamma X^4 + (A^2 + \Delta^2)X^2 - B^2 = 0. \qquad (22)$$

To examine the stability of the state (Φ, X), put $\phi = \Phi + \xi$, $x_0 = X + \eta$. Then from (19) and (20), by differentiation, the linearized equations take the form:

$$\dot{\xi} = 2\gamma X\eta + B\eta \cos \Phi/X^2 + B\xi \sin \Phi/X = \alpha_1\xi + \alpha_2\eta,$$

and

$$\dot{\eta} = -A\eta - B\xi \cos \Phi = \alpha_3\xi + \alpha_4\eta,$$

the coefficients α being expressed in terms of X alone by use of (21):

$$\alpha_1 = -A, \qquad \alpha_2 = 3\gamma X - \Delta/X, \qquad \alpha_3 = \Delta X - \gamma X^3, \qquad \alpha_4 = -A.$$

The trace $T = \alpha_1 + \alpha_4 = -2A$ and is always negative; it can only confer stability. The determinant, however, $D = \alpha_1\alpha_4 - \alpha_2\alpha_3 = 3\gamma^2 X^4 - 4\Delta\gamma X^2 + A^2 + \Delta^2$, and may be caused to go negative. The critical value of Δ, at which a given solution becomes unstable, is that which makes D equal to zero. It is easily verified, by differentiating (22), that this is also the value of Δ at which the response curve, $X(\Delta)$, has a vertical tangent. This is, of course, the same criterion as we took for granted in discussing fig. 3.

The instability of an anharmonic vibrator under the influence of too strong a force sets a practical limit to the validity of neglecting anharmonicity. So long as the slope of the resonance curve nowhere exceeds that of the line (e.g. the chain line in fig. 3) showing the variation of natural frequency with amplitude, the response is single-valued and the harmonic approximation may well prove adequate. The losses in the system, by broadening the response, help to save the harmonic approximation. For a given amplitude, however, the resonance curve can be made as steep as one wishes by reducing the losses, so that in principle it is always possible to achieve the regime of instability, however small the amplitude. It follows then that the ideal lossless vibrator has a single-valued response at all frequencies only if the force-constant is strictly constant – any departure, however small, introduces instabilities. The harmonic vibrator is indeed an idealization though in practice, as already remarked, the presence of loss mechanisms confers a limited validity on its use.

390, 395 400, 415 There is no need at this stage to give any more examples of the use of the stability criteria for systems governed by (4); they will turn up again on several occasions in chapter 12.

Stability of a system with more than two independent variables

More elaborate systems than the examples just mentioned may need more than two variables to specify their state, but in the vicinity of an equilibrium configuration the linearized equations of motion will be similar to (4):

$$\dot{\xi}_1 = \alpha_{11}\xi_1 + \alpha_{12}\xi_2 + \cdots,$$

or in general, for the ith coordinate:

$$\dot{\xi}_i = \sum_j \alpha_{ij} \xi_j.$$

If we substitute the trial solution, $\xi_i = \xi_{i0} \, e^{pt}$, the equations reduce to linear algebraic equations:

$$(\alpha_{ii} - p)\xi_i + \sum_{j \neq i} \alpha_{ij} \xi_j = 0.$$

There are as many equations as there are independent variables, and since there are no constant terms the equations can only be mutually compatible if the determinant of the coefficients vanishes. The possible values of p are thus those which satisfy the equation:

$$\begin{vmatrix} \alpha_{11} - p & \alpha_{12} & \cdots \\ \alpha_{21} & \alpha_{22} - p & \cdots \\ \vdots & \vdots & \end{vmatrix} = 0,$$

which on multiplying out gives an equation of the same order in p as the number of variables, n. The system is capable of n independent modes of motion, vibratory if p is complex; only if the real part of every p is negative or zero can the system avoid instability, for any mode having a positive real part of p describes a growing disturbance. The criteria for stability are those rules which enable one to determine from the set of coefficients α_{ij} that there are no growing modes; the general solution to this algebraic problem was given by Routh, and in modified form by Hurwitz.[8] It is sufficient here to indicate the nature of the general problem and the existence of a solution; we do not propose to tackle any examples where more than two variables are present. There are full acounts available in standard textbooks.

Stability testing by discrete sampling (stroboscopic method)[9]

The rotating rigid body, like the simple vibrator, is a system whose coordinates can readily be followed continuously, and represented by a trajectory on the appropriate phase plane. It may be inconvenient, however, to keep note of so much detail if the behaviour is more complicated, and it may be sufficient to sample the coordinates at regular intervals to see how they are evolving, whether towards or away from steady values. This is particularly helpful for testing the stability of a dynamical system whose motion is periodic or approximately so, and especially if it is driven by a periodic force, for one can test at a standard moment in each cycle and cease to worry about what may be quite a complex pattern of behaviour between each sampling moment. This is what we did in deciding that the subharmonic generator in the arrangement of fig. 4 could not be stable in one of its possible configurations. Similarly if a maintained oscillator is found to be capable of executing a closed trajectory in the phase plane, the stability of this periodic motion can be examined by displacing it from its stationary cycle and noting each subsequent passage of a line intersecting the trajectory; in this case the state is determined by a single parameter, y say, and the stability depends on the relation between y_n and y_{n+1}. It will be realized that, merely by concentrating attention on discrete samples, one cannot in a detailed theoretical analysis avoid consideration of intermediate states of the system, since only by following the trajectory in detail between y_n and y_{n+1} can one discover how they are related; but the presentation of the results of such analysis is reduced to something manageable, and thereby it is made easier to catalogue the different types of behaviour that are possible in principle. A specific example of a fairly complicated trajectory reduced to simplicity by discrete sampling will be found in fig 10.6.

297

Let us illustrate this by reference to a one-parameter system whose evolution is completely specified by the dependence of any sample y_{n+1} on the value of the previous sample y_n, $y_{n+1} = F(y_n)$. We shall not limit the discussion to the particular type of application mentioned above, for which the form of F is restricted to monotonic variation by the requirement that trajectories on the phase plane do not intersect. Much more variety is possible with biological applications, as for instance the annual variations in the population of an insect species which matures, breeds and dies in a single season. The population, y_{n+1}, in year $n+1$ is thus entirely the progeny of the previous population, y_n, and if (a very dubious assumption) we can neglect random disturbances such as weather we may hope to write $y_{n+1} = F(y_n)$. For small y_n one expects $y_{n+1} > y_n$, for otherwise the species can hardly survive to allow us to examine its ecology, but for denser populations y_n may saturate ($y_{n+1} \sim y_n$); alternatively, epidemic disease or other hazards of overgrowth may cause havoc and result in F decreasing at large values of y_n. Fig. 15 exhibits various cases of interest. Equilibrium

Fig. 15. The full curves are various hypothetical possibilities for the dependence of y_{n+1} on y_n. A stationary value is possible at an intersection with the line $y_{n+1} = y_n$, shown broken. Successive values shown here illustrate that A and D are unstable, while B and C are stable. The crosses on the line through D are mirror images in the 45° line and are a possible (not necessarily stable) alternating steady solution.

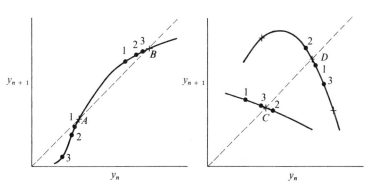

behaviour is possible only when $y_{n+1} = y_n$, at the intersection of the line representing F with a line of unit slope. The numbered points show successive values of y in the neighbourhood of an intersection. If the slope of F at the intersection is greater than unity, the point is unstable and divergence from the line is monotonic; if the slope is positive but less than unity it is stable, with monotonic convergence. These are the one-dimensional analogues of nodal stability and instability. If the slope is negative but not as steep as -1, there is oscillatory (analogous to focal) stability; if steeper than -1, oscillatory instability. The oscillation amplitude grows until two points are found on F which are mirror images in the line of unit slope; the solution may then become stationary with alternate values of y lying on each of these points. There is, however, no guarantee that this stationary solution is itself stable. If the negative slope at the intersection is too steep the stable behaviour may be a cycle of 4, 8 or more points; more remarkably there may be no periodic solution, and y may run almost at random through all points on finite segments of the curve. As we shall have no occasion, for the reason already explained, to treat cases in which the slope at intersection is negative, we shall go no further into this pathological behaviour, which has excited a certain amount of interest among mathematicians.[10]

Of more concern is the sampling of two-parameter systems. The device in fig. 4 can be characterized by the speed of the pin at the moment of contact with the base and the phase of oscillation of the base at this instant; given these two it is a matter of dynamics to calculate when the next contact will occur and what the pin speed will be. In general terms we may, as before, characterize a small variation from a stationary dynamical pattern by two parameters, ξ and η, which in the stationary configuration are found to vanish at every sampling. Close to such a configuration we expect linear behaviour:

$$\xi_{n+1} = a_1 \xi_n + a_2 \eta_n, \qquad \eta_{n+1} = a_3 \xi_n + a_4 \eta_n. \tag{23}$$

By the same transformations as before (23) may be expressed in terms of two coefficients only: **257**

$$X_{n+1} = b_1 X_n + b_2 Y_n, \qquad Y_{n+1} = \pm b_2 X_n + b_1 Y_n, \tag{24}$$

but for all the similarity between (7), with $\beta_1 = \beta_4$ and $\beta_3 = \pm \beta_2$, and (24), the stability criteria are different. We once more distinguish two cases, according to the sign in the second equation of (24) (note that the sign of b_2 is irrelevant – either sign may be generated in the transformation).

Case (a), negative sign. Put $b_1 = s \cos \theta$, $b_2 = s \sin \theta$, where $s^2 = b_1^2 + b_2^2 = D$. Then (24) takes the form:

$$X_{n+1} = s(X_n \cos \theta + Y_n \sin \theta), \qquad Y_{n+1} = s(Y_n \cos \theta - X_n \sin \theta).$$

This means that (X_{n+1}, Y_{n+1}) is obtained from (X_n, Y_n) by rotating the vector (X_n, Y_n) through θ and multiplying its length by s. The successive points therefore lie at equal angular increments along an equiangular spiral which converges if $s < 1$, but diverges if $s > 1$. These are the discrete analogues of focal stability and instability.

Case (b), positive sign.   As before, rotate the axes through $\pi/4$ to give the analogues of (12):

$$u_{n+1} = (b_1 - b_2)u_n, \qquad v_{n+1} = (b_1 + b_2)v_n.$$

Each coordinate evolves by geometrical progression and we distinguish three possible types of behaviour:

> nodal stability – convergence when $|b_1 - b_2|$ and $|b_1 + b_2| < 1$;
> nodal instability – divergence along the same types of trajectories when $|b_1 - b_2|$ and $|b_1 + b_2| > 1$;
> saddle-point instability – when one of $|b_1 - b_2|$ and $|b_1 + b_2|$ is greater and the other less than unity.

Since the determinant and trace, D and T, of the a-matrix are unchanged by the transformation of coordinates, it is convenient to express these criteria in terms of them.
We have that

$$D = a_1 a_4 - a_2 a_3 = b_1^2 \mp b_2^2, \tag{25}$$

and
$$T = a_1 + a_4 = 2b_1; \tag{26}$$

therefore $b_2^2 = \pm(b_1^2 - D) = \pm\frac{1}{4}(T^2 - 4D)$, the upper sign referring to case (b). All points inside the parabola, $T^2 = 4D$, belong to case (a) and all points outside to case (b). On the straight line, $T = 1 + D$, which is tangent to the parabola and wholly outside it, (25) and (26) show that $b_2^2 = (1 - b_1)^2$; since the sign of b_2 is irrelevant, we choose the positive root, so that $b_1 + b_2 = 1$ on this line. Similarly, on the line $T = -(1 + D)$, $b_1 - b_2 = -1$. From here on it is easy to label the various regions of the (D, T)-plane according to their stability behaviour, as in fig. 16. The criterion for stability is simple enough: (D, T) must lie in the triangle defined by the points $(-1, 0)$ and $(1, \pm 2)$.
We are now in a position to discuss the stability of the subharmonic generated by the bouncing pin.

The subharmonic generator revisited

To develop the model illustrated in fig. 4 quantitatively, let us assume the spring exerts constant pressure on the pin, so that it has constant acceleration f when free of the base, and let the coefficient of restitution at impact be ε. If the mass arrives with downward velocity V when the base is moving up with velocity v, the relative velocity after bouncing is $\varepsilon(V + v)$

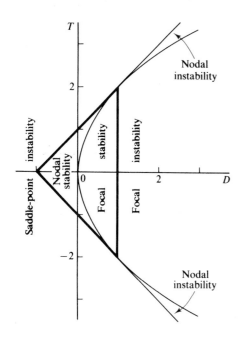

Fig. 16. Domains in which the solutions of (23) have various characters. The triangle defines the limits of stable solutions.

and the actual upward velocity of the mass is $\varepsilon V + (1+\varepsilon)v$. Since in a periodic solution this must equal V,

$$v/V = (1-\varepsilon)/(1+\varepsilon). \tag{27}$$

If the vertical position z of the base plate is $z_0 \sin \omega t$, and a bounce occurs at t_0,

$$v = \omega z_0 \cos \omega t_0. \tag{28}$$

Moreover, for the nth subharmonic the time between the bounces, which is $2V/f$, must be $2\pi n/\omega$. Hence the phase at the moment of contact is determined:

$$\cos \omega t_0 = \frac{\pi n f}{\omega^2 z_0} \frac{1-\varepsilon}{1+\varepsilon}. \tag{29}$$

This equation has two alternative solutions, as shown in fig. 5, and although we know already that one is unstable we shall keep both solutions in mind and test both for stability.

The two parameters characterizing the periodic solution are conveniently chosen to be V and t_0, so that a slightly perturbed solution has the mass arriving with velocity $V + \delta V$ at $t_0 + \delta t$. At the first bounce to be considered, when these parameters are δV_1 and δt_1, the velocity of the base is not v, as given by (28), but $v + \delta v_1$, where

$$\delta v_1 = -\omega^2 z_0 \sin \omega t_0 \cdot \delta t_1.$$

Consequently, the mass rebounds with velocity $V + \delta V_1'$, where

$$\delta V_1' = \varepsilon \, \delta V_1 - (1 + \varepsilon)\omega^2 z_0 \sin \omega t_0 \cdot \delta t_1. \tag{30}$$

If the base were at the same height when the mass returned at time $t_0 + 2\pi n/\omega + \delta t_2$, δV_2 would be the same as $\delta V_1'$. But the base has risen a further distance $\delta z = \omega z_0 \cos \omega t_0 \cdot (\delta t_2 - \delta t_1)$, and the velocity on arrival is thereby diminished by $f \delta z / V$. Hence

$$\delta V_2 = \varepsilon \, \delta V_1 - (1 + \varepsilon)\omega^2 z_0 \sin \omega t_0 \cdot \delta t_1 - f \omega z_0 \cos \omega t_0 \cdot (\delta t_2 - \delta t_1)/V. \tag{31}$$

Further, the mass after rebounding with velocity $V + \delta V_1'$, as given by (30), arrives $2\pi n/\omega + \delta t_2 - \delta t_1$ later with downward velocity $V + \delta V_2$, having suffered uniform acceleration in the interval. Hence.

$$2V + \delta V_1' + \delta V_2 = f(2\pi n/\omega + \delta t_2 - \delta t_1),$$

i.e.
$$\delta V_1' = f(\delta t_2 - \delta t_1) - \delta V_2.$$

From (30), by eliminating $\delta V_1'$, we have:

$$\varepsilon \, \delta V_1 + \delta V_2 = f \, \delta t_2 + [(1 + \varepsilon)\omega^2 z_0 \sin \omega t_0 - f] \, \delta t_1. \tag{32}$$

Rearranging (31) and (32), and making use of (27) and (28), we have

$$(f \, \delta t_2) = a_1(f \, \delta t_1) + a_2 \delta V_1$$

and
$$\delta V_2 = a_3(f \, \delta t_1) + a_4 \, \delta V_1,$$

where
$$\begin{aligned} a_1 &= 1 - (1 + \varepsilon)^2 B, \\ a_2 &= \varepsilon(1 + \varepsilon), \\ a_3 &= -\varepsilon(1 + \varepsilon)B, \\ a_4 &= \varepsilon^2, \end{aligned} \Bigg\} \tag{33}$$

and B is written for $\omega^2 z_0 \sin \omega t_0 / f$. The problem is now formulated in the same way as (23); by choosing $f \delta t$ for ξ and δV for η we have arranged that the as are dimensionless. From (29),

$$B^2 = \omega^4 z_0^2 / f^2 - (1 - \varepsilon)^2 \pi^2 n^2 / (1 + \varepsilon)^2, \tag{34}$$

and the positive and negative values of B define the two possible solutions shown in fig. 5.

As the amplitude of the base vibration is altered, B varies while ε remains constant; the as are functions of one parameter, so that the point on the (D, T)-plane defining the behaviour traces out a line, in this case the vertical straight line $D = \varepsilon^2$, as follows from (33). Since $T = 1 + \varepsilon^2 - (1 + \varepsilon)^2 B$, and B can take any value, we arrive at the following description:

(i) $\omega^2 z_0 / f \geq (1 - \varepsilon)\pi n / (1 + \varepsilon)$ for the nth subharmonic to have any chance of being generated; this follows from (34).

(ii) At the critical value of z_0, $B = 0$ and $T = 1 + \varepsilon^2 = 1 + D$; the system is in neutral equilibrium at the upper boundary between nodal stability and saddle-point instability in fig. 16.

(iii) As z_0 increases, negative values of B increase T and lead to instability (the broken curve of fig. 5), while positive values of B diminish T at constant D to allow first nodal and then focal stability. Further increase of B results in a short stretch of nodal stability giving place to saddle-point instability. The overall range of B that can confer stability is thus determined:

$$0 < B < 2(1 + \varepsilon^2)/(1 + \varepsilon)^2,$$

or in alternative form, by use of (34), in terms of the range of amplitudes of base vibration that can excite the nth subharmonic:

$$\frac{1 - \varepsilon}{1 + \varepsilon}\ \pi n < f_0/f < \frac{1 - \varepsilon}{1 + \varepsilon}\ \pi n\left[1 + \left(\frac{1 + \varepsilon^2}{1 - \varepsilon^2}\ \frac{2}{\pi n}\right)^2\right]^{\frac{1}{2}}, \qquad (35)$$

where $f_0 = \omega^2 z_0$ and is the maximum acceleration of the base. The ranges for different subharmonics may overlap but need not do so. For instance, if $\varepsilon = 0.8$, the fundamental excitation ($n = 1$) can occur for values of f_0/f between 0.35 and 1.07, the second subharmonic between 0.20 and 1.23, the third between 1.05 and 1.45, the fourth between 1.40 and 1.72, and the fifth between 1.75 and 2.01. The first four ranges overlap, but the rest do not. With ε rather smaller, none overlap, while with ε nearer unity several subharmonics may be stable with the same value of z_0. It will also be observed that stable bouncing motion is possible even when $f_0 < f$ and, left to itself, the mass would move with the base and not be thrown up. Different modes of response may therefore be brought about by different initial conditions.

At once a number of questions raise themselves, to indicate how superficial our analysis has been even of this elementary example. When several subharmonics can coexist, what determines which, if any, actually occurs? As z_0 is varied so that a subharmonic ceases to be stable, how does the system respond, especially if there is no other stable subharmonic to which it can switch? Are there other forms of response, fractional harmonics possibly, or integral subharmonics with more than one impact during the cycle?

To answer the last question first, undoubtedly other patterns can occur. Examination of the system pictured in fig. 4 with a stroboscope is enough to show this, even if it also shows that the base is significantly disturbed by the impact, so that the model is not very close to reality. Nevertheless, double impacts can certainly occur, and even multiple impacts, especially if ε is low. Fig. 17 shows in schematic form such a possibility, the mass rebounding so as to be caught again and again by the rising base, but eventually being flung off as the downward acceleration of the base exceeds f. The waveform picked up by a microphone placed near the loudspeaker

Fig. 17. Multiple bounces
in the experiment of fig. 4.

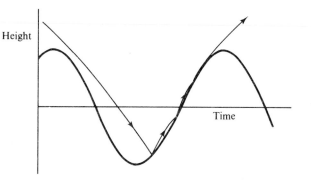

confirms the complexity of the behaviour, with many spikes and discontinuities to indicate the multiple impacts. In spite of this, careful
adjustment of the amplitude allows a given pattern to persist for a long
time, and the possibility unquestionably exists of stable locking of the
bouncing mass to a low subharmonic of the loudspeaker. At the same time,
one is left in no doubt that most locking patterns are complex and that they
are very numerous. Evidently an extended search of the possibilities would
be needed to satisfy oneself that all stable modes had been located. This,
however, would still be incomplete since there are almost certainly
aperiodic trajectories, even at values of z_0 for which stable periodic
behaviour is possible. The procedure for testing stability is now seen to be
of limited utility if a full picture of what can happen is required. Probably
nothing short of tracing out trajectories in detail, by computer for instance,
would satisfy this need, should it arise; for the stability test is only the last
stage – and the easiest – a critical examination of what has been discovered
by hard work. As it happens, this particular system is not in itself of any
great interest and there is no encouragement to undertake a full investigation. It is enough to recognize that as soon as one leaves harmonic or nearly
harmonic problems even the most elementary cases may prove extremely
formidable.

The cyclotron, and Azbel'–Kaner cyclotron resonance

A characteristic feature of the problem just discussed is that the applied
force is not independent of the coordinate of the oscillatory system, as with
all other cases of forced vibration examined so far. The equation of motion
takes the form:

$$\ddot{\xi} + f(\xi) = F(\xi, t),$$

instead of a simple $F(t)$ on the right-hand side, the change being enough to
generate almost unlimited possibilities for complication in the behaviour.
If the restoring force $f(\xi)$ is proportional to ξ, so that the free motion is
harmonic, the possible types of solution are severely restricted – the system
is unable to adjust its frequency to that of the excitation by changing

amplitude, so that stable excitation usually demands close synchronization of the force to the natural frequency. The cyclotron is an example of such a system, where the force is applied only while the particle traverses very short sectors of its orbit. A particle crossing the gap between the two dees when the phase of the oscillatory field between them is such as to accelerate it, is also accelerated at every subsequent crossing, twice per cycle, if the e.m.f. has the cyclotron frequency. In this case the localization of the accelerative force is an engineering convenience; the acceleration would take place with the electric field applied uniformly to the orbit, as in semiconductor cyclotron resonance. There would, however, be a significant difference if, for some reason, excitation at a higher frequency were desired. For with the dee arrangement an odd harmonic of the cyclotron frequency would also serve to accelerate particles; it would not matter that the e.m.f. ran through several cycles while the particles were inside the dees and isolated from electric fields. The point is the same as that illustrated in fig. 5.1˙ – if a sinusoid is sampled at equally spaced discrete points, π/ω_c apart in this case, the frequency remains indeterminate to the extent that a multiple of $2\omega_c$ may be added or subtracted.

237

Although excitation at a harmonic of the cyclotron frequency is not usual in particle accelerators, something very similar takes place in the arrangement used for the study of cyclotron resonance of conduction electrons in metals, originally proposed by Azbel' and Kaner.[11] In contrast to pure semiconductors, the skin effect in metals prevents a high-frequency field from penetrating deep into the interior. In a pure metal at low temperatures the free path between collisions is long and the conductivity proportionately high; a field at a microwave frequency of, say, 10^{10} Hz penetrates little more than 10^{-7} m into the metal, much less than the free path and also less than the radius of a typical cyclotron orbit of an electron. When a magnetic field is applied parallel to the surface some electrons are caught in helical orbits that just miss the surface (fig. 18), so that for a small fraction of each orbit they experience the electric field due to the microwave excitation. If successive entries into the surface layer catch the field in the same phase the electrons have their interaction with the field enhanced, while if there is phase reversal at successive entries the

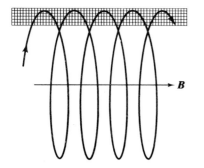

Fig. 18. Electron motion in Azbel'–Kaner cyclotron resonance. The shaded area represents the skin layer containing the high-frequency electric field.

interaction is diminished. With change of magnetic field, and therefore of ω_c, the effective conductivity of the metal alternates in magnitude, and this is detected as fluctuations in the surface resistance.

To make the argument more quantitative, let each entry into the surface layer result in an impulse $P\,e^{j\omega t}$ transmitted by the field to the electron. We define the response of the electron to this single impulse as the norm, the sum of all such individual responses being the surface current that would be set up by the field if the effects of resonance were negligible. In reality, however, each electron has a past history of similar impulses at previous entries, with the same phase difference θ between successive impulses. The average response is therefore as if to an impulse $P'\,e^{j\omega t}$, where

$$P'\,e^{j\omega t} = P\,e^{j\omega t} + \alpha P\,e^{j(\omega t - \theta)} + \alpha^2 P\,e^{j(\omega t - 2\theta)} + \cdots,$$

θ being $2\pi(1 - \omega/\omega_c)$; α is the probability that an electron will execute a cyclotron orbit without suffering a collision. If τ is the mean time between collisions, $\alpha = e^{-2\pi/\omega_c\tau}$. Summing the geometrical series, we have that

$$P'/P = (1 - \alpha\,e^{-j\theta})^{-1}. \tag{36}$$

189

This resonant denominator is the same as that already met with in the discussion of transmission line resonators (chapter 7). When the free path is long ($\tau \gg 2\pi/\omega_c$), so that α is very nearly unity, P' has sharp peaks whenever θ passes through zero, at regular intervals of $1/\omega_c$, i.e. of $1/B$, as the magnetic field is changed. The electrons can be considered as responding in an integral subharmonic series to the applied oscillatory field. An experimental curve showing the observation of Azbel'–Kaner resonance in tungsten is given in fig. 19.

A number of points need to be made briefly in connection with this curve. In a real metal, different groups of electrons have different cyclotron

Fig. 19. Recording of cyclotron resonance in a very pure single crystal of tungsten at a frequency of 35 GHz (W. M. Walsh[12]).

$\dfrac{\mathrm{d}R}{\mathrm{d}B}$

0   0.02   0.04   0.06

B/T

frequencies (more precisely, there is a continuous spread of ω_c) and their oscillations of conductivity mostly cancel each other out. Only dominant groups reveal themselves, and then as a comparatively minor oscillatory variation superimposed on a steady background. To observe the effect more readily it is common practice to record the derivative of the behaviour. For example, the surface resistance (the real part of the surface impedance Z, which is the value of $\mu_0 \mathscr{E}/\mathscr{B}$ for the microwave fields at the surface of the metal) changes with B, and causes changes in Q and hence in the output of a resonant cavity incorporating the sample as one of its walls. A small oscillatory variation of B then causes oscillation of the output with an amplitude proportional to the derivative of output with respect to B. This is picked up and amplified, and presented as a measure of dR/dB, the derivative of surface resistance of the sample. Slowly varying changes of background are largely eliminated by this procedure, and the oscillation left correspondingly more visible.

It is obvious that the curve of fig. 19 bears a closer resemblance to a succession of low-Q resonances, as in fig. 7.24, than to the differential of such a function. To understand why this is so one must recognize that under the conditions of the experiment, the free path l of the electrons being very long, Z is proportional to $(-\omega^2 l/\sigma)^{\frac{1}{3}}$, where σ is the conductivity; this is the theoretical expression of the *anomalous skin effect*.[13] If the variations in σ expressed in (36) are superimposed on a larger steady background, the form of Z is given by the expression

190

$$Z \sim [-1 - a/(1 - \alpha\, e^{-j\theta})]^{-\frac{1}{3}} \sim (-1)^{-\frac{1}{3}}[1 + \tfrac{1}{3}a/(1 - \alpha\, e^{-j\theta})]$$

when a, measuring the amplitude of the σ-oscillations, is much less than unity. As θ changes, $a/(1 - \alpha\, e^{-j\theta})$ runs round a circle, though at a non-uniform rate (see fig. 7.23 which illustrates a similar problem). In fig. 20 a typical trajectory of Z is plotted out, with points at equal intervals of θ marked. It is the orientation of the circle, lying along the 60° line instead of the axis, because of the term $(-1)^{-\frac{1}{3}}$ in Z, that gives the characteristic form to the variation of R and of $dR/d\theta$, which is nearly equivalent to dR/dB. The curve still matches the experimental curve only rather approximately, but the discrepancies are to be explained in terms of minor experimental details or features of the electronic behaviour in real metals not fully taken into account in our crude model. Further refinements may well be left to experts for, significant though they are to the understanding of electrons in metals, they add little to our appreciation of resonance phenomena.

Conditions for subharmonic generation

When a dissipative system is acted on by a periodic force the maintenance of periodic motion must be dependent on a steady flow of energy to make up for dissipative losses. A sinusoidal force can only continue to do work if the object on which it acts has a Fourier component in its motion at the same frequency as the force. In the cases already treated – a mass bouncing

Fig. 20. The origin of the line shape in cyclotron resonance. As *B* is changed the point representing *Z* runs round a circle (shown here considerably enlarged). With α taken as 0.9 the marks show successive values at equal increments Δ*B* – there are 110 increments in a complete revolution, and most of the trajectory is described in a small fraction of a cycle. The real parts of *Z* are marked on the real axis in (*a*), and the variation of *R* and its derivative exhibited in (*b*) and (*c*).

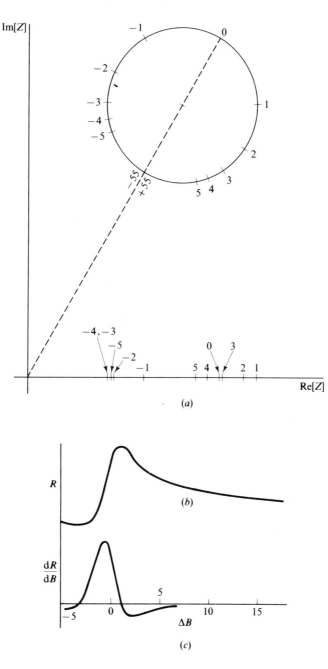

(*a*)

(*b*)

(*c*)

on a vibrating base, and cyclotron resonance – the applied force was intermittent at a frequency determined by the motion of the object affected. It was this that generated a component of the applied force at the frequency of the resulting motion even though, as with cyclotron resonance, the electric field variations and the orbital motion of the electrons were both harmonic and at different frequencies. When, however, a moving body is acted upon by a harmonic force at all times, it can only respond at a subharmonic frequency by possessing in its own motion a Fourier component to match the applied force. Slightly anharmonic vibrators (pendulum, vibrating string) with odd harmonics of their fundamental frequency present to a small degree in their motion, can in principle be set into vibration in this fundamental mode by applying a periodic force at 3, 5 etc. times the frequency; but the effect is weak, and energy conservation is no guarantee of stability. This system is difficult to analyse, but one can illustrate the process by means of a rather artificial example where the effect is strong and the analysis easy.

Consider a ball rolling without friction on a horizontal plane and bouncing to and fro between two walls. By tilting the plane periodically, as sketched in fig. 21, one can subject the ball to a harmonic force; can it be set into periodic motion at the driving frequency or a subharmonic of it? Clearly the energy argument allows the possibility – whatever the driving frequency there is a speed V for the ball to be projected such that its period is a subharmonic of this; if it bounces elastically its velocity has a square-wave variation with time, being alternately $\pm V$, and possesses all odd harmonics in its Fourier analysis. In principle, therefore, work can be done when the motion is synchronized to an odd subharmonic of the driving frequency, so that even when bouncing is inelastic there is the possibility of maintaining the periodic response. Let us discover some of the possible patterns of periodic response and then apply the tests of stability.

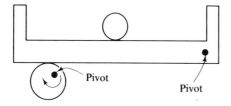

Fig. 21. A ball bouncing between end-plates is excited by periodic tilting.

Let the ball, of mass m, be confined to the region $-X < x < X$ and be acted upon at all times by a force $F \cos \omega t$, so that except at the moment of bouncing it obeys the equation of motion

$$m\ddot{x} = F \cos \omega t,$$

with general solution

$$x = -F \cos \omega t / m\omega^2 + vt + x_0,$$

where v and x_0 are constants during any one traverse of the range. In dimensionless form,

$$z = -\cos \theta + y\theta + c, \tag{37}$$

where $z = m\omega^2 x/F$, $\theta = \omega t$, $y = m\omega v/F$ and $c = n\omega^2 x_0/F$; $-Z < z < Z$, where $Z = m\omega^2 X/F$. The analysis is greatly simplified if only odd subharmonic solutions are considered. There are probably many other stable solutions, but we shall not search for them. The particular solutions chosen are those in which the range is traversed in time $n\pi/\omega$, i.e. in $\frac{1}{2}n$ periods, n being odd, so that the ball reaches Z when the phase is exactly opposite to that at which it left $-Z$, though one or more complete periods may have intervened. If the coefficient of restitution is such that at Z it rebounds with the same speed as it possessed on leaving $-Z$ it can repeat in the opposite direction the motion that took it from $-Z$ to Z, and this double traverse can then be repeated indefinitely, provided it is stable against small perturbations.

Let the phase be θ_0 at the moment the ball leaves $-Z$, and $\theta_0 + n\pi$ on arrival at Z; then (37) shows that

$$Z = \cos\theta_0 + \tfrac{1}{2}n\pi y. \tag{38}$$

The (dimensionless) velocity is $dz/d\theta$, which is $\sin\theta_0 + y$ on leaving $-Z$ and $-\sin\theta_0 + y$ on arriving at Z. If the former is to be ε times the latter,

$$y = -\frac{1+\varepsilon}{1-\varepsilon}\sin\theta_0. \tag{39}$$

These two equations, (38) and (39), fix the values of θ_0 and y needed to give synchronous motion at the nth subharmonic. The physical significance is made clear in fig. 22. When $\theta = n\pi$, $dz/d\theta = y$; with perfectly elastic bouncing ($\varepsilon = 1$) the points A and B would be the end-points of the traverse for the third subharmonic, y being adjusted so that the vertical

Fig. 22. Illustrating (38).

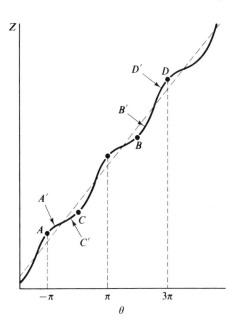

distance from A to B represented $2Z$. Equally well C and D are possible end-points for the same subharmonic when $\varepsilon = 1$, and both possibilities must be borne in mind. When $\varepsilon < 1$, the end-points must be shifted so that the initial velocity is ε times the final velocity. Thus A' and B' (at $-\pi + \Delta$ and $2\pi + \Delta$) or C' and D' (at $-\Delta'$ and $3\pi - \Delta'$) are possible end-points.

To investigate the stability of these two possible solutions, consider the ball leaving $-Z$ at a phase $\theta_0 + \delta\theta$, with such a velocity that y takes the adjusted value $y + \delta y_1$; c also will be changed, but we shall eliminate it from the argument. The effect of these perturbations on the position of the ball follows by differentiating (37):

$$\delta z = (\sin\theta + y)\,\delta\theta_1 + \theta\,\delta y_1 + \delta c. \tag{40}$$

Since these perturbations refer to changes in the description of the motion at $-Z$, we have that $\delta z = 0$ when $\theta = \theta_0$,

$$\text{i.e. } 0 = (\sin\theta_0 + y)\,\delta\theta_1 + \theta_0\,\delta y + \delta c. \tag{41}$$

At the next bounce, when $z = Z$, let $\theta = \theta_0 + n\pi + \delta\theta_2$; the same argument shows that

$$0 = (-\sin\theta_0 + y)\,\delta\theta_2 + (\theta_0 + n\pi)\,\delta y_1 + \delta c. \tag{42}$$

Eliminating δc from (41) and (42), we have that

$$\delta\theta_2 = [-n\pi\,\delta y_1 + (y + \sin\theta_0)\,\delta\theta_1]/(y - \sin\theta_0). \tag{43}$$

This is one equation relating the perturbations at the start of the second traverse to those at the start of the first. For the other we note that the velocity, being $\sin\theta + y$, is perturbed by $\cos\theta \cdot \delta\theta + \delta y$, and that at the moment of arrival at Z, when $\theta = \theta_0 + n\pi + \delta\theta_2$, δv before bouncing is $-\cos\theta_0 \cdot \delta\theta_2 + \delta y_1$. After bouncing the perturbation is reduced by a factor ε, and this must be described by a modified value of y,[†] such that

$$\cos\theta_0 \cdot \delta\theta_2 + \delta y_2 = \varepsilon(-\cos\theta_0 \cdot \delta\theta_2 + \delta y_1). \tag{44}$$

This is the second equation which, with (43), defines the a-matrix relating $(\delta y_2, \delta\theta_2)$ to $(\delta y_1, \delta\theta_1)$. Substituting for y from (39) we have

$$\left.\begin{aligned}
a_1 &= \varepsilon - B, \\
a_2 &= -C, \\
a_3 &= \varepsilon B/C, \\
a_4 &= \varepsilon,
\end{aligned}\right\} \tag{45}$$

where $B = \tfrac{1}{2}n\pi(1 - \varepsilon^2)\cot\theta_0$ and $C = \varepsilon(1 + \varepsilon)\cos\theta_0$.

---

† Since alternate traverses are in opposite senses, y changes sign at each bounce. To avoid this we describe each traverse as if it were in a positive direc-tion, taking care to reverse the phase of the applied force at each bounce. This accounts for the signs in (44).

This a-matrix is rather similar to (33); in particular, $D = \varepsilon^2$, independent of θ_0, so that as the strength of the applied force, and hence Z and θ_0, are changed, the locus on the (D, T)-plane is once more a vertical line in fig. 16. Since $T = 2\varepsilon - B$, the limits of stability, $T = \pm(1 + D)$, are expressed by the equation:

$$2\varepsilon - \tfrac{1}{2}n\pi(1 - \varepsilon^2)\cot\theta_0 = \pm(1 + \varepsilon^2),$$

i.e.
$$\cot\theta_0 = -\frac{2}{n\pi}\frac{1-\varepsilon}{1+\varepsilon} \quad \text{or} \quad \frac{2}{n\pi}\frac{1+\varepsilon}{1-\varepsilon}. \tag{46}$$

Four values of θ_0 satisfy these equations, but only two are admissible since (39) shows that θ_0 must lie in the third or fourth quadrants.

To interpret these critical conditions in terms of the limits within which the exciting force F can lie, it is necessary to proceed from $\cot\theta_0$ through (39) to (38), evaluating y and then Z, the latter containing the required information. Numerical evaluation is easiest at this stage, and some typical results are shown in fig. 23. In contrast to the mass bouncing on a vibrating plate, the subharmonics appear at weaker, rather than stronger, excitation; this is because the natural frequency increases with energy here, while in the other case it decreases. The shape of the upper limit of fundamental excitation results from the fact that y, as computed from (46) and (39), becomes less than unity when ε is small. The velocity can now become

Fig. 23. Conditions for exciting fundamental response and third subharmonic in the system of fig. 21.

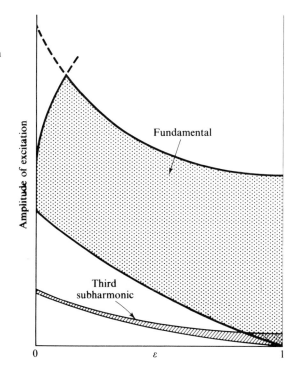

negative, and it is necessary to verify that it does not stray beyond the limits $\pm Z$. This additional feature precludes the whole of the stable range being realized.

It is seen from fig. 23 that the third subharmonic is stable only within a rather narrow range of force, especially if $\varepsilon \ll 1$. In particular, it should be noted that too strong a force precludes stability, though obviously not for energetic reasons. This has a consequence for the possibility of subharmonic excitation of systems whose anharmonicity is not so extreme as in this example. With correspondingly smaller terms in the Fourier expansion of the velocity–time behaviour, a given amplitude of response (and hence, roughly, of dissipative loss) requires a stronger force to supply the dissipated energy. If this force lies above the stability limit, no subharmonic excitation will occur, even though energetically it appears possible. We may therefore expect, for any given dissipation per cycle, that there will be a critical degree of anharmonicity needed to permit subharmonics, and that this critical anharmonicity will be greater in more lossy systems. So much we may aver fairly confidently; it is another matter to delineate the limits of the subharmonic regime and we shall take the matter no further.

10 Parametric excitation

It has long been known that a resonant system can be set into oscillation by periodically varying the parameters. A very extensive analysis was given by Rayleigh,[1] who drew attention to analogies with rather different physical processes – the perturbation of a planetary orbit by another planet having a quite different period in some near-integral relationship, and the propagation of waves in periodically stratified media. This latter example relates to a phenomenon of much greater physical interest nowadays than in Rayleigh's since it is basic to Bragg reflection of X-rays and to the motion of electrons in solids. Nevertheless, these are analogies only in the sense that the same type of differential equation is involved in all of them, as well as in the theory of vibrating elliptical membranes. It can hardly be claimed that the solution of one of these problems adds anything to the intuitive understanding of another. A mathematical framework, however, is well worth having, for these are not easy problems. The examples presented here earn a place in their own right as showing yet another aspect of the variety of oscillatory phenomena in nature. At the same time, there are important technological applications in the form of *parametric amplifiers*, as will be discussed in outline at the end of the chapter.

Probably the easiest way to demonstrate parametric excitation is to set up a simple pendulum whose length can be slightly varied at twice the natural frequency. If the amplitude of this variation is too small, nothing happens except that the bob vibrates up and down on the end of the string, but with a larger excursion the pendulum spontaneously begins to swing and its amplitude grows exponentially. A suitable demonstration apparatus is shown in fig. 1. The pendulum has a brass bob of about 100 g hanging on a fine steel wire, which passes through a small hole in the plate A and is hung by a loop on an eccentric pin fixed to the wheel C. Rotation of the wheel at, say, 1 revolution a second causes the effective length of the pendulum below A to vary with this frequency, and conditions for parametric excitation are achieved by adjusting the length of the pendulum so that its period is twice that of the wheel C. In practice the experiment works rather better if the string is pulled sideways by the light cord attached at B. This serves to hold the string against the side of the hole in A, thus defining the length of the pendulum rather precisely, but it also ensures that oscillations in and perpendicular to the plane of the paper have well-separated frequencies, so that only one mode is excited. Pendulum oscillations normal to the plane, with A as the effective point of support, are used to show the phenomena, and it is these whose frequency is carefully adjusted, by moving the bob on the threaded rod, to be exactly half that of the wheel. To get good results, it is worth being very patient in adjusting this frequency; the motor should run at a constant speed, for example synchronously with the a.c. mains. It is convenient to attach a side arm to the axle of C which can strike a metal spring once every turn; the ear and eye used together are very sensitive to slow departures of wheel and pendulum from synchronism. The pin on C is first adjusted to be central, so that C can turn without changing the pendulum length, and the frequencies are adjusted. Then the pin is moved to an eccentric position, say 5 mm off-centre, and the wheel started with the pendulum at rest. It will not be long before the pendulum begins to swing with ever-increasing amplitude, limited only by the breakdown of synchronism as its wide swing appreciably lengthens the period. With a smaller amplitude of excitation the growth will be found to be less rapid; if all is in good adjustment, the rate of growth (the reciprocal of the number of swings in which the amplitude is doubled) should be a linear function of the amplitude of excitation. We stress again the proviso about everything being well adjusted – it is in practice hard to obtain points on the straight line with weak excitation, when the rate of growth is slow. The reason will become clear in due course from the mathematical analysis. If the variation of pendulum length is insufficient to induce growth, the parametric excitation serves only to modify the rate of decay of any oscillation already present.

There is a certain critical excitation which allows the oscillation to persist at constant amplitude. It is virtually impossible to exhibit this in practice, but the reader is encouraged to try, since more can be learned about the physical processes by being frustrated than by accepting defeat at the bidding of the printed page.

An explanation of how the pendulum oscillation can be maintained follows from considering the energy supplied by the motor driving C. The tension, T, of the string is not constant but is slightly greater than mg as the pendulum swings through vertical, and slightly less as it pauses at the ends of its swing; T therefore runs through two cycles for each complete cycle of the pendulum. If the motion of C is such that the upward velocity of the pin, v, is in phase with the variations of T, energy is fed into the pendulum at all phases of the oscillation. To achieve this optimal condition, the vertical displacement, a, which determines the shortening of the effective length of the pendulum, must lag behind T, in phase quadrature.

To obtain a quantitative measure of the condition for spontaneous oscillation, let the pendulum have length l and angular displacement θ. The inward radial acceleration of the bob, $l\dot{\theta}^2$, results from the imbalance of T and the radial component, $mg \cos \theta$, of the weight of the bob.† For small amplitudes we approximate to $\cos \theta$, writing

$$T - mg(1 - \tfrac{1}{2}\theta^2) = ml\dot{\theta}^2.$$

Substituting $\theta_0 \cos \omega t$ for θ, and $-\omega\theta_0 \sin \omega t$ for $\dot{\theta}$, yields the result

$$T = mg(1 + \tfrac{1}{4}\theta_0^2 - \tfrac{3}{4}\theta_0^2 \cos 2\omega t),$$

of which only the oscillatory last term, $-\tfrac{3}{4}mg\theta_0^2 \cos 2\omega t$, concerns us here. Vertical displacement of the string according to $-a_0 \sin 2\omega t$ causes an upward velocity of $-2a_0\omega \cos 2\omega t$; the resulting instantaneous power input, Tv, is $\tfrac{3}{2}mga_0\omega\theta_0^2 \cos^2 2\omega t$, which is always positive and has mean value $\tfrac{3}{4}mga_0\omega\theta_0^2$. The energy E in the oscillating pendulum is $\tfrac{1}{2}ml^2\omega^2\theta_0^2$, and the equation for the increase in energy therefore takes the form

$$\dot{E} = E/\tau_e', \quad \text{where} \quad \tau_e' = \tfrac{2}{3}l/a_0\omega.$$

The build-up is exponential, with time-constant τ_e', and the system may be assigned a (negative) quality factor, $-Q'$, related to τ_e' in the usual way,

$$Q' = \omega\tau_e' = \tfrac{2}{3}l/a_0; \tag{1}$$

Q' is the number of radians of oscillator required for an increase in energy by a factor e.

Of course any real pendulum suffers internal damping which, if linear, is described by an intrinsic quality factor, Q. Spontaneous oscillation sets in if $Q > Q'$, the gain from the parametric process outweighing dissipative loss.

---

† No account need be taken of tension changes which result from vari- ations of l, since these make no net contribution to the energy transfer.

The critical amplitude a_c for parametric excitation therefore follows by setting $Q' = Q$ in (1):

$$a_c = \tfrac{2}{3}l/Q. \qquad (2)$$

The system described and illustrated in fig. 1 has typically $Q \sim 2500$ and $l \sim 1$ m, so that a_c is a little under $\tfrac{1}{3}$ mm. If a_0 is considerably greater than this the dissipation may be neglected without much harm, and the oscillations will build up exponentially at a rate governed essentially by Q'. Experimentally it was observed that with $l = 0.84$ m and $a_0 = 8.5$ mm the amplitude increased fourfold, and the energy sixteenfold, in 33 swings. The measured Q' is therefore $33 \times 2\pi/\ln 16$, i.e. 75. According to (1) it should have been 66, or 68 if the intrinsic Q is allowed for. The higher measured Q' means a slightly slower build-up, possibly explained as an error of observation but more likely due to imperfect tuning of the parametric excitation. We shall return to the question of tuning shortly.

There are many analogous examples of parametric excitation, since somewhere in most oscillators one can find a force (or its equivalent) oscillating at twice the frequency, which can be used to feed in energy:

(*a*) A trivial example is the toy illustrated in fig. 2, a disc suspended on a loop of string, which can be made to spin alternately clockwise and anticlockwise by pulling on the loops twice per complete cycle. The cycle involves many revolutions back and forth, and the system is highly anharmonic, but the parametric excitation does not depend on this; nor is it vitiated by variations in period, for the skilful operator adjusts his pulling frequency as required. The following two examples are more nearly harmonic, the second strictly so.

Fig. 2. Child's toy working by parametric excitation.

(*b*) A string vibrating transversely (fig. 3) exerts a different longitudinal component of force at the point of attachment S according as it is in the configuration A or in either of the equivalent configurations B and C. If the tension is constant the longitudinal force at S varies as $T \cos \theta$. More probably the tension increases with displacement and overcompensates the decrease due to $\cos \theta$, but this does not eliminate the essential property of the longitudinal component, that it oscillates at twice the natural frequency. Longitudinal motion of S at this double frequency may supply enough power to excite transverse oscillations parametrically. We shall return to this example in more detail.

Fig. 3. Three configurations of a stretched string in its lowest mode.

(c) If a current is oscillating in the circuit of fig. 4 and the capacitance C is varied at twice the resonant frequency, for example by changing the plate separation, energy can be fed into the circuit. For the attractive force between the plates is $\frac{1}{2}\varepsilon_0 \mathscr{E}^2$ per unit area when the electric field is $\mathscr{E}$, and takes a maximum value twice in each cycle as the current passes through zero and the charge on the plates reaches its peak positive or negative value. These are the moments when the separation should be increasing, the process being reversed when the charge is around zero.

Fig. 4. Parametrically excited resonant circuit.

Detailed analysis of parametrically excited systems

The energy argument gives valuable insight into the conditions necessary for parametric excitation, but for a full discussion direct analysis is best. The equation of motion for the circuit of fig. 4 is readily written down, but not so readily solved. Let the spacing of the capacitor plates vary as $1 + \beta \cos 2\Omega t$, β being much less than unity, so that the capacitance varies as $C_0(1 + \beta \cos 2\Omega t)^{-1}$. Then the charge on the plates obeys the equation:

$$L\ddot{q} + R\dot{q} + q(1 + \beta \cos 2\Omega t)/C_0 = 0, \tag{3}$$

which is linear but, in contrast to systems with fixed parameters, has one time-dependent coefficient. The principle of superposition applies, in the sense that two independent solutions of (3) can be combined by addition to yield all possible solutions. Translational invariance has, however, disappeared; a solution does not remain valid when shifted arbitrarily along the time axis, but only when the shift is a multiple of π/Ω.

It is not easy, except for a very practised performer, to follow the general behaviour of (3) by sketching solutions. Let us therefore start with a simplified version in which the spacing of the plates is changed by sudden steps, being alternately $1 \pm \beta$ times the mean, and let the changes take place every quarter of a cycle at the resonant frequency. Further, let $R = 0$, so that (3) can be written (in a shorthand notation):

$$\ddot{\xi} + \omega_0^2 \xi(1 \pm \beta) = 0. \tag{4}$$

In a phase-plane plot the trajectory of (ξ, v), v being defined as $\omega_0\dot{\xi}$, would be a circle if $\beta = 0$. If, however, $\beta \neq 0$, the natural frequency is increased to $\omega_0(1+\beta)^{\frac{1}{2}}$ at the wider spacing of the plates, and decreased to $\omega_0(1-\beta)^{\frac{1}{2}}$ at the narrower spacing. The trajectories in these circumstances are elliptical, with the major axis lying along the v-axis in the former case, and along the ξ-axis in the latter. Fig. 5 shows how the trajectory is changed by switching from positive to negative sign (a) at the moments when $\xi = 0$ and (b) when $v = 0$. In (a) the response grows exponentially, in (b) it decays, and the general form of behaviour when the system is started at some arbitrary point on the plane is a mixture of both types, with the growing mode (a) ultimately dominating.

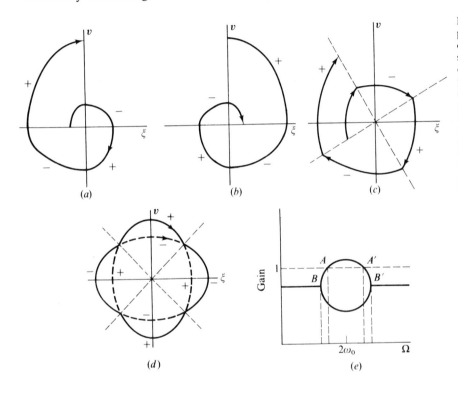

Fig. 5. Trajectories in phase-plane when capacitor in fig. 4 has its spacing increased and decreased suddenly twice per cycle (a) with oscillation so phased as to give maximum amplification, (b) phased to give maximum attenuation. The + and − signs indicate the quadrants in which the spacing is large and small. In (c) switching occurs at an intermediate phase, and the amplification is less than in (a). In (d) the oscillation has steady amplitude, and there are two modes – slow (full line) and fast (broken line). The graph (e) shows schematically how the gain (amplitude ratio of successive swings) changes with tuning. On the circle the modes are phase-locked to the excitation; outside the range BB' they are not and their mean gain, over many cycles, is independent of the excitation.

Considered as a stability problem like those treated in the last chapter, with the value of (ξ, v) noted only once each cycle of the excitation, the solution in terms of two exponentials is seen to have the characteristics of saddle-point instability. If a large enough resistance is inserted in the circuit, the parametric excitation will prove inadequate to sustain oscillation, and the behaviour will be described by two decaying exponentials, such as characterize nodal stability. Another way of stopping oscillation, even if the resistance is not large enough to do so in a perfectly tuned system, is to detune the excitation. For small changes in excitation frequency the system will remain within the domain of saddle-point instability, its response being locked in phase with the excitation at exactly half the frequency, but as the detuning is increased a point will be reached where phase-locking cannot be maintained. At this stage, if the analogy with other systems is sound, one may expect a transition to nodal stability and thence to focal stability. This can be illustrated diagrammatically with the simplified model, though detailed analysis involves rather tedious algebra and will be omitted. When the switching frequency for the capacitor is not exactly $2\omega_0$ the oscillation may still be able to adjust itself to synchronism by suitable phasing, as in fig. 5(c), but the rate of growth or decay is reduced thereby. In the absence of losses, detuning may be carried to the point where steady oscillation can just persist in one of the modes shown in fig. 5(d). The switch-overs then occur at phase angles shifted by $\pi/4$ from those in (a) and (b); for the circuit to be made to follow the outer, rosette-shaped trajectory the switching speed must be lower than $2\omega_0$, and to follow the inner, cushion-shaped trajectory it must be higher. It is easy enough to calculate the time taken to execute one cycle of either, and to show that the tuning range for excitation is $2\omega_0(1 \pm 2\beta/\pi)$. Schematically the amplification per cycle varies with the excitation frequency as shown in fig. 5(e), which has been drawn so as to include a small amount of dissipation. Between A and A' an amplified mode exists and the system is unstable. In the narrow ranges AB and $A'B'$ there is nodal stability – the oscillation is phase-locked to the excitation so that sampling once per cycle of the latter reveals nothing of the oscillation but only the two modes of exponential decay of the amplitude. Beyond B and B' phase-locking fails, and sampling shows up an oscillatory decay to rest, i.e. focal stability. It would be a mistake to make too much of the distinction between nodal and focal stability. Mathematically the two types of critical point, A and B, are clearly marked; to the observer the difference between stability and instability is more important than the precise category to which the stability should be assigned.

The general features revealed by this model are substantiated by detailed calculation of the behaviour when the capacitance is varied sinusoidally. Since we shall ultimately compare the calculation with measurements on a pendulum, we shall analyse this case rather than the circuit with variable capacitor. The same method applies to both and the general behaviour of the solutions is very similar. The equation of motion of the pendulum of variable length is readily set up. If the length l is $l_0(1 + \beta \cos 2\Omega t)$, and the deflection θ is small, the tangential force on the bob is $-mg\theta - \lambda l\dot{\theta}$, λ being the viscous drag coefficient. Now a particle whose position is defined by l and θ has a tangential component of acceleration $l\ddot{\theta} + 2\dot{l}\dot{\theta}$; the equation of motion therefore takes the form

$$ml_0(1 + \beta \cos 2\Omega t)\ddot{\theta} - 4m\beta l_0 \sin 2\Omega t \cdot \dot{\theta} + \lambda l_0(1 + \beta \cos 2\Omega t)\dot{\theta} + mg\theta = 0.$$

$$(5)$$

We may restrict the analysis to systems of low loss, not too strongly excited, without losing any essential results; this allows second-order terms (e.g. $\lambda\beta$) to be neglected. If, further, we write θ as $y(t)\,e^{-\frac{1}{2}\lambda t/m}$, the intrinsic loss term disappears, and the simpler form results:

$$\ddot{y}(1+\beta\cos 2\Omega t)-4\beta\Omega\dot{y}\sin 2\Omega t+\omega_0^2 y = 0, \tag{6}$$

where $\omega_0^2 = g/l_0$. For comparison the same treatment converts (3) into the form

$$\ddot{y}+\omega_0^2 y(1+\beta\cos 2\Omega t)=0. \tag{7}$$

Both (6) and (7) are linear equations with coefficients that are periodic functions of t, having π/Ω as period. It is a general property of such equations (*Floquet's theorem*)$^{(2)}$ that their solutions can be cast in the form $e^{ipt}P(t)$, where p is a constant (possibly complex) and $P(t)$ is a periodic function with period π/Ω, i.e. having the property that $P(t+\pi/\Omega)=P(t)$. Since the equations are of second order, there will be two independent solutions of this form. We now express $P(t)$ as a Fourier series, writing y in the form

$$y=\sum_{n=-\infty}^{\infty} A_n\,e^{i(p+2n\Omega)t}. \tag{8}$$

On substituting in (6), and writing the trigonometrical terms in exponential form, we see that

$$\sum_{-\infty}^{\infty}\{-A_n(p+2n\Omega)^2(1+\tfrac{1}{2}\beta\,e^{2i\Omega t}+\tfrac{1}{2}\beta\,e^{-2i\Omega t})$$

$$-2\beta\Omega A_n(p+2n\Omega)\,(e^{2i\Omega t}-e^{-2i\Omega t})+\omega_0^2 A_n\}\,e^{2in\Omega t}=0.$$

This equation contains terms in $e^{2in\Omega t}$ with n taking all integral values, but by choosing a particular value of n, multiplying by $e^{-2in\Omega t}$ throughout and integrating over one period, π/Ω, everything can be made to disappear except the coefficient of the chosen $e^{2in\Omega t}$. It follows that the coefficient of each $e^{2in\Omega t}$ must vanish separately. Hence for all n,

$$[(p+2n\Omega)^2-\omega_0^2]A_n+\tfrac{1}{2}\beta[(p+2n\Omega)^2-4\Omega^2](A_{n-1}+A_{n+1})=0. \tag{9}$$

A recurrence relation involving three coefficients normally presents difficulties, and (9) is no exception. For a full discussion the reader should consult treatments of Mathieu's equation, of which (7) is an example.$^{(3)}$ It will be enough here to treat the case $\beta\ll 1$, for which simplifying approximations are available. If one arbitrarily fixes the ratio of two coefficients, A_1/A_0 say, the rest may be systematically calculated by repeated application of (9). As one proceeds to larger $|n|$ less error is occasioned by neglecting the difference between ω_0^2 and $4\Omega^2$ in the square brackets, and we can write the asymptotic form of (9):

$$A_n+\tfrac{1}{2}\beta(A_{n-1}+A_{n+1})\approx 0. \tag{10}$$

When the A_n follow a geometrical progression, $A_{n+1} = -cA_n$, (10) is satisfied, and when $\beta \ll 1$, c is either $\frac{1}{2}\beta$ or $2/\beta$. Although one may discover an initial value of A_1/A_0 that makes the values of A_n tend, as $n \to +\infty$, towards a series having $c = \frac{1}{2}\beta$ and converging to zero, in general one cannot expect the same choice to impose convergence as $n \to -\infty$. For convergence at both extremes the value of p must also be adjusted correctly. If this has somehow been achieved in the case where β is small, convergence is very rapid and only a few values of A_n will have any significant size; physically speaking, a very gently excited pendulum, with no damping, will execute nearly sinusoidal oscillations, possibly growing or decaying in amplitude, possibly showing beats between its free and its driven vibrational modes, but free from such irregularities of waveform that announce the presence of high Fourier terms.

With this in mind we concentrate attention on the behaviour of (9) when n is small. For example, when $n = -1$,

$$[(p-2\Omega)^2 - \omega_0^2]A_{-1} + \tfrac{1}{2}\beta[(p-2\Omega)^2 - 4\Omega^2](A_{-2} + A_0) = 0. \quad (11)$$

Now if the excitation is very nearly tuned to twice the natural frequency, $\omega_0 \sim \Omega$, and if $p \sim \omega_0 \sim \Omega$, the coefficient of A_{-1} is much smaller than that of $(A_{-2} + A_0)$; in spite, therefore, of the smallness of β we may expect A_{-1} to be sizeable. Putting $n = -2$ in (9) shows, on the other hand, that A_{-2} is much smaller than $\frac{1}{2}(A_{-1} + A_{-3})$ and can be neglected in (11) without much harm. Similarly by putting $n = 1$ we can see that A_1 and terms above this are of minor importance, so that the solution is dominated by A_0 and A_{-1}. If all other terms are assumed to vanish, only two equations of the type (9) retain any interest, those that result from putting $n = -1$ and 0:

$$[(p-2\Omega)^2 - \omega_0^2]A_{-1} + \tfrac{1}{2}\beta[(p-2\Omega)^2 - 4\Omega^2]A_0 = 0, \quad (12)$$

and
$$[p^2 - \omega_0^2]A_0 + \tfrac{1}{2}\beta[p^2 - 4\Omega^2]A_{-1} = 0. \quad (13)$$

Here then is the explicit demonstration that arbitrary values of p are not permitted – (12) and (13) together demand that

$$[p^2 - \omega_0^2][(p-2\Omega)^2 - \omega_0^2] = \tfrac{1}{4}\beta^2[p^2 - 4\Omega^2][(p-2\Omega)^2 - 4\Omega^2]. \quad (14)$$

This quartic equation looks less formidable when we realize that the square brackets on the right are very close to $-3\Omega^2$ each, and that if p, ω_0 and Ω are nearly equal we may replace $(p^2 - \omega_0^2)$ by $2\omega_0(p - \omega_0)$ and $[(p-2\Omega)^2 - \omega_0^2]$ by $2\Omega(2\Omega - p - \omega_0)$. Let us take Δ, defined as $(\Omega - \omega_0)/\omega_0$, to measure the extent to which the parametric excitation is not tuned exactly to $2\omega_0$, and ε, defined as $(p - \omega_0)/\omega_0$ to measure the relative change in frequency caused by the parametric excitation. On making these and other equivalent approximations, after substituting the new variables, we arrive at a more transparent expression for the behaviour of the system:

$$\varepsilon(2\Delta - \varepsilon) = \Delta_c^2, \quad \text{where } \Delta_c = \tfrac{3}{4}\beta. \quad (15)$$

This has the solution

$$\varepsilon = \Delta \pm (\Delta^2 - \Delta_c^2)^{\frac{1}{2}},$$

i.e.
$$p = \Omega \pm \omega_0 (\Delta^2 - \Delta_c^2)^{\frac{1}{2}}. \tag{16}$$

If the excitation is perfectly tuned ($\Omega = \omega_0$ and $\Delta = 0$)

$$p = \Omega \pm i\omega_0 \Delta_c, \tag{17}$$

and the full solution is the superposition of the oscillations corresponding to both choices of p. From (13) it follows, with the same approximations as led to (15), that $A_{-1}/A_0 = \varepsilon/\Delta_c = i$ if the positive sign is chosen in (17), $-i$ if the negative sign is chosen. The character of the oscillation follows from (8):

$$y = A_0 e^{\mp \omega_0 \Delta_c t} (e^{i\Omega t} \pm i e^{-i\Omega t}),$$

and A_0 must be chosen so as to make y real, i.e. A_0 must be a real constant times $e^{\mp i\pi/4}$. Hence

$$y = C e^{\mp \omega_0 \Delta_c t} \cos (\Omega t \mp \tfrac{1}{4}\pi). \tag{18}$$

The physical meaning is immediately clear; the pendulum, whose length has been taken to vary as $l_0(1 + \beta \cos 2\Omega t)$, is being shortened most rapidly when $\Omega t = \pi/4$, $5\pi/4$ etc. If the lower signs are chosen in (18) these are the moments when y vanishes and the tension is greatest, while with the upper signs y is at its peak as the tension is least. The former choice leads, as expected, to growth of the oscillation, the latter to decay. All this, of course, is in addition to the natural decay $e^{-\frac{1}{2}\lambda t/m}$ which was removed by the change of variables leading to (6). The full expression for the swing of the pendulum takes the form

$$\theta(t) = C_1 e^{-(\frac{1}{2}k + \omega_0 \Delta_c)t} \cos (\Omega t - \tfrac{1}{4}\pi) + C_2 e^{-(\frac{1}{2}k - \omega_0 \Delta_c)t} \cos (\Omega t + \tfrac{1}{4}\pi), \tag{19}$$

in which $k = \lambda/m$, and C_1 and C_2 are arbitrary real constants. The two independent solutions represent oscillations locked in different phases to the excitation, the second being locked so that energy is fed in and the natural decay counteracted, the first being locked so as to enhance the natural decay. However, the pendulum may have been swung initially, and this is what determines C_1 and C_2, after a long enough interval the second term will dominate the behaviour; and if β exceeds the critical value $\frac{2}{3}k/\omega_0$, i.e. $\frac{2}{3}/Q$, the second term represents a growing oscillation. This is the same result as obtained at the outset by energy considerations.

The word 'locked' in the last paragraph is perhaps not justified, since with $\Delta = 0$ there is no inherent tendency for the pendulum to change phase relative to the excitation. It is when we try to excite the pendulum with a frequency other than $2\omega_0$ that the question of locking becomes significant. As (16) shows, when the detuning is not excessive (to be precise, when $|\Delta| < \Delta_c$), the solution (17), for $\Delta = 0$, is modified in the imaginary part only, by replacing Δ_c by $(\Delta_c^2 - \Delta^2)^{\frac{1}{2}}$. The full solution corresponding to (19) can be derived in the same way without difficulty:

$$\theta(t) = C_1\,e^{\alpha_1 t}\cos\,(\Omega t - \gamma) + C_2\,e^{\alpha_2 t}\cos\,(\Omega t + \gamma), \qquad (20)$$

where $\alpha_{1,2} = -\frac{1}{2}k \mp \omega_0(\Delta_c^2 - \Delta^2)^{\frac{1}{2}}$ and $\tan 2\gamma = (\Delta_c^2/\Delta^2 - 1)^{\frac{1}{2}}$. The oscillations are still locked to the excitation, but at phases that are not optimal for transferring energy to or from the pendulum; the modifications of the natural decrement are correspondingly less. This agrees with the simple picture presented in fig. 5. Amplification sets in only when the excitation is rather larger than is required in the state of perfect tuning; the critical value of β is $\frac{2}{3}(1 + 4\Delta^2 Q^2)^{\frac{1}{2}}/Q$ instead of $\frac{2}{3}/Q$. We now appreciate how carefully the system must be tuned to obtain quantitative agreement between theory and experiment. The pendulum mentioned earlier, having $Q \sim 2500$, needs its parametric excitation to be better tuned than 1 in 5000 (i.e. $|\Delta| < 1/5000$) if the measured critical amplitude for spontaneous oscillation is to be within $\sqrt{2}$ of the theoretical minimum. This analysis also explains why it is hard to set the system up so that the rate of growth is a linear function of excitation; α_1 and α_2 are linear in Δ_c, and therefore in β, only when the excitation is perfectly tuned.

If the amplitude of excitation is held constant while the frequency is progressively detuned, two critical points are passed. The first marks the transition from saddle-point instability to nodal stability; when $\omega_0(\Delta_c^2 - \Delta^2)^{\frac{1}{2}}$ is less than $\frac{1}{2}k$ (i.e. $\Delta^2 > \Delta_c^2 - \frac{1}{4}/Q^2$) the gain due to parametric excitation is too small to compensate the intrinsic loss. The second critical point occurs when Δ reaches the value Δ_c and, as (16) shows, p becomes real. Instead of the response being locked in frequency and suffering exponential gain or decay of amplitude, frequency locking now breaks down and, apart from the intrinsic loss, the amplitude is constant. This is the transition from nodal to focal stability. The behaviour when Δ exceeds Δ_c deserves elucidation, for it is not simply an arbitrary superposition of two oscillatory modes – the requirement that y shall be real ensures that the two modes coexist in a well-defined relationship.

Let us first define quantities $\delta \equiv \omega(\Delta^2 - \Delta_c^2)^{\frac{1}{2}}$ and $X \equiv \Delta/\Delta_c + (\Delta^2/\Delta_c^2 - 1)^{\frac{1}{2}}$, and note that $\Delta/\Delta_c - (\Delta^2/\Delta_c^2 - 1)^{\frac{1}{2}} = 1/X$. From (16) and (13) it follows that $p = \Omega \pm \delta$, with $A_{-1}/A_0 = X$ for the positive sign, and $1/X$ for the negative sign. The general solution then takes the form

$$y = A_0'\,e^{i\delta t}(e^{i\Omega t} + X\,e^{-i\Omega t}) + A_0''\,e^{-i\delta t}\!\left(e^{i\Omega t} + \frac{1}{X}e^{-i\Omega t}\right),$$

in which A_0' and A_0'' are arbitrary complex constants. It is convenient, for the purpose of writing y in its most general real form, to pair the first term in the first bracket with the second term in the second bracket, and take the other two terms as a second pair:

$$y = [A_0'\,e^{i(\Omega+\delta)t} + (A_0''/X)\,e^{-i(\Omega+\delta)t}] + [A_0''\,e^{i(\Omega-\delta)t} + XA_0'\,e^{-i(\Omega-\delta)t}]$$

If now we put $A_0' = C\,e^{i\zeta}$ and $A_0'' = XC\,e^{-i\zeta}$, where C and ζ are arbitrary real constants, y assumes a real form with the required complement of two independent constants:

$$y = C\{\cos[(\Omega+\delta)t+\zeta] + X \cos[(\Omega-\delta)t-\zeta]\}, \qquad (21)$$

and consequently:

$$\theta = C e^{-\frac{1}{2}kt}\{\cos[(\Omega+\delta)t+\zeta] + X \cos[(\Omega-\delta)t-\zeta]\}. \qquad (22)$$

The requirement that y and θ be real prevents any single oscillatory mode occurring alone; the characteristic response (22) consists of two beating sinusoidal oscillations compounded with an exponential decay. The term in brackets has a beat period of $\frac{1}{2}\Omega/\delta$ swings between successive maxima, and the ratio of maxima to minima is $(X+1)/(X-1)$. In consequence, the amplitude may not decrease exponentially but may show a periodic rise and fall imposed on a steady decrease, a natural consequence of the inability of the excitation to lock the pendulum to exactly half its frequency; for, as time goes on, the phase of the pendulum drifts steadily with respect to the excitation, so that it passes through periods of being helped and being hindered by the parametric amplification, but with no mitigation in the long run of the intrinsic decay.

It may appear at first sight that this beating character, which may have a long period, is inconsistent with the relatively simple approach to equilibrium described as focal stability. It should be remembered, however, that the characteristic parameters, $(\theta, \dot\theta)$ say, are not observed continuously but only at regular intervals dictated by the period of the excitation. All that happens in between reveals the beats, but the moments of observation pick up a steady progression of discrete values $(\theta, \dot\theta)$ round a converging elliptical spiral. This is an immediate consequence of Floquet's theorem, as expressed in (8); successive values of y (and hence of θ), taken at intervals of π/Ω, reflect the geometric progression by a factor $e^{i\pi p/\Omega}$ at each step, but altogether miss the interesting structure described by the Fourier series. Of course there are two values of p, but direct substitution of $n\pi/\Omega$ for t in (21) shows that both terms in the braces oscillate as $(-1)^n \cos(n\pi\delta/\Omega\pm\zeta)$, having the same frequency and differing only in amplitude and phase, so that they compound to give a sequence of points lying evenly spaced on a simple sinusoid. The same holds for $\dot\theta$ as for θ, and the result is, as stated above, a sequence of points converging on a spiral path. This system, with its apparently more complicated behaviour, lies well within the stability formalism of the last chapter. An example is shown in fig. 6.

In an experiment designed to illustrate the foregoing analysis, the pendulum of fig. 1 was modified by replacing the brass bob by one of plastic which, being lighter, conferred a lower Q on the system. The pendulum was not of the correct length for exact tuning, and with a rather weak excitation amplitude a_0 of 3.9 mm ($\beta = 4.7\times10^{-3}$, or $\Delta_c = 3.5\times10^{-3}$) its swing, after being set going, varied in the way shown in fig. 7(a). The logarithmic plot 7(b) reveals a steady beat pattern imposed on an exponential decay which corresponds, as one would expect from (20), with

Fig. 6. Phase-plane
trajectory for a typical
solution of (21).
Observations made at the
same phase in successive
cycles of the excitation lie
on an ellipse, as indicated
by the marked points.

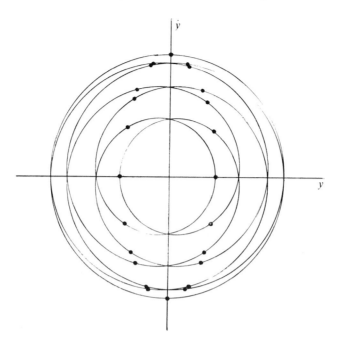

the natural decay when there is no parametric excitation. To see how well this curve agrees with (22), we shall use it to evaluate the relevant parameters. Since the pendulum makes 103 swings during one beat period, δ/ω_0 is $1/206$; as this is $(\Delta^2 - \Delta_c^2)^{\frac{1}{2}}$ and $\Delta_c = 3.5 \times 10^{-3}$, we find $(\Delta^2/\Delta_c^2 - 1)^{\frac{1}{2}}$ to be 1.39 and Δ/Δ_c to be 1.71. Hence $X = 3.1$ and the ratio of maxima to minima in the beat pattern should be 4.1/2.1 or 1.95. The straight lines in fig. 7(b) are drawn to exhibit this ratio, and the comparison between theory and experiment leaves one in little doubt of the essential correctness of the theory.

By way of contrast, when the amplitude of excitation was increased above the critical value and the pendulum set swinging in an unfavourable phase, the amplitude initially decreased and then rose again, as in fig. 8, without any superimposed oscillations, since both modes were now locked to the excitation. The asymmetry of the curve is obvious; in the initial stages, dominated by the first term of (20), the excitation enhances the natural decrement and the fall is more rapid than the subsequent rise, dominated by the second term, where the natural decrement partially counteracts the parametric amplification.

To take a last look at the simple pendulum, let us remove the string tying it back at B (fig. 1), and so convert it into a conical pendulum, which is then set swinging in a circular orbit. The tension in the string now remains constant, so that no net energy is communicated to the pendulum by periodically changing its length. Moreover there are no tangential forces on the bob and in the absence of friction its angular momentum must stay

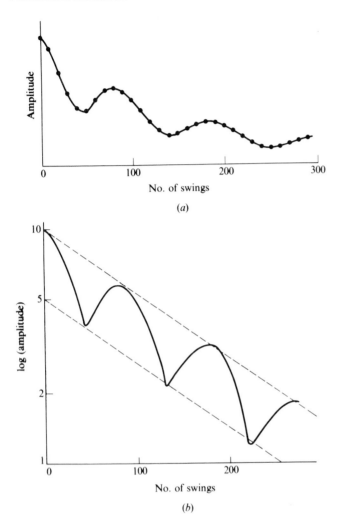

Fig. 7. Variation of amplitude of pendulum of fig. 1 when excitation is detuned so far as to prevent phase-locking; (*a*) direct observations, (*b*) logarithmic plot.

constant. It is tempting to think that no change could occur in the motion of the pendulum, but this is not so. As the bob is raised, energy is communicated to the pendulum, to be removed once more as the bob is lowered, and if this process is tuned to twice the orbital frequency, or near enough, a progressive deformation of the orbit occurs. The circular motion may be analysed into two linear harmonic oscillations, one phased so as to grow in purely exponential fashion and the other so as to decay exponentially, until in the end the motion is almost confined to the plane in which the phase is most favourable to growth. Since this means separating the two terms in (20), which are not necessarily in phase quadrature, the two linear components are not at right angles nor do they vibrate in phase. There is no need to work out precisely what combination is needed to synthesize circular motion since the conservation of angular momentum, which we

Fig. 8. As in fig. 7 but
with excitation increased
to allow phase-locking and
parametric amplification.

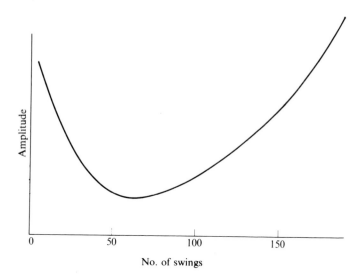

desire to verify, does not depend on such detail. Let us start with two linear vibrations of amplitude A and B with an angle θ between them; treating the horizontal plane as a complex plane the displacement corresponding to the sum of the two motions is given by

$$r = A \cos \omega t + B\, e^{i\theta} \cos (\omega t + \phi), \tag{23}$$

ϕ being the phase difference between the vibrations. The angular velocity is the imaginary part of $r\dot{r}^*$, which evaluates to $AB\omega \sin \theta \sin \phi$. Now let the first term in (23) grow by a factor $e^{\alpha t}$ while the second decays by the same factor (note that if $k = 0$, $\alpha_1 = -\alpha_2$ in (20)). Then Im $[r\dot{r}^*]$ is unaffected, which shows that angular velocity is conserved. As for the total energy, which is proportional to $(rr^* + \dot{r}\dot{r}^*/\omega^2)$, this on evaluation is found to vary as $\cos \phi \cos \theta + \cosh 2\alpha t$, if t is taken as 0 at the moment when $A = B$. Since the derivative of $\cosh 2\alpha t$ vanishes when $t = 0$, the energy of a pendulum in circular orbit is indeed momentarily unchanged by the excitation, but as the orbit becomes elliptical the energy grows steadily. Observation confirms this analysis.

The pendulum, by virtue of its slow vibration, is easy to observe though tedious if much information is to be gathered. A more rapidly vibrating system which is readily excited parametrically is a string under tension, as in fig. 3. This can be realized by use of the arrangement shown in fig. 9.1, the impeller now being placed above the end to excite the string longitudinally. While it is not so easy with this as with the pendulum to verify the theory quantitatively, it is particularly well suited to illustrate how the process of parametric excitation is limited by non-linearity. The behaviour of a strictly linear system, as expressed in (19) and (20), for example, must continue as it started, growing exponentially without limit. A real system, however, will saturate either because the driver cannot deliver unlimited

249

power or, before this limit is reached, because non-linearity of one sort or another reduces the rate of growth to zero. With the vibrating string, increased air loss and shift of resonant frequency with amplitude both play a part, as is illustrated by the experiment now to be described.

The impeller is set to vibrate vigorously, and the frequency slowly raised; parametric excitation starts as a well-defined frequency, and within a second or two (the natural decay time for the string) the amplitude stabilizes. With further increase of frequency the amplitude rises, as is shown in fig. 9, until suddenly the oscillation collapses entirely and the string comes to rest. When the frequency is lowered, nothing happens (unless the string is violently plucked) until the point of collapse is well passed but there is a lower frequency at which oscillation begins again spontaneously, and with further lowering of the frequency the original curve is retraced. With smaller excitation a similar curve is traced on a reduced scale.

Fig. 9. The wire of fig. 9.1 excited parametrically.

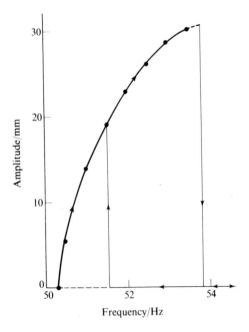

Frequency/Hz

The explanation of the shape of the curve follows naturally from the anharmonic behaviour exhibited and discussed in chapter 9 (see fig. 9.2). **250**
The increase of tension as the string is displaced is the dominant mechanism for parametric excitation, easily overcoming the $\cos\theta$ reduction in the longitudinal component of tension at the point of excitation. Yet the motion is nearly enough harmonic for the assumption to be made that the linear theory need only be modified by allowing ω_0 to vary slightly with amplitude. Now the detuning Δ_1, at which the vibration need neither grow nor decay is given by $(\Delta_c^2 - \frac{1}{4}/Q^2)^{\frac{1}{2}}$, independent of amplitude if the excitation is kept constant (i.e. Δ_c fixed) and if Q does not change. We therefore

draw two lines on an amplitude–frequency graph, separated in frequency by $\pm\Delta_1$ from the line of the resonant frequency, to show the conditions in which vibration at a steady amplitude can occur (fig. 10). The lower line, however, is unstable since a small increase in amplitude brings the system to a point where the amplitude increases still further. If this were the whole story, as the frequency was raised nothing would happen until ω_1 was reached, when parametric excitation would begin to be effective. With further increase of frequency the amplitude would grow, following curve A. There would be no reason for the sudden collapse shown in fig. 9. This must arise from the lowering of Q at large amplitudes, already discussed in chapter 9, which diminishes the critical detuning and brings curves A and B together, as shown by the broken curve. The sequence of events shown in fig. 9 is now clear.

252

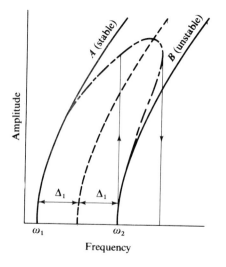

Fig. 10. Illustrating the explanation of fig. 9.

If this interpretation is correct, the observed frequency of 53.5 Hz just below the point of collapse must be very close to twice the natural frequency of the string when its amplitude is 30 mm, as against 50.9 Hz at low amplitudes. On the assumption that the change of natural frequency is proportional to the square of the amplitude, at 17.5 mm it should be 0.44 Hz, which is very close to what can be inferred from the results in fig. 9.2.

Parametric amplifiers[4]

The periodic variation of a circuit element, a capacitor for instance, produces distortion of any oscillatory current already present, generating sum and difference frequencies which may excite resonant response in a suitable tuned circuit. It is this effect which is employed in parametric amplifiers, particularly at the microwave frequencies used in satellite

communication and radio astronomy, where incoming signals are very
weak and easily swamped by extra noise generated at the first stage of
amplification. At this stage what is needed above all is a modest degree of
amplification with minimal noise; from then on the signal strength will be
sufficiently far above the noise level of conventional amplifiers for the
danger of losing information to be past. It is the merit of parametric
amplifiers that their circuit elements can be cooled, so that Johnson noise is
low, and that there are no elements generating shot noise. Precisely how
little the noise can be, and how to minimize it, are matters that have been
extensively studied. Here, however, we shall confine attention to the one
basic question of how the device works. It is unnecessary to go into details
about the physical processes underlying the operation of the *varactor*, the
variable capacitor which lies at the heart of the amplifier.[5] Let it suffice to
remark that it depends on the *depletion layer* at the junction of a semiconducting p–n rectifier; when it is reverse-biased only a very small current
flows through the rectifier, since between the regions of p-type and n-type
material is a thin layer from which virtually all mobile electrons have been
swept by the applied field. It is this layer that acts as the dielectric of the
capacitor, and whose thickness can be changed by changing the applied
field. When the varactor, which is a tiny semiconductor chip, is mounted in
a microwave cavity shaped so as to concentrate the electric field across
it, its capacitance can then be modulated (*pumped*) at the resonance
frequency of the cavity, and the range of modulation controlled by adjusting the pumping power.

In fig. 11 the input circuit on the left is supposed resonant or in some
other way isolated by filters so that its impedance is prohibitively large at

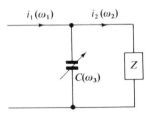

Fig. 11. Currents at
different frequencies in a
parametric amplifier
circuit.

any other frequency than ω_1. In the same way the *idler* circuit on the right is
resonant near ω_2 and is unaffected by anything happening elsewhere at ω_1
(input) or ω_3 (the pumping frequency). Now let the currents and capacitance
vary as follows:

$$
\left.
\begin{aligned}
i_1 &= a_1 \cos (\omega_1 t + \phi_1), \\
i_2 &= a_2 \cos (\omega_2 t + \phi_2), \\
1/C &= [1 + \beta \cos (\omega_3 t + \phi_3)]/C_0.
\end{aligned}
\right\}
\qquad (24)
$$

The current flowing to the capacitor is $i_1 - i_2$, from which the charge follows by integration and hence the potential difference across it:

$$C_0 V = [1 + \beta \cos (\omega_3 t + \phi_3)] \left[\frac{a_1}{\omega_1} \sin (\omega_1 t + \phi_1) - \frac{a_2}{\omega_2} \sin (\omega_2 t + \phi_2) \right]$$

$$= \frac{a_1}{\omega_1} \sin (\omega_1 t + \phi_1) - \frac{a_2}{\omega_2} \sin (\omega_2 t + \phi_2)$$

$$+ \frac{\beta a_1}{2\omega_1} \{ \sin [(\omega_1 + \omega_3)t + \phi_1 + \phi_3] - \sin [(\omega_1 - \omega_3)t + \phi_1 - \phi_3] \}$$

$$- \frac{\beta a_2}{2\omega_2} \{ \sin [(\omega_2 + \omega_3)t + \phi_2 + \phi_3] - \sin [(\omega_2 - \omega_3)t + \phi_2 - \phi_3] \}. \quad (25)$$

This contains terms in ω_1, ω_2 and the combination frequencies $\omega_1 \pm \omega_3$ and $\omega_2 \pm \omega_3$. When the idler circuit is tuned so that $\omega_2 = \omega_3 - \omega_1$, four separate frequencies are present in V: ω_1, ω_2, $2\omega_1 + \omega_2$, $2\omega_2 + \omega_1$, of which the last two generate no current in the sharply tuned circuits.† In fact the only terms relevant to the individual circuits are the following:

$$\text{circuit 1:} \quad C_0 V_1 = \frac{a_1}{\omega_1} \sin (\omega_1 t + \phi_1) - \frac{\beta a_2}{2\omega_2} \sin (\omega_1 t + \phi_3 - \phi_2). \quad (26)$$

$$\text{circuit 2:} \quad C_0 V_2 = - \frac{a_2}{\omega_2} \sin (\omega_2 t + \phi_2) + \frac{\beta a_1}{2\omega_1} \sin (\omega_2 t + \phi_3 - \phi_1). \quad (27)$$

It is only by virtue of V_2 that any current i_2 flows in the idler circuit. If its resistance is R_2 and reactance X_2 the required form of V_2 is given by

$$V_2 = R_2 a_2 \cos (\omega_2 t + \phi_2) - X_2 a_2 \sin (\omega_2 t + \phi_2). \quad (28)$$

Since (27) and (28) must be identical the amplitude and phase of i_2 take such values that

$$\frac{\beta a_1}{2a_2} \sin \phi = \omega_1 C_0 R_2 \quad \text{and} \quad \frac{\beta a_1}{2a_2} \cos \phi = \frac{\omega_1}{\omega_2} (1 - \omega_2 C_0 X_2), \quad (29)$$

where $\phi = \phi_3 - \phi_2 - \phi_1$. When the idler circuit, including the varactor in its average state, is exactly tuned to resonance, $X_2 - 1/\omega_2 C_0 = 0$, and therefore $\cos \phi = 0$; then

$$a_2 / a_1 = \beta / 2\omega_1 C_0 R_2. \quad (30)$$

Off resonance a_2 / a_1 is less and (29) shows, not unexpectedly, that a significant response is only to be obtained if the idler circuit is tuned so that $\omega_3 - \omega_1$ (or $\omega_1 + \omega_3$) lies within its resonant peak.

---

† If the idler circuit is tuned so that $\omega_2 = \omega_1 + \omega_3$ the behaviour is very similar, and we shall not examine this case separately.

To evaluate the input impedance of the circuit at frequency ω_1, we note that with the idler on resonance (the only case we shall treat) (26) takes the form

$$C_0 V_1 = \frac{a_1}{\omega_1} \sin(\omega_1 t + \phi_1) - \frac{\beta a_2}{2\omega_2} \cos(\omega_1 t + \phi_1)$$

$$= \frac{a_1}{\omega_1} \sin(\omega_1 t + \phi_1) - \frac{\beta^2 a_1}{4\omega_1 \omega_2 C_0 R_2} \cos(\omega_1 t + \phi_1), \quad \text{from (30).}$$

The first term is in phase quadrature with the input current $a_1 \sin(\omega_1 t + \phi_1)$ and expresses the reactive part of the input impedance, which can be tuned out by adding an opposite reactance in series – this is just the operation of adjusting the input circuit to resonance. When this is done the circuit appears as a negative resistance

$$R_{\text{in}} = -\beta^2 / 4\omega_1 \omega_2 C_0^2 R_2. \tag{31}$$

The negative resistance allows a larger voltage signal to be extracted than is present in the input. With the circuit of fig. 12 an input voltage V_{in} from a source of resistance R_1 (e.g. a transmission line of characteristic impedance R_1) produces a current $V_{\text{in}}/(R_1 + R_{\text{in}})$. Across the capacitor there is therefore a voltage V_1, at ω_1, of amplitude $V_{\text{in}} R_{\text{in}}/(R_1 + R_{\text{in}})$. The voltage amplification is $R_{\text{in}}/(R_1 + R_{\text{in}})$, which can be made equal to 10, say, giving a power gain of 20 dB, by adjusting $|R_{\text{in}}|$ to the value $\frac{10}{11} R_1$. If $|R_{\text{in}}|$ exceeds R_1 the circuit is unstable, and oscillates spontaneously even without an input signal. This is another form of parametric excitation of an oscillator, with two resonant systems excited simultaneously by an input at the sum (or difference) frequency, and is clearly to be avoided in an amplifier. The more precisely the pumping power can be controlled, the nearer can β be held to the critical value for instability, and the amplification benefits correspondingly.

Fig. 12. Equivalent circuit of tuned parametric amplifier, fed by a transmission line of characteristic impedance R_1.

An amplified signal may also be extracted at frequency ω_2. The voltage across R_2 is $R_2 a_2$, i.e. $\beta a_1 / 2\omega_1 C_0$ from (30), when the idler circuit is well tuned. Hence

$$V_2 / V_{\text{in}} = \beta / 2\omega_1 C_0 (R_1 + R_{\text{in}}),$$

and this is the voltage amplification for this arrangement. To compare the two arrangements, note that

$$V_2 / V_1 = -\beta / 2\omega_1 C_0 R_{\text{in}},$$

and that β is adjusted so that $-R_{in}$ is very nearly R_1. Hence

$$V_2/V_1 \sim (\omega_2 R_2/\omega_1 R_1)^{\frac{1}{2}}.$$

Whether it is advantageous to pick off V_2 or V_1 for the next stage of amplification is a technical matter to be determined by the particular application. The analysis so far presented offers little guidance.

The design of real parametric amplifiers has progressed far beyond this point, and extended to wave–propagation devices in which parametric pumping causes progressive amplification as the wave proceeds. These are advanced matters which would be inappropriate here.

11 Maintained oscillators

This chapter is concerned only with simple systems in steady oscillation, such as were classified in chapter 2 under the heading of negative resistance **39** devices, feedback oscillators and relaxation oscillators. Here we shall attempt to refine the description and classification of the different types though, as is common in such attempts, firm divisions are hard to find. At the same time we shall analyse a number of examples so as to understand what conditions must be satisfied for them to oscillate spontaneously, how the amplitude of oscillation is limited by non-linearity, and what determines the ultimate waveform. No attempt will be made to establish rigorously the general conditions for oscillation to occur. This is an important and well-studied problem, but one which deserves the fuller treatment that will be found in specialized texts.

It has already been remarked, in chapter 2 and elsewhere, that a resonant system governed by a second-order equation such as (2.23) will **30** oscillate spontaneously if k is negative. A source of energy is required to overcome inevitable dissipative effects, and among the many examples of how the energy may be injected probably the commonest is by feedback. Let us start, then, with a general survey of the feedback principle, with particular reference to the influence feedback may have on the performance of a resonant system, and not solely in setting it in spontaneous oscillation. The argument will be conducted in terms of electrical circuits, which account for the overwhelming majority of applications.

The feedback principle

Baldly stated, feedback consists of taking a signal from a system, subjecting it to some processing (e.g. amplification, filtering) and injecting the result back into the system. Since the injected signal combines with what is already present, it has an influence on what is extracted for processing, and in this sense feedback establishes a cyclical flow pattern. This view of the matter is not, however, especially useful or, indeed, always valid. It is usually better to think of the electrical circuit, or mechanical system (or even the economic or social system to which the concept is also usefully applied) as having cross-links between different elements and as behaving in the only way that is consistent with the relationships that are dictated by the linkages. The idea of a sequence of events – extraction, processing, reinjection – derives from the non-reciprocal character of an amplifier, whose output is its response to its input, but not vice versa. It is thus natural to follow the process as if it were a time sequence, and it may be helpful to do so if the processing introduces a time delay. But the advantage is usually slight, and we shall have little occasion to discuss problems in these terms.

Consider a passive LCR circuit, modified so that the e.m.f. between any two points can be picked up and fed back, after amplification, across any two other points. As shown in fig. 1(a), pick-up and feedback are both across resistors, but the arrangement may be more general, as in fig. 1(b). Fig. 1(c) shows the idealized form of amplifier that is assumed, an input ε generating $A\varepsilon$ at the output stage. The input resistance R_{in} is supposed large enough to prevent any significant current being diverted from the circuit; the resistance R_{out} is similarly taken to be much larger than any impedance in the circuit, so that the current fed back, $A\varepsilon/R_{out}$ $(=g\varepsilon)$, is proportional to the input voltage. In addition, the input and output stages are considered to be insulated from one another, so that, among other things, there is freedom to reverse the sign of g by reversing the output connections; this is not impossible with modern devices, but is unusual and not to be taken for granted in any actual example – indeed, some of the difficulties experienced by electronic designers in the days of thermionic tubes sprang directly from the electrical connection between stages. So long as we are concerned with basic principle, however, we may disregard the practical problem, returning to reality later with a clearer idea of the distinction between essential features of the behaviour and those arising from practical limitations.

Fig. 1. (a) The e.m.f. across a resistor in a resonant circuit is picked up, amplified and fed back through another resistor; (b) generalized form of (a); (c) equivalent circuit of the amplifier.

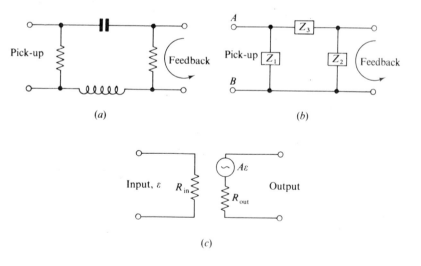

To analyse the circuit of fig. 1(b) imagine the currents i_1 and i_2 to have complex frequency ω. If the impedance between the tapping points AB is $Z_1(\omega)$, $i_2 = gZ_1i_1$ and the extra e.m.f. across the resistor, due to i_2, is $gZ_1Z_2i_1$. The effect of feedback is therefore to add another term to the equation for the circuit. Whereas without feedback the natural frequency is that at which Z_t, defined as $Z_1 + Z_2 + Z_3$, vanishes, with feedback the equation takes the form

$$Z_t - gZ_1Z_2 = 0. \qquad (1)$$

For the circuit of fig. 1(c), the additional impedance is resistive, gR_1R_2. With the amplifier connected so that g is negative, the extra resistance is positive and the resonant properties of the circuit are damped (*negative feedback*). With the connections reversed, however, to make g positive (*positive feedback*) R_1 and R_2 may be chosen so that the total resistance in the circuit becomes negative, and the circuit goes into spontaneous oscillation, with an amplitude that rises exponentially until saturation of the amplifier lowers the effective value of g and prevents further growth. Systems of this sort will occupy much of the space in this chapter, and before becoming involved in them it is as well to appreciate that positive and negative feedback are not the only choices possible. One may, for example, choose a capacitor for Z_1 and feed back through a resistor, in which case $Z_1 = 1/j\omega C_1$ and the extra impedance is reactive, being equivalent to a capacitance C_1/gR_2. The resonant frequency, rather than Q, is now the property directly affected.

A rather early and ingenious realization of this sort of feedback, which shows how the restoring force constant in a resonant system can be manipulated to advantage, is the galvanometer amplifier invented and developed by Gall.[1] Light reflected from a galvanometer mirror is focussed on a photocell, which is part of a thyratron circuit (fig. 2(c)) running directly off the a.c. mains, without rectification. The details of this very economical circuit are left to the interested reader to work out or learn from the original account; it is enough to know that the thyratron passes no current so long as the photocell is in the dark, but that as soon as illumination of the photocell causes it to conduct, the phase relationships of the grid and anode of the thyratron are disturbed so that it passes a heavy current pulse every cycle. This is smoothed and fed back to the input circuit of the galvanometer in either of the ways shown in fig. 2(a) and (b).† In (a) the feedback current i_f sets up a potential difference across R_1 that adds to that across G, so that the input p.d. for a given current input is considerably increased and the system has large input resistance. In (b) the feedback current through G is opposed to the input current, so that the p.d. across G is reduced and the input resistance reduced by the same factor. In different circumstances both arrangements have merit, the former where a sensitive voltmeter is required, the latter when the need is for an ammeter. The feedback current, being an amplification of the input, and rather accurately proportional to it,

---

† It will be observed that the use of an optical lever enables the output and input stages of the amplifier to be electrically isolated. Much more recently 'cal stages have been incorporated into electronic circuits for the same purpose, a light-emitting diode serving as transmitter of information which is picked up by a phototransistor, all built into the same module.

Fig. 2. (*a*) and (*b*) are two forms of feedback network for the galvanometer amplifier. Current injected through the input terminals deflects the galvanometer, and the amplifier injects the feedback current through the output terminals; the latter current is measured by the ammeter A and is proportional to the input current. (*c*) is the thyratron amplifier circuit (D. C. Gall[1]).

as we shall see when working through the circuit in detail, provides a very sensitive measure of the input. These matters, however, while explaining the utility of the device, are only incidental to the present discussion which is concerned with the effect of feedback on the dynamical behaviour of the suspended coil, a mechanical analogue of a resonant circuit. It will become clear that the changes due to feedback are wholly beneficial – the instrument is indeed a rare (if minor) example of an invention with virtually no disadvantages.

To develop a quantitative theory, we define the overall amplification α as the ratio of feedback current to current through the galvanometer; this may be 10^5. To consider only the arrangement 2(*b*), the input current i_{in} and feedback current i_{f} are accompanied by a current i' in the feedback loop; application of Kirchhoff's law to this loop shows that

$$(G + R_2 + R_3)i' - Gi_{\text{in}} - R_2 i_{\text{f}} = 0.$$

Moreover, the current through the galvanometer is $i_{\text{in}} - i'$, and therefore

$$i_{\text{f}} = \alpha(i_{\text{in}} - i').$$

Elimination of i' between these two equations yields the result:

$$(R_2 + R_3)i_{in} = \left(R_2 + \frac{R_2 + R_3 + G}{\alpha} \right)i_f,$$

or $i_f \doteqdot \beta i_{in}, \quad \text{where } \beta = 1 + R_3/R_2,$ (2)

if α is large enough to make the last term on the right negligible.

This equation expresses the essential result that makes the feedback principle so attractive. With a very large amplification factor, α, it is possible to choose β to be large and still have i_f/i_{in} sensibly independent of α. There is then no need to worry about the stability of the amplifier; a small input current i_{in} is magnified by β, whose constancy depends only on the stability of the resistors, and i_f itself can be measured or recorded by a comparatively crude instrument. The Gall galvanometer amplifier allowed currents much less than $1\,\mu$A to be recorded directly on a moving pen chart recorder. With modern electronics this may not seem a great achievement, but one may still admire the economy which achieves this with a single photocell and a thyratron, without even needing a power pack. Moreover this greater current sensitivity is accompanied by a substantial reduction in the input resistance. The current through the galvanometer is i_f/α, or $\beta i_{in}/\alpha$ from (2); the input resistance is therefore $\beta G/\alpha$, which may be many times less than G.

The virtual elimination of α from the ratio i_f/i_{in} is easily understood. With very large amplification, the smallest deflection of the galvanometer mirror sets up a huge feedback current which returns it very nearly to zero; in equilibrium the mirror deflects much less than it would without the feedback, and the feedback current is such that, after partitioning between R_2 and $R_3 + G$ (with the lion's share going to R_2), there is in effect no current through G. Appreciation of this point enables one to see why the dynamical properties are also greatly affected. The feedback current provides a strong restoring torque for any deflection from equilibrium, and therefore enhances the value of the torque constant of the suspension. The periodic time is thereby shortened, and so is the time required to reach equilibrium deflection after a current is applied – another significant gain in performance.

Let us analyse this in terms of the equation of motion of the galvanometer mirror. The free motion is governed by an equation of the form (2.29); when a current i flows through the suspended coil its contri-   **33**
bution to the torque adds an extra term:

$$I\ddot{\theta} + (\lambda_0 + A/R)\dot{\theta} + \mu\theta = S\mu i,$$ (3)

in which S, the current sensitivity, is the steady deflection produced by unit current.† From the definition of the amplification α it follows that a

---

† The damping term, $A\dot{\theta}/R$, arises from the e.m.f. $A\dot{\theta}/S\mu$ induced by movement of the coil. The value to be assigned to R is the resistance seen by a ͵rator inserted in series with the galvanometer, i.e. the resistance of the network formed by G, R_2, R_3 and any resistance connected across the input terminals. It is unaffected by the feedback.

deflection Si of the galvanometer produces a feedback current αi, so that in general the relation between deflection and feedback current takes the form:

$$i_f = \alpha\theta/S.$$

Hence $$i = i_{in} - \alpha\theta/\beta S,$$

which, when substituted into (3) gives the modified equation of motion:

$$I\ddot{\theta} + (\lambda_0 + A/R)\dot{\theta} + \mu(1 + \alpha/\beta)\theta = S\mu i_{in}.$$

This confirms the qualitative view of feedback as enhancing the torsion constant. The natural frequency is increased by $(1 + \alpha/\beta)^{\frac{1}{2}}$, and the decrement per cycle reduced. The condition for critical damping is

$$\lambda_0 + A/R = 2[\mu I(1 + \alpha/\beta)]^{\frac{1}{2}},$$

so that R can be much smaller without causing overdamping. This allows the instrument to be used to measure small currents in a low resistance circuit, and marks a considerable improvement over the unaided galvanometer for which current sensitivity is bought at the price of high resistance.

Negative resistance and spontaneous oscillation

Much of the rest of this chapter is concerned with the performance of maintained oscillators once they have attained a steady state. This, as already remarked, takes us out of the realm of linear processes, and it is therefore convenient to deal at the outset with that part of the discussion that belongs properly to linear theory; that is to say, the critical behaviour as the feedback parameters are changed infinitesimally so as to take the system from a stable equilibrium state to one in which oscillations grow spontaneously, but are still small in amplitude – the crossing of the line between focal stability and focal instability in fig. 9.6, for example. In the simplest cases, as with the circuit of fig. 1(a), g being independent of frequency, the behaviour can be seen at a glance; (1) takes the familiar form

$$j\omega L + R(1 - gR_1R_2/R) + 1/j\omega C = 0, \tag{4}$$

and the critical value of g is R/R_1R_2, at which value the second term vanishes. The matter is not so obvious when g is a (complex) function of frequency, as happens with amplifiers when internal capacitances, electron transit times etc. degrade the amplification at higher frequencies as well as introducing frequency-dependent phase shifts. Even if g is expressible as a polynomial in ω (4) may turn into an algebraic equation of high order, with a multitude of roots, any one of which having a positive imaginary part can be responsible for a growing spontaneous oscillation, making the circuit unstable. The same problem arises, even when g is constant, if the circuit

256

containing the feedback loop is more complicated than fig. 1(a), having a number of linked networks and being described by a high-order differential equation. The resolution of this problem by application of the Routh–Hurwitz tests is one of the oldest successes in the theory of stability. It will be realized, however, that it is not a panacea, since in general the frequency-variation of g will not be representable by a polynomial. Fortunately an alternative procedure exists, and one which is far more closely related to experiment, in that it operates on $g(\omega)$ as a known function, whether measured or calculated, but does not assume any particular analytical form for g. This is the procedure developed by Nyquist[2], but while going just far enough into the argument to make it plausible, we shall not go to the considerable lengths of developing a rigorous proof, nor shall we multiply examples of its use.

The concept of *loop gain*, that plays a central role in Nyquist's theorem, may be understood from the circuit of fig. 1(b), in which ε, the e.m.f. across Z_1, is the input to the amplifier whose output is the source of i_2. Let us disconnect the input leads at AB, and with an external source supply an input ε to the amplifier. Then if R_{in} is large the output current $g\varepsilon$ is divided between Z_2 and $Z_1 + Z_3$, so that the e.m.f. across Z_1 is $g\varepsilon Z_1 Z_2/(Z_1 + Z_2 + Z_3)$. The loop gain, G, of the circuit is $gZ_1 Z_2/(Z_1 + Z_2 + Z_3)$, being the ratio of output to input e.m.f. when the loop is open.† At one or more values of ω (in general complex) the loop gain is unity, and the loop can be closed without perturbing the currents and e.m.f.s anywhere in the circuit. These values of ω are the natural frequencies of the circuit with feedback; indeed, equating $gZ_1 Z_2(Z_1 + Z_2 + Z_3)$ to unity simply reproduces (1).

Our concern, however, is not with complex frequencies but with the trajectory of the loop gain on an Argand diagram as ω runs through all real values between $\pm\infty$. When, to take a specific example, Z_1 and Z_2 are resistances and Z_t ($= Z_1 + Z_2 + Z_3$) is $j\omega L + R + 1/j\omega C$, the impedance of the resonant circuit in fig. 1(a), the loop gain takes the form:

$$G(\omega) = gR_1 R_2/(j\omega L + R + 1/j\omega C).$$

If g is real and independent of ω, the trajectory of G is a circle with the points 0 and $gR_1 R_2/R$ at opposite ends of a diameter, as shown in fig. 3. At a certain value of g the trajectory passes through the point U at which G is real and equal to unity; it is then possible for the circuit to oscillate with real ω, that is to say it is neutrally poised between stability and instability. This condition is of course the same as that derived from (4), $g = R/R_1 R_2$. For smaller g the circuit is stable, for larger it is unstable, and the onset of instability is marked by the trajectory expanding to enclose U. There are obviously many other feedback circuits with trajectories other than circular, especially if g is a complicated function of frequency, to

† Care must be taken in general that the open loop is terminated by the impedance which would lie ahead of it when it is closed. We have avoided the problem here by opening at a point of high impedance (the input to the amplifier).

Fig. 3. Loop gain (Nyquist diagram) for circuit of fig. 1(a). In increasing order of g and radius of circle, the circuit is stable, neutral and unstable. U is the point, G = 1.

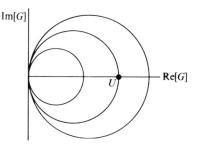

Fig. 3. Loop gain (Nyquist diagram) for circuit of fig. 1(a). In increasing order of g and radius of circle, the circuit is stable, neutral and unstable. U is the point, G = 1.

which the same argument could be applied. For if the circuit is stable, as it should be, when g is very small and the trajectory of G confined to a region close to the origin, then if g is steadily increased so that the trajectory expands, a time will come when it passes through U and is in neutral equilibrium with respect to one mode of oscillation. A small increase in g will set it off into instability, and at the same time the point G = 1 will have been enclosed within the trajectory. Although this does not constitute a proof of Nyquist's theorem, it does make it plausible that *a feedback circuit is stable if the trajectory of G does not encircle U.*

364

The argument also makes plausible the further statement that if the trajectory encircles U n times, there are n distinct modes of instability. For each increase in the value of n is achieved by the trajectory crossing U and allowing a new mode to pass through neutral equilibrium; though this can only be observed, of course, if the already existing unstable modes can be prevented from building up by the most careful choice of initial conditions. In the case of the circuit of fig. 1(a) the trajectory describes the circle twice as ω runs between $\pm\infty$, and this is consistent with there being two solutions of the equation of motion, Ω_1 and Ω_2; whether they are really different, seeing that $\Omega_2 = -\Omega_1^*$, is a matter of definition, but if we allow ω to take negative as well as positive values, then in relating the number of encirclements of U to the number of unstable modes they must be counted separately.

In the practical application of Nyquist's theorem occasional difficulties arise, for example when the system is unstable in the absence of feedback but stabilized by the feedback itself. Again, it may happen that a system is stable when g is small, becomes unstable as g is increased, but then restabilizes at still higher g. In this case the first critical value of g marks the entry of U into the interior of the trajectory, and the second its re-emergence. Each special case serves to test, and ultimately to confirm, the great generality of the theorem.

At this point we leave the general and turn to particular examples of self-maintained oscillators, where more interest attaches to the steady oscillatory state than to diagnosis of the initial instability. It will be seen, indeed, that the latter is normally a trivial matter with the elementary examples we shall treat. The former may by contrast present challenging problems.

Negative resistance by means of feedback; class C oscillators

The idea summed up in (1) found practical realization at the beginning of electronic history, when the basic triode oscillators operating on the feedback principle were invented.[3] The triode (and still more the pentode) approximates rather closely to the idealized amplifier postulated in the last section, in that a voltage applied to the grid causes a change of anode current, the input and output impedances both being high enough, at moderate frequencies, for the analysis to give a satisfactory account of the observed behaviour. The quantity g is the *transconductance* of the triode, the change in output current due to unit change of grid voltage. Some typical primitive oscillator circuits are shown in fig. 4, from which various circuit components, inessential to understanding the mode of operation, have been eliminated. The capacitor C', resistor R' and choke L' in the Colpitts circuit are essential to the operation, in the sense that the grid and cathode cannot be allowed to float, but must have d.c. contact to the rest of the circuit to ensure that they are held at the correct mean potential. At the same time the oscillatory current must be channelled towards the right electrodes. Thus C' must be large enough to present no significant impedance at the oscillating frequency while R' and L' must be effective obstacles ($RC' \gg 1/\omega$ and $L'C \gg 1/\omega^2$). Then these components play little part in determining the character of the oscillation.

Fig. 4. Four early types of oscillator using triode amplifiers and feedback.

In the Hartley and Meissner circuits Z_1 and Z_2 are both inductive, so that gZ_1Z_2 is real and of order ω^2. In the Colpitts circuit gZ_1Z_2 is also real but of order ω^{-2}. In both cases the equations for ω are of third order; thus for the Meissner circuit the differential equation for a series resonant circuit must be supplemented by a term of the form $G\ddot{q}$, where $G = gM_1M_2$:

$$G\ddot{q} + L\ddot{q} + R\dot{q} + q/C = 0.$$

If we seek solutions of the form e^{pt}.

$$Gp^3 + Lp^2 + Rp + 1/C = 0. \tag{5}$$

A similar equation governs the Hartley circuit, but the Colpitts circuit has its new term at the other end,

$$Lp^3 + Rp^2 + p/C + G' = 0, \tag{6}$$

where $G' = g/C_1C_2$. In both (5) and (6), G and G' are positive. With all coefficients positive, and R small, the equations have one real and two complex roots as is easily verified by sketching the forms of (5) and (6). The real root is negative and therefore harmless, but if G or G' is large enough the other roots describe growing oscillations. The critical condition for spontaneous oscillation to start is that p shall be purely imaginary:

i.e. $\qquad\qquad p = \pm j\omega_0, \quad \text{where} \quad \omega_0^2 = 1/LC,$

and $\qquad\qquad G = R/\omega_0^2 \quad \text{or} \quad G' = R\omega_0^2.$

For larger values of G or G' the effect of feedback is to introduce a negative resistance greater than R.

The Tickler feedback circuit differs from all these, in that the pick-up is from a capacitor and the feedback across an inductor; the pick-up e.m.f. is q/C and the feedback e.m.f. gMq/C. Either sign of feedback can be chosen (as also with the Meissner circuit), and for oscillation the feedback must counteract the resistor's action:

$$L\ddot{q} + (R - gM/C)\dot{q} + q/C = 0. \tag{7}$$

This is the simplest case of all those in fig. 4, feedback providing a frequency-independent resistive term of either sign and virtually unlimited magnitude. As with the other oscillators, at the critical feedback for initiating oscillation the frequency is exactly ω_0. With very strong feedback focal instability may give way to nodal.

In view of the simplicity of this oscillator we shall use it to illustrate the effect of saturation of the amplifier in limiting the amplitude of oscillation. When the amplitude is low the amplifier responds linearly, but the anode current, defined in terms of the rate of arrival of electrons, cannot be driven negative nor can it exceed the rate of emission from the cathode. A typical characteristic is shown in fig. 5, together with a convenient approximate form which is helpful in discussing saturation behaviour. The d.c. bias of the triode is irrelevant to the analysis; for treating oscillatory effects the origin is chosen to be the working point of the characteristic. We therefore write the anode current as a function of grid potential

$$i_a = i_a(v_g),$$

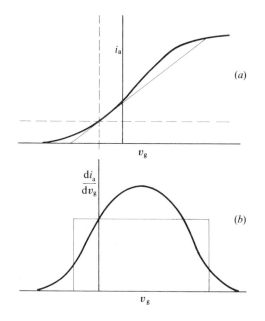

Fig. 5. (a) Schematic triode characteristic (heavy line) relating anode current, i_a, to grid voltage, v_g. The light line shows a simplified model characteristic, used in approximate analysis of the behaviour. The broken lines are the displaced axes with the operating point of the characteristic at the origin. (b) The derivative of the characteristics in (a), the top-hat being the approximation.

where $i_a(v_g) = gv_g$ when v_g is small. Instead of (7) the equation for q now has the form (since $v_g = q/C$)

$$LC\ddot{v}_g + RC\dot{v}_g + v_g = M\,di_a/dt = M\frac{di_a}{dv_g}\dot{v}_g. \tag{8}$$

The form of di_a/dv_g is sketched in fig. 5(b), together with the 'top hat' which is the derivative of the approximate characteristic.

Assuming this approximate characteristic we can readily appreciate what happens. Once M has been increased above the critical value to initiate oscillation, there is nothing to prevent the oscillation growing until the amplitude of v_g extends to one limit of the amplifying range. It will continue to grow beyond this value, but the amplifier will not be operative over part of the cycle and the rate of growth will begin to diminish, until a steady state is reached in which the resistive dissipation and the power input from the amplifier are balanced. This is the *limit cycle* already introduced in general terms in chapter 9. If M is considerably greater than the critical value the final amplitude may reach far beyond the limits of the amplifying range. Then over much of the cycle $di_a/dv_g = 0$ and, as (8) shows, the circuit is in free oscillation. Provided its intrinsic Q is high, the crests and troughs of its waveform are almost pure sinusoids, very slightly modified by resistive loss. These sinusoidal sections are joined by sections lying within the amplifying range, and which may be parts of a growing sinusoid or even, if the gain is very high, exponential in form. But since these sections are only a minor part of the whole waveform, it hardly matters what their exact shape is – they are very nearly straight lines. Thus

259

although the cycle is composed of sections which are mathematically distinct, to the eye of the observer there is little difference between the actual waveform and a pure sinusoid. During some of each cycle a small fraction of the energy is dissipated, and during the rest it is restored, and one may calculate the steady–state amplitude as that which allows the two effects to balance.

Alternatively one may calculate the effective amplification at the oscillating frequency and find at what amplitude it is just sufficient to neutralize the resistive effects. For this purpose we imagine a sinusoidal signal $v_{g0} \cos \omega t$ applied to the grid; the anode current is truncated by saturation and the induced e.m.f. therefore contains harmonics of ω which, however, we suppose to have a quite negligible influence on the current in the circuit, being so far removed from the resonant frequency. The effective transconductance, g_{eff}, on this assumption is determined by the amplitude of the fundamental component of the anode current.

If, for the purpose of illustration, the approximate characteristic is assumed symmetrical about the working point, the anode current is $g v_{g0} \cos \omega t$ so long as it is less than $g v_{sat}$, and $g v_{sat}$ otherwise. The fundamental component in the Fourier transform is easily calculated and the details need not be given here:

$$g_{eff} = g \times \frac{2}{\pi}[\sin^{-1}(1/r) + (r^2 - 1)^{\frac{1}{2}}/r^2], \quad \text{if } r > 1, \qquad (9)$$

r being written for v_{g0}/v_{sat}. This function is shown in fig. 6; a more realistic characteristic would give a smooth curve but the limiting behaviour at large

Fig. 6. Effective amplification for a sine wave limited by a top-hat characteristic.

values of r would be the same, being the result for a square wave of amplitude $\pm g v_{sat}$, whose fundamental has amplitude $4 g v_{sat}/\pi$;

$$g_{eff} \approx 4g/\pi r \quad \text{when } r \gg 1. \qquad (10)$$

The oscillation builds up until, as follows from (7), $g_{eff} = RC/M$. Thus when there is strong enough feedback for (10) to apply,

$$v_{g0} = 4gMv_{sat}/\pi RC. \qquad (11)$$

To reach the same conclusion by an energy argument, note that the feedback e.m.f. is $Mg\dot{v}_g$ while v_g runs between $\pm v_{sat}$, and that in this short interval the current in the resonant circuit is near its maximum i_0, which is equal to $\omega_0 v_{g0} C$. Each short burst of power introduces energy $Mgi_0 \int \dot{v}_g \, dt$, i.e. $2Mgi_0 v_{sat}$, and there are two bursts per cycle. The energy dissipated in the resistor during one cycle is $\pi R i_0/\omega_0$, equal to the energy fed in when (11) is satisfied. An oscillator excited in this way by short bursts of power is said to operate as a class C oscillator; the historical origin of this term, deriving from a classification of thermionic tube amplifiers, is irrelevant to our discussion. One merit of this type of oscillator is its intrinsic frequency stability, arising from minimal interference with the resonator circuit. The escapement of a pendulum clock makes use of the same principle, though the details are slightly different, as we shall discuss in the next section.   **323**

It is of interest to see how the oscillator approaches its final amplitude; when feedback is strong enough the energy argument may be extended without difficulty. It is enough to note that the dissipation by the resistor varies as the square of the amplitude while the power input varies as the first power. If y denotes the amplitude of oscillation of v_g at any instant, energy conservation determines the differential equation for y,

$$\frac{d}{dt}(y^2) = (v_{g0}y - y^2)/\tau_e. \tag{12}$$

The constants have been supplied by inspection; if there is no feedback the first term on the right vanishes, and τ_e is the time-constant for energy dissipation in the free vibrator, while with feedback the stable amplitude, v_{g0} in (11), must make the right-hand side vanish. The solution of (12) has the form

$$y = v_{g0} - A\,e^{-\frac{1}{2}t/\tau_e} = v_{g0} - A\,e^{-t/\tau_a} \tag{13}$$

showing that the approach to the final amplitude is governed by the time-constant of the free resonant ciruit, once the feedback is strong enough for v_{g0} to be given by (11).† Stronger feedback enhances the final amplitude and speeds up the initial growth, but does not alter the time scale of the last stages. A high-Q resonator used as the basis of a class C oscillator has the same sort of insensitivity to short-term disturbances as it has when running free. This point will be amplified later in this chapter, when we discuss the effects of noise, and in the next chapter, where we   **361** shall discuss how readily an injected oscillatory signal, at a different frequency, can entrain the oscillator, i.e. pull it into synchronism.   **392**

Although we have described the behaviour of a triode oscillator, it is more convenient for the purpose of checking the theory experimentally to use a transistor circuit or, better still, an integrated circuit designed as an

---

† An alternative approach is to note that twice per cycle the anode current switches rapidly between $\pm g v_{sat}$, producing in the resonant circuit an impulsive e.m.f. whose strength, $2Mg v_{sat}$, is independent of the oscillatory amplitude. The circuit responds like a free vibrator driven by a succession of constant impulses. This is only true once class C operation is fully developed and the impulses are much shorter than the period.

operational amplifier.[4] This can be treated as a 'black box' with certain characteristics, and for the low frequency application here the d.c. characteristics are sufficient. With the amplifier represented in fig. 7(a) the output voltage V_{out} is determined by the difference between the voltages of the two inputs, $V_{out} = f(V_+ - V_-)$. The input impedance is very high, and the output impedance very low, any current required to maintain V_{out} in the presence of a load being supplied through the earth lead (which is usually

Fig. 7 Operational amplifier; (a) the basic device in schematic form; (b) with feedback to limit and stabilize the amplification.

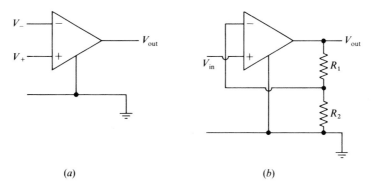

(a)   (b)

omitted in circuit diagrams). So long as V_{out} is not too close to the supply voltage, between ± 10 V, say, the amplifier is linear, with extremely high gain, perhaps 10^5. Thus when $|V_+ - V_-| < 10^{-4}$ V it is linear, but for higher imput differentials it saturates. The high gain is normally not required, but it enables feedback to be introduced to lower the gain to a convenient level, which is then extremely well-defined, for the same reason as was explained in connection with the galvanometer amplifier. In the present case it is clear from fig. 7(b) that the resistor chain fixes V_- in relation to V_{out}:

$$V_- = V_{out}/\beta, \quad \text{where } \beta = (R_1 + R_2)/R_2.$$

In the linear range of response, $V_{out} = A(V_+ - V_-)$, where $A \sim 10^5$, and these two equations show that

$$V_- = V_+/(\beta/A + 1)$$

and $\qquad\qquad V_{out} = \beta V_+/(\beta/A + 1) \doteqdot \beta V_+ \quad \text{if } \beta \ll A.$

Provided the value chosen for β is much less than 10^5, the gain is β and is determined by the resistors, being extremely insensitive to the value of A; amplification is thus linear until saturation reduces the effective value of A to something approaching β. The range of V_+ is now of the order of $10/\beta$ V, much larger than the range before feedback is introduced, since feedback causes V_- to follow V_+ almost exactly. A measured characteristic is shown in fig. 8, with a slope of 50.5 instead of the expected 51, and a knee so sharp as to give confidence in the comparison of experimental behaviour with the theory developed above on the basis of the approximate characteristic in fig. 5.

310

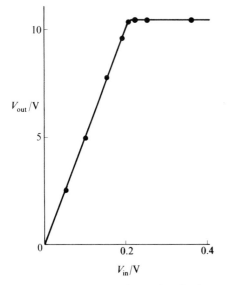

Fig. 8. Measured characteristic of the circuit of fig. 7(b), with $R_1 = 5000\,\Omega$, $R_2 = 100\,\Omega$.

The oscillator configuration is shown in fig. 9. Since the amplifier is a low impedance voltage source, a resistor R_3 ($10^4\,\Omega$, much greater than ωL_2) is inserted to ensure that the current in L_2 is proportional to the p.d. across the capacitor of the resonant circuit; β/R_3 plays the part of g in (7), and the negative resistance is $\beta M/R_3 C$, i.e. $(R_1 + R_2)M/R_2 R_3 C$. To eliminate possible non-linear complications in ferrite cores the inductors L_1 and L_2

Fig. 9. Tickler feedback oscillator (see fig. 4) used to test theory.

were plain copper coils originally made as windings for an electromagnet. They were large and ungainly, but had the merit of high enough inductance ($L_1 = 69\,$mH) combined with low resistance ($5.2\,\Omega$ at the operating frequency) to give a circuit of good Q (48) at so low a frequency (581 Hz) that the amplifier responded almost precisely in accordance with its d.c. characteristic. The two coils were spaced to give a value of M that permitted oscillations to begin when g was $3.78 \times 10^{-4}\,\Omega^{-1}$; if the theory is correct this means that $M = 15\,$mH. The value of g was then increased to $50 \times 10^{-4}\,\Omega^{-1}$ and the spontaneous build-up of oscillations recorded, as shown in fig. 10(a), which is the p.d. across the capacitor. Only the first few cycles after switching on fall entirely on the linear amplifier characteristic and are therefore suitable for determining the rate of exponential growth. To compare experiment with theory note that unless Q is very small indeed (see (2.27)) the relative change in amplitude per cycle is a linear function of resistance, and hence of g. A change of g from 0 to 3.8×10^{-4} reduced the decrement from its free value to zero; the addition of 12 times as much should therefore give rise to as much increase in one cycle as the free oscillation loses in 12, a factor of 2.2 according to the trace of the decrement in fig. 10(b). The trace in fig. 10(a) is not very suitable for determining the increase in the first few cycles; it appears to be nearer 2.0 per cycle than 2.2, but a more careful study would be needed to see if this small discrepancy is real. Certainly there is no major disagreement here.

32

Fig. 10. (a) Record of the build-up of oscillation in the circuit of fig. 9 from the moment of connecting the amplifier in circuit; (b) free decrement of the resonant circuit with amplifier disconnected. The curves have been disposed so as to show the same time-constant governing the decrement in (b) and the approach to the steady state of oscillation in (a).

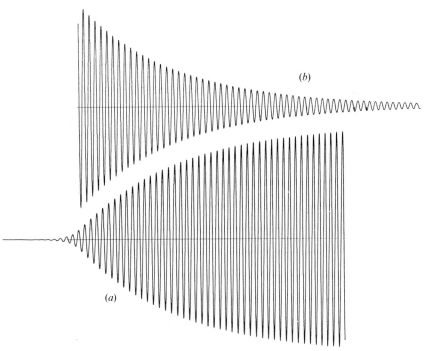

Turning to the behaviour near saturation, we see immediately from fig. 10 that the time-constant governing the approach to the maximum amplitude matches very closely that for free decay – the curves can be aligned so as to make their envelopes run parallel. This confirms (13), and a further simple check is provided by the amplitude in the steady state. The voltage amplitude across the capacitor is more than 3 V, at least 15 times what is needed to saturate the amplifier, whose output voltage therefore has an almost square waveform as it switches rapidly between the positive and negative limits. At the same time the oscillation in the resonant circuit is an almost perfect sinusoid; this is true class C operation. The amplitude of the fundamental component of the output is $4/\pi$ times the amplitude of the square waves. Measurement on an oscilloscope showed the square wave to have an amplitude 3.1 times that of the sine-wave input, so that the effective value of β at the fundamental frequency was 3.95, i.e. $g_{eff} = 3.95 \times 10^{-4}\,\Omega^{-1}$. Since this maintains the oscillations at constant amplitude it should be the same as what is needed at a low level when the amplifier is linear, and this is the value $3.78 \times 10^{-4}\,\Omega^{-1}$ already quoted. Once more agreement is close enough to show that the theory is at least a very good approximation. We may be satisfied in this case that the oscillator is in essence a simple device.

Further confirmation is provided by the behaviour when the resistance R_3 is decreased to the point where the combination of R_3 and L_2 has a substantial reactive component. In the linear range of amplification the impedance introduced into the resonant circuit is no longer purely resistive, having the form $-B/(R_3 + j\omega L_2)$ instead of $-B/R_3$, where B is $(R_1 + R_2)M/R_2 C$. In addition to the negative resistance $-BR_3/(R_3^2 + \omega^2 L_2^2)$ there is a reactance $j\omega L_3 B/(R_3^2 + \omega^2 L_2^2)$ which supplements the inductive reactance and lowers the resonant frequency, though not by a great deal if M is not too large.† The critical condition for oscillation to be maintained now takes the form

$$BR_3/(R_3^2 + \omega^2 L_2^2) = R, \tag{14}$$

and when B is adjusted to satisfy this condition the extra inductance is $L_2 R/R_3$. Hence

$$\omega_0^2/\omega^2 = 1 + E/R_3, \tag{15}$$

where $E = L_2 R/L_1$, and ω_0 is the limiting frequency for large R_3. This variation is readily verified with the circuit used for the earlier measurements, and the magnitudes of L_1, L_2 and R, as determined from the oscillatory behaviour, yield a value of E within 1% of what is needed to describe the frequency variation.

If ω is assumed constant, (14) predicts a hyperbolic variation of the critical value of B with R_3, the asymptotes being the lines $B = RR_3$ and $B = 0$, both passing through the origin. To verify this the value of M was reduced considerably to provide a convenient working range of amplifier gain, and the curve shown in fig. 11 was obtained, having the expected form. In particular the diameter D, constructed graphically, passes very close to the origin, as predicted. The oscillatory regime lies within the hyperbola; as R_3 is reduced from a large value, the gain being kept constant, oscillation starts at one critical value and dies out again at another. The frequency varies with R_3, as is indicated by the values attached to the critical curve, which agree well with (15).

---

† By contrast, when R_3 is large and the impedance introduced by feedback is almost purely resistive, the frequency is (as expected) sensibly unaffected by the value of M, and does not alter by more than 0.1% even when the oscillator is in the extreme class C condition.

Fig. 11. The oscillator of fig. 9 with R_3 reduced to provide a significant inductive contribution to the feedback. The boundary between oscillatory and non-oscillatory conditions is hyperbolic, with the vertical axis and A as asymptotes, and the frequency varies along the critical curve as indicated by the arrowed values.

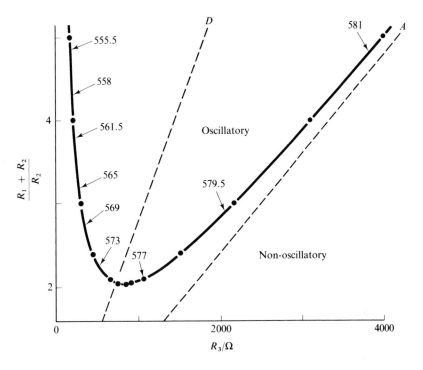

The pendulum clock, a non-linear class C oscillator

The traditional long-case pendulum clock has the typical characteristic of the class C oscillator in that its resonant element, the pendulum, is only stimulated during a small fraction of each vibration, more or less at the phases of maximum velocity so as to perturb the natural frequency as little as possible. In more recent designs (e.g. the Synchronome clock)[5] the stimulus is much more infrequent, perhaps twice a minute, and the time-keeping correspondingly dissociated from disturbing influences, apart from those affecting the pendulum's period directly. Here we shall discuss only the oldest type of escapement still in common use, the recoil escapement of Hooke, such as is illustrated in fig. 12. The eighteenth-century long-case clock in the author's possession has an escapement very much like this, and provides experimental data for comparison with the theory which we now present.

The swing of the pendulum rocks the pallets to and fro while the escapement wheel, which is gently impelled in a clockwise sense by the clock weights, through a gear train, keeps a light pressure on one or other of the pallets.† As the diagram is drawn, the pendulum has just swung far enough to the right to release the wheel from the right-hand pallet, so that

---

† The details of the suspension of the pendulum and the crutch which transmits the motion to the pallets are not essential for our discussion, though their correct design is important for the working of the clock.

Fig. 12. The recoil escapement.[6]

it has jumped forward and been caught on the left-hand pallet. Further swinging of the pendulum to the right pushes the wheel back a little (this is the recoil which gives the escapement its name), but as the pendulum swings towards the centre the wheel moves forward and does work on the pendulum. As the pendulum moves to the left, the left-hand pallet disengages from the wheel and the process is repeated with the right-hand pallet receiving the force of the wheel.

There is a certain minimal amplitude of swing needed to actuate the escapement, but once this is exceeded the mechanism is such that, if friction could be neglected, the same amount of energy would be delivered during each swing, irrespective of the amplitude of the swing. For there is a certain angle at which one pallet disengages and the other is brought to bear, and another angle at which the process is reversed. Between these two instants in the swing the wheel has a well-defined net forward movement; with a larger swing, all that happens is that the wheel recoils further back, but the larger amount of work done on it by the pendulum is recovered on the return swing. The net work is just fixed by the forward movement of the wheel while in contact with the pallet. Indeed, if the escapement had no 'drop' – if it were so perfectly made that there was no movement of the wheel as the escapement switched over – the rate at which work was done by the wheel on the pendulum would be simply the product of the tangential force exerted by the wheel on the pallet and the mean tangential velocity of the wheel.

In detail, then, this maintained oscillator differs considerably in its mode of action from that in fig. 9, requiring a certain amplitude to get the process started, and thereafter being automatically self-limiting since the energy delivered per cycle does not increase with amplitude, while the dissipation does. A further difference lies in the dissipation being mostly due to friction at the escapement, so that effectively the motion of the pendulum is resisted by a constant force, not one that is proportional to its velocity. This reveals itself in the behaviour at an amplitude too small to actuate the escapement; the vibrations decay linearly, not exponentially (cf. fig. 2.22).   **36**

It is reasonable, therefore, to consider the escapement-driven pendulum as an oscillatory system in which the energy communicated per swing is constant and the energy dissipated per swing is proportional to the amplitude, a. And since the stored energy is proportional to a^2, the differential equation for the variation of amplitude with the number of swings takes the form

$$a\,\mathrm{d}a/\mathrm{d}n = \varepsilon - \lambda a, \qquad (16)$$

in which ε represents the energy delivered by the escapement and λa the frictional loss. The general solution has the form

$$n = -a/\lambda - (\varepsilon/\lambda^2) \ln (\varepsilon - \lambda a) + \mathrm{const.} \qquad (17)$$

By choosing the constant of integration to be $(\varepsilon/\lambda^2) \ln \varepsilon$, we elect to count n from the (hypothetical) starting point when $a = 0$. The limiting value of a, as n goes to infinity, is ε/λ to make the argument of the logarithm vanish. With f defined as $a\lambda/\varepsilon$, i.e. the amplitude measured as a fraction of its limiting amplitude, and N as ε/λ^2, (17) is converted to the form

$$n/N = -\ln (1-f) - f. \qquad (18)$$

This function is plotted in fig. 13, but it should be remembered that the initial steep rise is not observable, since the escapement does not work at small amplitudes. The experimental points were taken with the clock weight increased to about half as much again in order to make the final amplitude considerably larger than was intended by the maker and so provide a larger range for observation. No great sophistication was attempted, the amplitude being read by means of a pointer on the pendulum swinging past a paper scale.

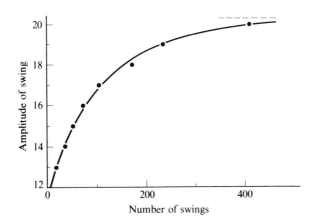

Fig. 13. Build-up of escapement-driven pendulum towards its steady state.

To fit the experimental points to the curve it was necessary to take N to be 150 and the limiting value of a to be 20.3 on the paper scale. From these values λ is found to be 0.136. If the same value applies also at low amplitudes when the escapement is inoperative and the equation of motion is like (16) but with ε put equal to zero, we expect the linear decay of amplitude to involve a reduction of 0.136 per swing. This is somewhat less than the observed value of 0.160, but near enough to indicate that the essence of the behaviour of the escapement is described by the simple model embodied in (16). One should not in such a rough experiment look for too close an agreement; apart from the crudities of observation the escapement is assumed in the theory to be free from eccentricity, something that is not very probable after a long working life. Any eccentricity in the pivot of the escapement wheel will cause the energy fed to the pendulum to vary periodically with a period of one minute, which is the rotation period of the wheel; this may well affect the details of the observations.

Further examples of negative-resistance oscillators

(a) *A resistive feedback network.* We have analysed a class C oscillator in some detail to show both the origin of its negative resistance and the final state of oscillation. The present section exhibits a varied selection of arrangements whose oscillatory instability may be attributed to negative resistance or an analogous property, but whose final states may differ considerably among themselves. It will be appreciated that these are only a few out of a vast range of choices, and that the selection is to a considerable degree a matter of personal whim. They have in common that they are based on resonant systems so that, as with the class C oscillator, the condition for oscillations to build up is that the loss mechanism must by some means be more than compensated; in a circuit the resistance must be made negative, in a mechanical vibrator the viscous drag must be reversed, in a electromagnetic cavity oscillator the dielectric loss must be replaced by dielectric gain, etc. We shall discuss examples in each of these categories, first elucidating the mechanism for introducing gain, and then proceeding to the non-linear effects that govern their approach to the steady state. Some of the saturation effects are conveniently disposed of in passing, where the oscillatory system is not particularly important in itself. The general question of the saturation of negative-resistance circuits is, however, so basic to the appreciation of various types of oscillator that we shall devote a separate section to it.

In the feedback circuit of the oscillator in fig. 9 the e.m.f. across the capacitor was fed back, after amplification, across the inductor. It is just as easy to construct a negative resistance by taking off and feeding back across resistors, since the extra term in (1), gZ_1Z_2, is real if both Z_1 and Z_2 are real. A simple circuit having this property is shown in fig. 14(a), together with two current–voltage characteristics. It is easy to show that for small currents, when the amplifier is operating in its linear range, the resistance between A and B is $-R_1R_3/R_2$. What happens after saturation depends on the values of the circuit components. If they have high resistance, so that the amplifier is not called upon to supply excessive current, the

Fig. 14. Negative resistance device based on operational amplifier: (*a*) connected in series with resonant circuit; (*b*) connected in parallel; (*c*) two typical characteristics for the device.

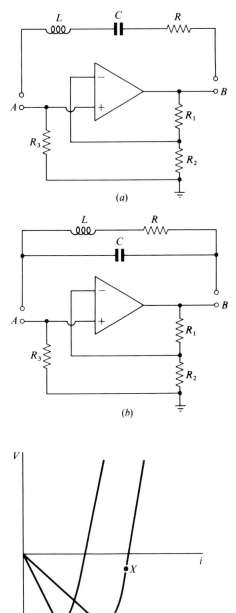

potential at B saturates and thereafter any further current alters the potential of A relative to earth but leaves that of B unchanged. The gradient of the characteristic then reflects the value of R_3. With lower resistances in the circuit the current needed to saturate the amplifier is enough to modify its behaviour considerably; the positive slope is then not related solely to the values of the resistors. We may take it as an empirical parameter when we come to discuss its influence on the steady state of oscillation. With the oscillatory circuit connected in series, as in fig. 14(*a*), and the negative resistance increased in magnitude, spontaneous oscillation sets in when the circuit resistance *R* is compensated. This aspect of the behaviour needs no further discussion.

364

One might expect that the same negative resistance device could also be used in the parallel circuit of fig. 14(*b*), which at resonance has a high resistance between *A* and *B*, approximately equal to $\omega^2 L^2/R$. A negative resistance less than this should cause oscillation, but in fact the arrangement is always unstable, usually jumping, immediately after switching on, to a point such as *X* in fig. 14(*c*), where the ratio *V/i* is equal and opposite to the resistance *R*; a steady current circulates through *L*, *R* and the negative resistance. It is worth devoting a little space to this behaviour which is not predicted by an elementary analysis of the circuit. To appreciate this point, consider a frequency-independent negative resistance, $-R'$, connected across *AB*, instead of the actual device. Since an e.m.f. may be established between *A* and *B*, but no net current may flow from one point to the other, the sum of the three admittances in parallel must equal zero:

$$1/(R+j\omega L)+j\omega C-1/R'=0;$$

i.e. $$LC\omega^2-j(CR-L/R')\omega-(1-R/R')=0. \qquad (19)$$

This equation is reduced to the equation for the plain resonant circuit by removing the two terms in R' (i.e. putting $R'=\infty$). It is clear that if *R* is small ($Q\gg1$ in the absence of R') as R' is reduced nothing important happens until $R'<L/RC$, which is very nearly $\omega^2 L^2/R$, when the second term representing the damping changes sign, and one expects oscillations to grow, as already remarked. But this is not what really happens, and the mistaken prediction is a consequence of neglecting the frequency-variation of the negative resistance device, arising from the time delay (perhaps 10^{-4} s) between injecting current into the circuit and the establishing of the steady-state voltage at all internal parts of the amplifier circuit.

Imagine that a current is suddenly switched through *AB* at time $t=0$, the resonant circuit being disconnected. Probably at the very outset most of the current follows the path through R_3, R_2 and R_1, and a positive potential difference is set up. Only as the amplifier responds is this replaced by a negative potential difference, the final state for the negative-resistance circuit. The step-function response therefore has something like the form shown in fig. 15(*a*) – the exact shape is of no great consequence, only the crossing of the axis. Correspondingly the response to a short impulsive current injected at time $t=0$ is the derivative of this, having a positive instantaneous δ-function response, followed by a negative delayed response. Because the asymptote of fig. 15(*a*) is negative, the integral of the negative part of fig. 15(*b*) is greater than the δ-function. From the

Fig. 15. (*a*) Schematic
diagram of step-function
voltage response of the
device in fig. 14 when a
steady current is switched
into it at time $t = 0$;
(*b*) the corresponding
impulse response function,
being the derivative of (*a*).

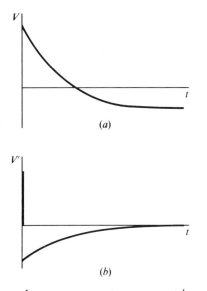

(*a*)

(*b*)

voltage response to a current impulse the impedance at frequency ω may
be derived as the Fourier transform. It is more convenient here, however,
since we are concerned to demonstrate exponential instability, to consider
the response of the circuit to a growing exponential current of the form
$i_0 e^{pt}$, p being real and positive. Since the voltage at any time t is the
resultant of the responses to earlier current impulses, the impedance $Z(p)$
is seen to be derived from $V'(t)$ by forming the integral (*Laplace transform*):

$$Z(p) = \int_0^\infty V'(t) \, e^{-pt} \, dt.$$

Instead of being constant and negative for all p, as would be the
consequence of neglecting the transient properties of the amplifier, $Z(p)$ is
negative only when p is small enough for the exponentially reduced
contribution of the negative tail of $V'(t)$ at large t to outweigh the positive
δ-function. As p increases, $Z(p)$ goes through zero and is thenceforward
positive, as in fig. 16.

Fig. 16. Impedance of
negative-resistance device
as a function of p.

Let us apply this result to the parallel resonant circuit of fig. 14(*b*). If it is
to be exponentially unstable there must be a positive value of p at which

the admittances total zero. When p is real the impedance of an inductance is pL and of a capacitance $1/pC$. Hence (19) may be rewritten in the form:

$$1/(R+pL)+pC = -1/Z(p). \tag{20}$$

The left-hand side of (20) takes a minimal value of about $2(L/C)^{\frac{1}{2}}$, as shown in fig. 17. If $Z(p)$ were negative and took the same value at all p as it does when $p=0$, the resulting positive horizontal line representing the right-hand side of (20) would not necessarily intersect the curve for the left-hand side, but in practice the presence of a zero in $Z(p)$ ensures intersection. The circuit is therefore always unstable. By contrast, the series arrangement of fig. 14(a) is governed by the condition that the impedances, not the admittances, sum to zero; there is no infinity in $Z(p)$ to guarantee trouble.

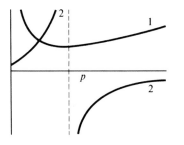

Fig. 17. Graphical solution of (20); curve 1 shows the variation with p of the left-hand side, and curve 2 the variation of the right-hand side.

It might be asked why the foregoing argument has been developed on the basis of the voltage-response to a step-function current, rather than vice versa – the current-response to a step-function voltage. The answer is simple – the circuit which enables the former to be determined is stable, but for the latter it is unstable, and the current-response function is not definable. To generate a constant current whose magnitude is unaffected by the behaviour of the load demands a high voltage source with a high resistance in series. So long as this resistance overwhelms the negative resistance of the load under test the current flowing is uniquely specified. But to generate a constant voltage across the load demands a very small series resistance, and the negative resistance of the load then causes the current to grow until non-linearity allows a stable configuation to be found.

(b) *The tunnel diode*[7]. Of the physical devices that exhibit intrinsic negative resistance behaviour, as distinct from the feedback artefact just discussed, the tunnel diode is probably the most straightforward, in that its behaviour is well represented by its current–voltage characteristic. By contrast, the Gunn diode and the discharge tube owe their negative resistance to spatially extended phenomena that result in complicated changes in internal structure (e.g. moving domains of reversed field). The tunnel diode shows none of these complexities. Unfortunately its behaviour is properly understandable only in the light of a basic course in solid-state physics, and it would be out of place here to attempt an adequate explanation. The reader is referred to one of the texts cited in the references, which provide justification for the highly abbreviated and over-simplified description given in the square brackets that follow.

[In a semiconductor like germanium, electrons can only move freely if their energy lies within one of a series of bands which are separated by *energy gaps*, bands of forbidden energies. Moreover each band can only accommodate a definite number of electrons, and in pure germanium the number present is exactly sufficient to fill all bands up to what is called the *valence band*, leaving the next higher *conduction band* empty. In a filled band there are as many electrons moving in any one direction as in the opposite direction, and the band as a whole cannot be made to carry current. The only reason why pure germanium conducts at all is that a few electrons are thermally excited into the conduction band. If one dissolves in the germanium enough (~0.1%) *donor* impurity (Sb, As), of higher valency than Ge, the extra electrons occupy the lowest levels of the valence band, as shown schematically in fig. 18. If, on the other hand, an *acceptor* impurity (Ga, In) of lower valency is added, the valence band is depleted.

Fig. 18. Allowed and forbidden energy bands in heavily doped n- and p-type semiconductors. When electrons can be exchanged between them, through the tunnel barrier, the allowed bands fill to the same level, shown by a broken line; the higher electron concentration in n demands that some overflow into allowed band C, while in p some states in allowed band *B* remain empty.

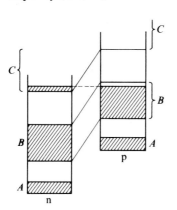

When a thin wafer of Ge has donors diffused into it from one side, acceptors from the other, a very narrow junction between the two types of material can be formed, through which electrons may pass (though with difficulty) by the process of quantum-mechanical tunnelling. The junction region can support a moderate potential difference and, left to itself, the wafer settles down in equilibrium with such a potential difference as will bring the highest filled states to the same total energy, as shown in the diagram. If an external p.d. is now applied so as to change the field in the junction layer and raise the electron energy in the n-material relative to that in the p-material, electrons will tunnel through the layer from n to p, finding unoccupied states of the same energy into which they can go. A simple ohmic current results from this process. But as the external p.d. is increased further, a state of affairs is reached when the occupied part of the conduction band in n is opposite the forbidden band in p; when this

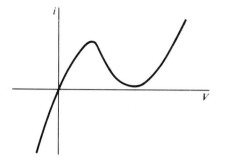

Fig. 19. Typical i–V characteristic of a tunnel diode.

happens tunnelling can no longer occur, for there is no adequate mechanism in the junction layer for electrons to lose energy as they tunnel and thus fall into the valence band of p. This is the reason for the drop of current as the p.d. is increased, as shown in the characteristic of fig. 19. The fact that it does not fall to zero, and increases again as the p.d. is further raised is due to the presence of thermally excited electrons which can pass between conduction bands on both sides.]

To understand the behaviour of the tunnel diode as a circuit element, almost all the required information is contained in the characteristic of fig. 19. In contrast to the negative-resistance circuit of fig. 14, for which V is a single-valued function of i, here i is a single-valued function of V. Further, though this is only a minor detail, the diode needs a steady bias to bring its mean operating point onto the negative resistance range of the characteristic. The arrangement shown in fig. 20 provides the bias and sets the negative resistance in parallel with the resonant circuit, which is now the

Fig. 20. Tunnel diode biased through a resistor to bring the operating point into the negative resistance range. The lead inductance and the intrinsic capacitance of the diode are shown explicitly.

appropriate configuration for exciting oscillation without incurring intrinsic instability of the circuit. In view of the preceding discussion this must not be taken as obvious, but demands an examination of the response-function of the diode. It is important to recognize that the diode has an intrinsic capacitance; the potential drop occurs across a very thin layer, perhaps 10^{-7} m thick, through which the electrons tunnel as indicated above, and this layer behaves as a parallel-plate capacitor with a negative leakage, i.e. a capacitor and negative resistance in parallel. In addition the leads connecting the diode to other parts of the circuit have a certain inductance, so that the equivalent circuit for small departures from the operating point is as shown in fig. 21. The cell is not shown, since it serves only to

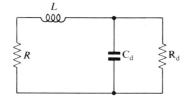

Fig. 21. Equivalent circuit of tunnel diode for small displacements from its operating point.

fix the operating point, where the value of dV/di is represented by R_d; the resistance R represents the resistance of the cell and other portions of the external circuit, while L is the lead inductance. This is the same circuit as is described by (19), and it is stable in the frequency-independent approximation provided that R_d, if negative, has a magnitude greater than R, and that $|L/R_d| < CR$. Keeping the lead inductance small obviates the spontaneous oscillations which will occur if the latter condition fails. The meaning of the former condition is seen by putting $L = 0$; the circuit consists then of a capacitor across which are connected two resistors in parallel, one positive and one negative; the time-constant of the combination is dominated by whichever is smaller, and for stability the positive resistance must dominate.

There appears to be no bar, therefore, to setting up a low-resistance, low-inductance circuit to apply a step-function voltage to the diode and determine its current-response, which has no tendency to diverge as time goes on. Initially, when the biasing voltage is increased by a very small amount to provide the step-function excitation, a positive incremental current flows to change the charge on the capacitor, but this current decreases and eventually the negative resistance takes over so that the final incremental current is negative. The response is similar in form, therefore, to fig. 15(a) but for i rather than V. Correspondingly it is the admittance $A(p)$ that is well-behaved and remains finite for all real positive p, and there is no danger of exponential instability when the diode is connected across a parallel resonant circuit.

The tunnel diode can be used to maintain low-frequency oscillations, but it is a much more delicate device than a transistor amplifier and its real value lies in its ability to work up to much higher frequencies than the latter. In recent years, however, it has been superseded for this purpose by other devices, especially the Gunn diode, where mode of operation is too complicated for it to be described as a simple negative resistance element, even if ultimately that is the significant feature of its behaviour.

(c) Mechanical analogues of negative and non-linear resistance; Liénard's construction. The phenomenon of whirling discussed in chapter 8 provided an example, quite different in kind from the circuits treated in the earlier sections, of a resonant system in which the damping term had had its sign reversed. Mechanical examples as simple as this are not encountered very frequently, most spontaneous mechanical vibrations

being more readily interpreted as feedback processes in which the driving force is so phased as to cause the amplitude to increase. As we have seen, particularly with class C oscillators, the distinction between these two interpretations is not easily drawn, being mainly a matter of mathematical formulation. The next examples are, however, fairly clear cases of negative resistance, complicated by non-linearities which make it desirable to preface their discussion with an account of a useful graphical technique, due to Liénard.[8]

Consider the standard form (2.23) for a simple vibrator, and let us **30** choose such a unit of time that the natural frequency $\omega_0 = 1$. Then

$$\ddot{\xi} + k(\dot{\xi}) + \xi = 0, \tag{21}$$

and we have indicated that the dissipative force, $-k(\dot{\xi})$, is now to be regarded as a more general function of velocity than mere proportionality. In terms of the variables ξ and v, defined as $\dot{\xi}$, (21) may be rewritten:

$$dv/d\xi = -[\xi + k(v)]/v. \tag{22}$$

The slope of the trajectory on the (ξ, v)-plane (the phase plane) is defined at every point by (22), and may be determined graphically very easily, as illustrated in fig. 22(a). The curve K is $-k(v)$; to find the tangent to the trajectory at any point P, draw a horizontal line to meet K in Q, and drop a vertical line from Q to cut the ξ-axis at N. Then T, drawn normal to NP, is the tangent to the trajectory at P. The proof follows immediately from the fact that $PQ = \xi + k(v)$, while $QN = v$.

The argument may be extended to provide a construction, not just for the slope, but for the curvature as well, so that the trajectory may be synthesized as a succession of circular arcs. The construction of C, the centre of curvature of the trajectory S at P, is shown in fig. 22(b). Draw PQ, QN and NP as before. Construct the normal to PN at P (i.e. the tangent to S) and the tangent to K at Q, and find their intersection, G. Join GN and let it cut PQ, produced if necessary, in F. Then C lies on PN vertically below F.

Proof: Fig. 22(c) shows two neighbouring points on S, P and P', with their associated Q and N; C is the intersection of PN and $P'N'$ in the limit as P' is moved towards P. It is required to show that FC is a vertical line. GL is the tangent to K at Q, GM the tangent to S at P. Since the triangles $EQ'Q$ and $HP'P$ are respectively similar to NLQ and NMP, with the same scaling factor, it follows that $NN'/HP' = NL/NM = FQ/FP$. But in the limit, as $P' \to P$, $CN/CP = NN'/HP'$. Therefore $CN/CP = FQ/FP$ and FC is parallel to QN, i.e. is a vertical line.

Given a set square, and squared paper to speed up the determination of Q and N, one can find C in a very short time and hence draw a short arc of S through P. The trajectory that results from this process carries more conviction than one constructed solely from the tangents as found by Liénard's method.

Fig. 22. (*a*) Liénard's construction for the tangent T at the point P on the phase-plane trajectory of a point obeying (22). (*b*) Construction for the centre of curvature, C, of the phase-plane trajectory at P. (*c*) Diagram used in proof of (*b*).

(*a*)　　　　　(*b*)

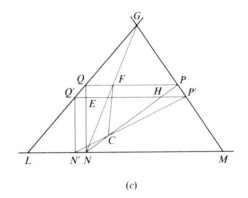

(*c*)

Apart from the graphical value of the construction, certain general points emerge, which the reader may readily prove, and which indicate why a trajectory has its particular shape:

(1) When S cuts the ξ-axis it does so vertically, and the centre of curvature C is the intersection of K with the ξ-axis.

(2) When S cuts K it does so horizontally, and the centre of curvature lies on the ξ-axis vertically below.

(3) C also lies on the ξ-axis when the corresponding point on K has a vertical tangent.

(4) If the tangents at corresponding points on K and S intersect on the ξ-axis, S has zero curvature. If they intersect at a vertical level between the ξ-axis and PQ, S is concave; otherwise it is convex.

Let us apply the graphical process to a few examples, starting with the simplest. For a lossless harmonic vibrator, $k = 0$, K coincides with the

v-axis and N is always the origin; the trajectory is a circle centred on O. When $k \propto v$ and positive, but less than $2v$, the trajectory is a converging spiral as in fig. 23(a) (focal stability). Convergence is forced by the necessity for the trajectory to run horizontally across K and vertically across the ξ-axis. Conversely, if $k \propto v$ and negative, the spiral diverges, as in (b) (focal instability). If, however, $k > 2v$ the vibrator is overdamped (nodal stability); (c) shows the trajectory in such a case, the system having been started (like a ballistic galvanometer) from its rest position by an impulse. These cases, of course, are easily treated analytically – the value of the construction reveals itself when k is not proportional to v.

Fig. 23. Trajectories for the case $k \propto v$ in (21); (a) damped oscillation for $0 < k < 2v$, (b) growing oscillation for $-2v < k < 0$, (c) overdamped response for $k > 2v$.

The model system sketched in fig. 24 may oscillate spontaneously if the frictional force varies with slipping speed in the appropriate way, as the slipping speed rises. In fig. 24(b) the speed of the belt is represented by v_0. If $v < v_0$ the mass M is dragged forward, i.e. $k < 0$, while if $v > v_0$ it is dragged back; the frictional force is depicted as lessening with increase of $|v - v_0|$. Where K cuts the ξ-axis, at W, there is a point of unstable equilibrium, the spring being stretched so that its force just balances the frictional force at a slipping speed of v_0. But the slope of K, as in fig. 23(b), shows that oscillations will grow until M achieves a speed of v_0 at some moment in the cycle. The trajectory cannot cross the horizontal part of K, i.e. M cannot move faster than the belt. For if the trajectory were to continue above v_0, Liénard's construction gives a negative slope (shown as

Fig. 24. (*a*) A mass *M* sliding on a moving frictional belt and restrained by a spring; (*b*) trajectory when the frictional force decreases with velocity.

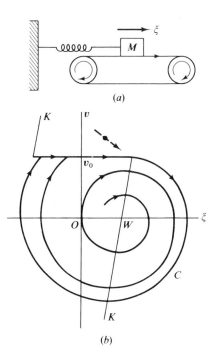

(*a*)

(*b*)

a broken line); but v is positive and ξ must be increasing – motion on this part of the trajectory, as shown by the arrow, is towards the horizontal part of K, not away from it. Of course, what actually happens is that the trajectory follows K, the frictional force increasing as M moves at the same speed as the belt, until the maximum friction is being exerted. From then on the vibration repeats itself regularly, as shown by the limiting form C.

A commoner form of this type of process is the stick–slip vibration, arising when the frictional force is sensibly constant once slipping has begun, but less than the force needed to start slipping.[9] Real examples of this phenomenon were mentioned in chapter 2. The phase-plane trajectory is shown in fig. 25(*a*). For small vibrations about W there is no gain or loss in this idealized system, but if the vibration is started with enough amplitude to reach v_0 it is then stable against extra dissipative effects, if not too large. For example, if there is a viscous drag on M in addition to the friction of the belt, K takes the form shown in fig. 25(*b*). Small amplitude vibrations die away, but larger amplitude vibrations may survive thanks to the extra kick they get from the stickiness of the belt. The form of the vibration, $\xi(t)$, is easily constructed when viscous drag is neglected, since it consists of a pure sinusoidal curve interrupted by straight sections. In fig. 26 the sine curve is drawn about a base line at ξ_0, which is the displacement that causes the spring to balance the sliding friction; its amplitude is such that at ξ_1, which corresponds to the sticking friction, the slope $\dot{\xi}$ is just equal to the speed of the belt. Then the portion AB, where the sine curve has a greater slope, is excised and replaced by a straight line. Clearly the

42

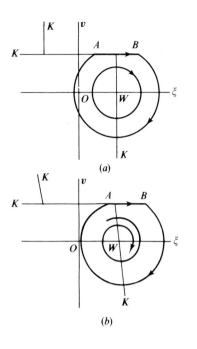

Fig. 25. Phase-plane
trajectories for stick–slip
oscillation (*a*) without
other damping effects,
(*b*) with additional
viscous damping.

Fig. 26. waveform of
stick–slip vibration,
corresponding to fig.
25(*a*).

period of vibration is slightly increased over the natural period; this is
equally clear from fig. 25(*a*), the stretch *AB* being traversed at a rather
smaller value of *v* than along an arc of the circle connecting these points.

A quite different case of negative resistance may be observed when a
metal block, suspended from a torsion wire, is rotated in a transverse
magnetic field, as shown in fig. 27. When the block turns slowly, the motion
of its parts in the field **B** induces eddy currents which create in the block a
magnetic moment **M** at right angles to **B**; the interaction of **M** with **B**
results in a drag on the rotation, so that the torsion wire acquires a twist. As
the angular velocity, Ω, of the torsion head is increased, **M** and the drag
are at first proportional to Ω, so that the block behaves as though immersed
in a viscous medium. The effect is surprisingly strong – a copper bar 5 cm in
diameter held transversely in the field of a good electromagnet can only be
turned very slowly by hand. At higher values of Ω the magnetic field of the
eddy currents is strong enough to cause a marked diminution of the field

Fig. 27. A conducting cylinder supported in a horizontal magnetic field by a wire attached to a steadily rotating torsion head.

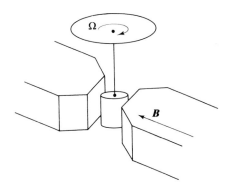

strength inside the block, and the value of M ceases to rise as fast as Ω. Indeed, the drag rises to a maximum at a certain Ω and at faster speeds decreases, in the way shown in fig. 28. At very high speeds the magnetic field is excluded from all but a thin layer at the surface of the block; this is the same as the well-known skin effect in metals, manifested here in a slightly different context.

Fig. 28. Eddy-current torque G on cylinder as a function of rotational angular velocity, Ω.

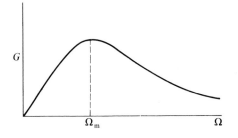

So long as the drag $G(\Omega)$ is a function solely of Ω – and it is only fair to give warning that this proves an overoptimistic assumption – we can write the equation of rotary motion in a form analogous to (21). Let ξ be the angle of twist of the torsion wire, and let the torsion head have uniform angular velocity Ω_0. Then the angular velocity of the block is $\Omega_0 + \dot{\xi}$, and

$$I\ddot{\xi} + G(\Omega_0 + \dot{\xi}) + \mu\xi = 0,$$

where I is the moment of inertia of the block and μ the torsion constant of the wire. If $(I/\mu)^{\frac{1}{2}}$ be taken as the unit of time, and $\dot{\xi}$ written as v, this is reduced to the same form as (22), with

$$k(v) = G(\Omega_0 + \dot{\xi})/\mu.$$

The phase-plane trajectory is now constructed as before, and fig. 29 shows how oscillations die out so long as Ω_0 is lower than the value Ω_m at the maximum of G, but grow to a limiting form if $\Omega_0 > \Omega_m$. If the torsion wire is weak, the unit of time is long and Ω_m is correspondingly large. However,

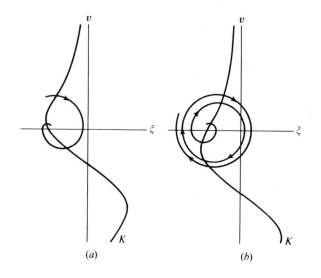

Fig. 29. Phase-plane trajectories for the arrangement in fig. 27: (*a*) damped oscillation when $\Omega < \Omega_m$; (*b*) maintained oscillation when $\Omega > \Omega_m$.

$\Omega_m \propto \mu^{-\frac{1}{2}}$, while $k(\Omega_m) \propto \mu^{-1}$, so that weakening the suspension causes the curve K to spread out horizontally, and the form of the oscillation is correspondingly distorted. The same effect is produced by strengthening the magnetic field. An example is given in fig. 30, which is worth comparing with fig. 23(*c*). When v is positive the block is turning faster than the torsion head, and the damping torque is relatively low – the motion is almost free. As soon, however, as it slows down and runs retrograde it gets caught up in the range of angular velocity where the torque varies rapidly with speed, in such a way as to reproduce the conditions for heavy over-damping. The trajectory hugs the curve K until the block has caught up

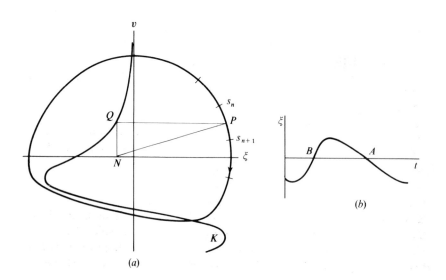

Fig. 30. Non-sinusoidal oscillation of conducting cylinder when damping is heavy; on the trajectory (*a*) is shown the construction by which the waveform (*b*) is derived.

enough speed to pass the maximum torque condition, when it begins to turn more freely again.

To translate the trajectory on the phase plane into a graph of ξ as a function of time, a simple graphical construction may be used, as illustrated in fig. 30. Divide the trajectory into equal arcs, each of length s, a typical arc being that between S_n and S_{n+1}. At the centre of each arc perform Liénard's construction. Then since QN is the ξ-component of the representative point's velocity along the trajectory, PN has the magnitude of the velocity itself. The time taken to traverse the arc s is therefore s/PN. In this way one may find the interval between each pair of points S_n and hence construct the waveform, as in fig. 30(b). It should be stressed that this construction, like the construction of the trajectory itself, is a qualitative graphical process that can only be made quantitatively accurate with very great care and skill. Its value lies in the insight it provides into the mechanisms at work and the possible forms the solution of the equation of motion may take. If one does not want insight but only a rather exact plot of the waveform, it is much better to integrate the equation of motion step-by-step with a computer.

To return to the physical system, experiment shows that the waveform is not quite like fig. 30(b). The observed forms in fig. 31 were recorded for a very pure copper cylinder cooled in liquid helium, having so high a conductivity that large eddy currents were induced by slow rotation and the maximum torque was generated at a speed of only $\frac{1}{10}$ revolution per second. So long as Ω was not much greater than Ω_m the oscillations, of

Fig. 31. Experimental record of oscillations of a conducting cylinder (unpublished observations by Dr. P. Martel).

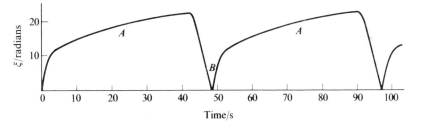

rather small amplitude, were not far from sinusoidal, but when more violent oscillations occurred the 'overdamped' regions A between segments of fairly free motion B were elongated and opposite in slope to those of fig. 30(b). The assumption that G is determined by Ω alone is in fact not valid. The eddy currents take a long time ($\gg 1/\Omega_m$) to change their character, and when the block is turning through a large angle in such a time they and their magnetic moment are carried round with it, creating all manner of changes in the torque. The value of G at any instant, therefore, is determined by the entire past history and not by the instantaneous value of Ω. Instead of a non-linear differential equation, the motion is governed by a non-linear integral equation, so complicated indeed as to justify the effort of computation only if the solution were needed for some important purpose. Since this condition does not apply here we make no apology for shirking a fuller explanation of the observations. The phenomenon has been introduced here partly to show how to apply Liénard's construction, but partly also as an illustration of how difficult a mathematical problem may be posed by a very simple physical arrangement. Non-linear integro-differential equations are so troublesome that one tends to put out of mind physical problems whose mathematical formulation leads in this direction. The very next example to be discussed is just such a one, but fortunately, like the rotating block, of no importance. It is a measure of the task that so much remains to be accomplished in the theory of musical instruments, including the violin and all wood and brass wind instruments, in which the excitation of the sound involves a non-linear time-dependent interaction between (on the one hand) bow, reed or lips and (on the other) the strings or an acoustic resonator. The same could be said of the human voice, and it can hardly be contended that these are unimportant vibratory systems. It is well, then, to remember that the problems of classical mathematical physics are far from exhausted – if it sometimes appears that they are, the blame is more likely to lie with limitations of mathematical technique than with the problem-setting resources of the natural world.

Two other oscillatory systems are worth mentioning rather briefly, being rather easy to make and experiment with, though otherwise not especially notable. They will serve as an interlude before turning to the far more significant *maser*, which unfortunately is not suitable for do-it-yourself demonstrations. The water-driven rocker in fig. 32 is particularly worth playing with, for it shows a remarkable variety of behaviour. The precise dimensions hardly matter. If the weight W is small so that the period of oscillation is rather long, a thin trickle of water serves to set the channel swinging and a large amplitude builds up quickly. The reason is plain – when one end of the channel is down, water collects near the septum on the other side, and runs down as soon as the channel rocks back beyond the level position; with a slow period there is plenty of time for the water to run

Fig. 32. Rocking channel driven by water stream. An aluminium channel, bent out of sheet, about 60 cm long and with sides of 4 cm, is suitable; a septum, cemented by epoxy resin into a saw-cut, serves also as pivot. The weight W can be used to adjust the rocking frequency. The behaviour can be studied with the gap below the septum blocked or unblocked.

off the end before the channel reaches its lowest point, and the weight of the water therefore exerts a couple in the same sense as the angular velocity, transmitting energy to the channel. If, however, W is heavy enough to make it rock fairly rapidly, it may start to rise before the water has reached the end, and the maximum torque is exerted against the angular velocity, causing damping of the vibration. This is particularly obvious when the amplitude is small and the flow of the water down the channel slow. At large amplitudes, however, there may still be time for the water to escape before the channel rises. Thus the vibration may be damped at small amplitudes and encouraged at large. This is, of course, a highly non-linear system, in which the time variation of the torque due to the water does not scale in amplitude alone, but suffers compression in time as the amplitude rises. It is worth constructing the rocker so that the septum can be removed, and also to provide for closing the ends of the channel. A number of quite surprising types of behaviour can be observed by making such minor changes, interpreting some of which offers a considerable challenge.

Thermally driven acoustic vibrations are not uncommon; almost every glass-blower, however inexperienced, has found a glass tube, after having one end sealed off, singing spontaneously. Rayleigh, in his *Theory of Sound*,[10] discusses several related instances, including one that is fairly easy to make and which was discovered by Rijke in 1859 – a large vertical pipe, open at both ends, and excited to emit a loud tone by a heated gauze placed about one-quarter of its length from the bottom. To understand what goes on we must first examine the dissipative effects present and then see how the heated gauze reverses their sign. The essential mechanism for loss arises from the oscillations of temperature of the vibrating gas column, which suffers alternately very nearly adiabatic compressions and rarefactions. In themselves these oscillations cause no dissipation, but if they lead to a transfer of heat to the walls of the resonator or other bodies in thermal contact with the air, there develops a phase difference between the pressure and volume oscillations of the air which has as its consequence the dissipation of acoustic energy as heat in the air.

This point is perhaps best appreciated in terms of the step response function of the pressure after a sudden small contraction in volume, as illustrated in fig. 33(a). At $t = 0$ the gas is compressed adiabatically by ΔV and thereafter held at its new volume; the instantaneous response of the pressure is in accordance with the adiabatic equation $PV^\gamma = $ constant, so that $\Delta P_0 / P = -\gamma \Delta V / V$. If there were no heat transfer to the walls or other bodies, the pressure would remain at ΔP_0, but if the gas can cool to the ambient temperature the subsequent change of ΔP is a (not necessarily exponential) relaxation to the value ΔP_∞ appropriate to isothermal compression, $\Delta P_0 = \gamma \Delta P_\infty$. The time-derivative of this curve, shown as fig. 33(b), is the impulse response function, which in addition to its instantaneous component at $t = 0$ shows a delayed response. It should be clear from the discussion of chapter 5 that the susceptibility (compressibility)

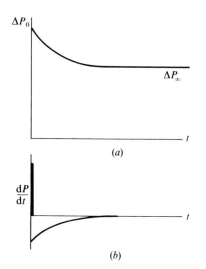

(a)

(b)

Fig. 33. (a) Excess pressure ΔP in a gas after a small step-function decrease in volume at time $t = 0$; (b) the corresponding impulse response function.

$\chi(\omega)$ derived from this by a Fourier transform will in general be complex, with dissipation as an inevitable consequence.

For the effect to be reversed the gas must receive heat, not lose it, when it is hot, and/or lose heat when it is cold. Let us see how this is achieved by the heated gauze. When the air in the pipe is in vibration in its fundamental resonance all parts suffer oscillations of pressure in the same phase, the amplitude being greatest at the centre. Except for the effect of end-corrections, no fluctuations of pressure occur at the open ends; the air moving in and out, to provide for the fluctuations inside, has its greatest speed at the ends, and does not move at all at the very centre, except for a steady upward convective flow past the heated gauze. The gauze, about one-quarter of the way up, lies in a region where there are substantial oscillations both of pressure and of velocity. Let us now show diagrammatically, as in fig. 34, how various parameters change with time in the vicinity of the gauze. Curve (a) shows the fluctuations of pressure, ΔP being the increase above atmospheric pressure; the same curve serves for the temperature fluctuations since in adiabatic compression temperature and pressure march together. In curve (b) is shown the upward velocity, v, consisting of a sinusoidal fluctuation superposed on a steady convective movement; the phase of the fluctuating component is easily deduced from the fact that it must be greatest at the moment when the pressure is increasing most rapidly, as, for example, at A. By integrating curve (b) we are enabled to sketch the variation with time of the height, h, of an element of air, and this must clearly, as in (c), be a sinusoidal fluctuation around the uniform rise due to convection. Different elements of air will show the same pattern, but displaced vertically. It will be seen that there is a certain fraction of each cycle, drawn more heavily, during which the air reaches a height it has not previously attained. The first passage of the air through the heated gauze occurs during one or other of these periods, and if

Fig. 34. Oscillations in Rijke tube of (a) ΔP and ΔT, (b) particle velocity, v, in lower half of tube, (c) particle height, h, being the integral of (b).

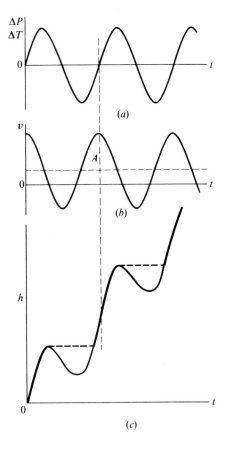

afterwards it falls back through the gauze and rises again, very little more happens since it does not have time to cool appreciably – it is the first passage that is important, and marks the moment when it receives a substantial input of heat. As the diagrams show, this moment is normally a little before the peak in the temperature fluctuation, and certainly not during that part of the cycle when the air is cooled by adiabatic expansion. The required conditions are present, therefore, for reversing the dissipative process, the heat being provided at the time the temperature is greatest, and we can understand in the light of this argument why the tube goes into spontaneous oscillation. As Rayleigh points out, if the gauze were to be put in the top half of the tube, it would communicate heat during the cold part of the cycle, and no spontaneous oscillation should occur – a prediction which is verified in practice.

(d) *The maser*[11]. Although it is often implied that maser action can only be understood as a quantum-mechanical process, this is an over-statement; there are processes which can be described equally well in classical or quantal terms, yet whose quantal description reveals the maser principle.

Nevertheless an attempt to base the development of the theory on classical ideas alone would be misguided, and we must therefore defer a full discussion until the second volume. This does not mean, however, that a qualitative discussion is ruled out, for this can be presented with no more than the most general ideas of quantum physics, and indeed is well worth attempting, in view of the great important of masers and their optical counterparts, lasers, in modern physics.

The original maser invented by Townes[12] consists of a microwave cavity maintained in spontaneous oscillation at about 2.5 GHz by passing through it, *in vacuo*, a stream of ammonia molecules in an excited state – separated from the unexcited molecules by means of an electrostatic lens. We need not spend time on the theory of this separation or on its practical realization; it is enough to recognize that the excited molecules provide the cavity with a dielectric filling having the property of negative dielectric loss, sufficient (in spite of the rather small number present at any time) to overcome the resistive losses of the walls, etc. Granted this mechanism, the oscillation in the cavity is seen essentially to be governed by the standard differential equation for a resonant system with the loss term reversed in sign, and there is no need to say more on this point. We concentrate therefore on why the excited molecules have negative loss, and for this purpose we consider how the normal process of dielectric loss is described in quantum-mechanical terms. Essentially, the oscillating energy in the cavity being quantized in units of $\hbar\omega$, dielectric loss involves the *irreversible* transfer of quanta from the oscillatory field to the molecules. If the molecules in their passage through the cavity start in a quantized state from which they can be excited to a higher state by absorbing $\hbar\omega$ of energy, they may make this transition and leave the cavity carrying their extra energy away with them.† Conversely, and this is the essence of the maser principle, molecules entering in the higher state may be stimulated by the oscillatory field to fall to the lower state, communicating one quantum to the field if they then leave the cavity before re-excitation occurs. It is not enough to pass an unseparated stream of ammonia molecules through the cavity, since in thermal equilibrium at room temperature there are very nearly equal numbers in the two states, the energy difference between the states, $\hbar\omega$, being much less than $k_B T$. In any case, the Boltzmann distribution ensures that there is an excess, amounting to $\hbar\omega/k_B T$ or something less than 1%, in the lower state, so that as a whole the stream of gas would be slightly lossy rather than the reverse. It is the separation of the

† This is a rather special process of dielectric loss. The more usual process, in dielectric materials as distinct from molecular beams, is that the quantum first excites internal motions of the molecule; subsequently this energy is transferred to other modes of motion which we characterize as heat, for example, translational kinetic energy or vibrations of the solid as a whole. But the end-result is the same from the point of view of the oscillatory field – the energy is irretrievably lost.

excited molecules that creates a stream of molecules capable of maintaining oscillation.

It should be noted that the negative-loss mechanism, unlike a negative resistance, is extremely frequency-sensitive; the natural frequency of the cavity must lie very close to the frequency $\Delta E/\hbar$, where ΔE is the level separation, if the quantum $\hbar\omega$ is to cause real transitions. How close may be estimated by a simple argument which may not carry conviction, standing alone, but in fact can be fully justified by a complete quantum-mechanical analysis. From the point of view of an ammonia molecule in its passage through the cavity, the radiation field is applied only for a time Δt equal to L/v, where L is the length of the cavity and v the molecular velocity. Correspondingly it experiences an oscillatory stimulation that is not perfectly sharp in frequency, but spread into a narrow range $\Delta\omega \sim 4\pi/\Delta t$ (cf. (4.21)). The cavity frequency must agree with the natural frequency of the ammonia excitation within this range for the process to work – if the natural frequency is not present in the field experienced by the molecule, no transition will occur. Since Δt is typically 10^{-4} s, $\Delta\omega$ is no more than 10^5 s^{-1}, a million times less than ω itself.

73

The foregoing description of maser action in terms of quantum jumps ignores the question of the dielectric response of the assembly of molecules – what we might call the classical description of the medium as a dielectric whose polarization is so phased relative to the exciting electric field as to transmit energy rather than absorbing it. A normal lossy dielectric has its oscillatory polarization lagging behind the field, but in the maser one must suppose it to lead. A quantum-mechanical calculation confirms this, and enables one to give a more precise account. An ammonia molecule in either of its two stationary states has no dipole moment, but after entering the cavity, and experiencing the electric field tuned exactly to synchronism with it, its moment grows steadily in magnitude, and is phased in quadrature with the field, lagging if it starts in the lower state, leading it in the upper state. With a weak field the rate of growth is constant and proportional to the field strength, and if the molecule leaves the cavity while in this linear regime the average dielectric response during its sojourn is linear and an effective dielectric constant ε can be assigned. But with stronger fields the dipole moment has time to reach its maximum and begin to fall; for the perfectly synchronized field causes it to shuttle continually between the two stationary states, and the dipole moment rises and falls in magnitude, sometimes leading and sometimes lagging. This is too pictorial a description to please a quantum physicist, but it is a good representation of the average behaviour of a large number of molecules, even if each separately cannot be described in such classical detail. With a very high field strength in the cavity the molecule makes so many transitions that it may emerge in either state. Each excited molecule then on the average communicates $\frac{1}{2}\hbar\omega$ to the cavity. If we define an effective (imaginary) dielectric constant in such a way that the rate of energy loss is $\frac{1}{2}\varepsilon''\mathscr{E}^2$ when the field amplitude is $\mathscr{E}$, the fact that the energy gain saturates means that

ε'' is negative and falls off as $1/\mathscr{E}^2$ at high field strengths. A schematic diagram showing how ε'' varies is given in fig. 35; the broken lines show the low and high field asymptotes, and the actual curve rises to something like twice the height of the latter when $\mathscr{E}$ is such as to cause each molecule to make a complete transition. In practice, of course, the amplitude grows until the effective value of ε'' is just right to supply as much energy as is dissipated elsewhere.

Fig. 35. The effective mean value of ε'' for ammonia molecules moving at a given speed through a microwave resonator, as a function of the oscillatory field-strength $\mathscr{E}$. The wide range of molecular velocities in practice smooths out the oscillatory tail so that it conforms to the asymptotic behaviour shown by the broken curve.

　　With this very sketchy account we must leave the maser, to return again when the time is ripe for a quantum-mechanical analysis. It may be remarked, however, that this analysis will establish an extraordinarily close correspondence between a two-level quantized system and the gyromagnetic top described in chapter 8 as a model of proton resonance (a proton in a magnetic field being itself a two-level system). Indeed, the account in the last paragraph of the growth of the oscillatory dipole to a maximum and its subsequent decay is simply an account of the way the transverse component of the moment of the top varies with time as its cone of precession expands and contracts under the influence of a perturbing field at the precessional frequency.

218

Saturation of negative-resistance oscillators

The maser reaches a steady level of oscillation when the source of energy just matches the losses, which are in any case very small. The molecules providing the compensating gain do so by being stimulated to develop dipole moments whose oscillations lead the oscillations of electric field. Both mechanisms of loss (resistance of walls or extraction of useful energy) and gain (dielectric) operate throughout the cycle of oscillation, cancelling each other's effects and leaving the cavity to a very good approximation in an ideally lossless state, executing pure sinusoidal vibrations. The class C oscillator obtained by inductive feedback also oscillates very nearly sinusoidally, but not quite so ideally as the maser, for it is a free-running oscillator, with decay, over most of its cycle and gets an occasional kick to keep each cycle up to the amplitude of its predecessors.

The approach to the steady state is different in these two cases. We have already considered the class C oscillator and now compare it with a high-Q resonant circuit just maintained by a slightly non-linear negative resistance; this is a convenient model of a maser. We assume the p.d. across the negative resistance to contain a positive cubic term, so that the current through this and the linear positive resistance of the circuit, taken together, generates a p.d. according to the equation:

$$V = -\alpha i + \beta i^3, \tag{23}$$

α and β being small enough for the oscillation to build up very gently and equally gently approach its steady amplitude. It is then permissible to assume a pure sinusoidal waveform for the purpose of calculating the mean rate of dissipation. If the current is $i_0 \sin \omega t$, the dissipation in one cycle is ΔE, where

$$\Delta E = 4 \int_0^{\frac{1}{2}\pi/\omega} (-\alpha i_0 \sin \omega t + \beta i_0^3 \sin^3 \omega t) i_0 \sin \omega t \, dt$$

$$= (2\pi/\omega)(-\tfrac{1}{2}\alpha i_0^2 + \tfrac{3}{8}\beta i_0^4). \tag{24}$$

The steady-state value of i_0 is $(\tfrac{4}{3}\alpha/\beta)^{\frac{1}{2}}$, such as to make ΔE vanish. If i_0 is increased by δi_0, the energy of oscillation is increased by $Li_0 \, \delta i_0$, and the dissipation per cycle by $(\mathrm{d}/\mathrm{d}i_0) \Delta E \cdot \delta i_0$, i.e. $2\pi\alpha i_0\delta i_0/\omega$. Hence the amplitude approaches its stationary value at a rate that may be expressed by a Q-value of $\omega L/\alpha$, which is the same as the Q for the initial build-up. This is in contrast to the class C oscillator, which has a Q for approaching the steady state of the same value as the resonant circuit without feedback, and a Q for initial build-up that can be very small indeed, if the amplifier is powerful.

Different again are the negative-resistance oscillators discussed earlier in which the current oscillations may, like the class C oscillator, have a rapid build-up, but as they approach the steady state may be subject to a resistance that alternates between large negative at low amplitudes and large positive at high. The amplification per cycle at low amplitude, and the decrement per cycle at high, are now such that it is quite unrealistic to pretend that the waveform approximates to a sinusoid. We shall not attempt to analyse the rate of approach to the steady state, but shall concentrate on the waveform in the steady state itself, and shall devote close attention to this because it will lead us to the important class of *relaxation oscillators*. There is a great similarity in the behaviour of a tunnel diode across a parallel resonant circuit and the series resonant circuit of fig. 14(a), and it is sufficient to treat the latter only. Let us consider, therefore, the general problem of a series resonant circuit consisting of an inductor, a capacitor, and a non-linear resistor across which the potential drop is an arbitrary function of current, $V(i)$, but with no time-dependence. The differential equation for the charge on the capacitor takes the form

$$L\ddot{q} + V(\dot{q}) + q/C = 0. \tag{25}$$

Now put $\xi = q(L/C)^{\frac{1}{2}}$, $\tau = t/(LC)^{\frac{1}{2}}$, $v = \mathrm{d}\xi/\mathrm{d}\tau = L\dot{q}$, and $k(v) = (LC)^{\frac{1}{2}} V(v/L)$. Then (25) takes the same form as (21) and can be analysed by the same methods. Fig. 36 shows an example of Lienard's construction applied to a non-linear characteristic corresponding in general terms to the negative-resistance feedback circuit whose characteristic is exhibited in fig. 14. The circuit oscillates with an amplitude that increases rapidly at first, and then reaches a steady state fairly quickly. The rapid expansion implies that the Q-value for the negative-resistance regime is low; in fact when $V = R\dot{q}$, the slope $\mathrm{d}k/\mathrm{d}v$ is just $R(C/L)^{\frac{1}{2}}$, or $1/Q$ if Q is not too small (cf. (2.27)). The Q-value chosen here is very close to unity, corresponding to an increase in amplitude by a factor $\mathrm{e}^{\pi/2}$, i.e. 4.8, each half-cycle.

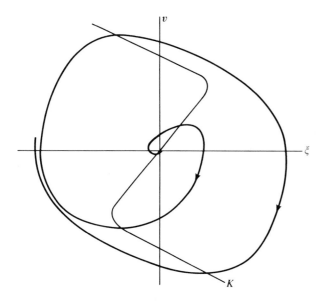

Fig. 36. Build-up to the steady state of the oscillations of the negative-resistance circuit in fig. 14(*a*), when feedback is not very strong and the oscillations are nearly sinusoidal.

The graphical procedure described earlier can be used to construct the waveform of the oscillation. In this particular case it is hardly worth exhibiting the final waveform, since it is only slightly distorted from a sinusoid. This is attributable to the choice of $k(v)$ with a small enough derivative everywhere for the circuit to remain oscillatory. Thus, as we have seen already, the value of Q at low amplitudes is about -1, while at high amplitudes, where the characteristic has also been drawn straight and with a slightly deeper gradient, Q is correspondingly slightly lower and positive. The fact that this linear stretch of the characteristic, at higher values of v, does not pass through the origin, does not prevent Q being defined, since by a change of variables the circuit equation may be reduced to the standard linear form. For if $V(\dot{q}) = V_0 + R\dot{q}$, (25) takes the form:

$$L\ddot{q}_1 + R\dot{q}_1 + q_1/C = 0, \quad \text{where } q_1 = q + CV_0. \qquad (26)$$

So long as $|R| < 2(L/C)^{\frac{1}{2}}$, i.e. $|dk/dv| < 2$, the solutions of (26) are oscillatory about an offset mean value of $-CV_0$ (only q shows an offset; $\dot{q}$ oscillates about zero). But if, say, L is decreased so that the slope of $k(v)$ rises to high positive and negative values, the trajectory hugs the K-curve as close as possible and thus assumes a much more angular form, as illustrated in fig. 37 together with the wave form of the oscillation of current, v.

Fig. 37. The same as fig. 36 but with stronger feedback, to show (b) the squaring of the current waveform.

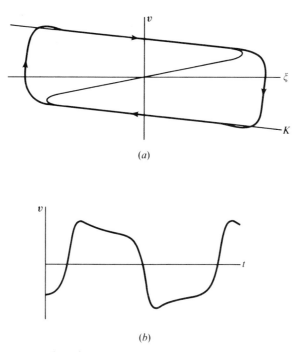

(a)

(b)

The general shape of the waveform follows from the solutions of (26). In the heavily overdamped case, the two time-constants for exponential growth or decay are L/R and RC. If R is negative, both exponentials are positive and that which grows faster dominates the growth; thus the low-current portions of the cycle are controlled by the short time-constant L/R. Conversely, if R is positive the rate of decay is limited by the slowest process, so that the high-current parts of the cycle are controlled by the long time-constant RC. With a sharply kinked characteristic $- V_0$, towards which q/C relaxes exponentially, may be well below the range of V traversed in the cycle; this part of the cycle will then be very nearly linear. When the inductor is removed altogether, the fast passage of the current through zero is limited only by the amplifier response or by stray inductances. The circuit of fig. 38 then shows a 'multivibrator' waveform,

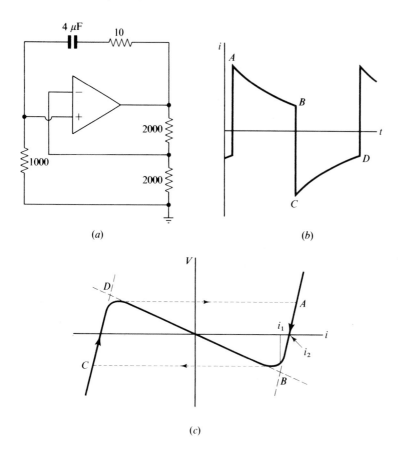

Fig. 38. (*a*) The circuit of fig. 14(*a*) with the inductor eliminated: (*b*) the 'multivibrator' waveform; (*c*) the trajectory on the *i–V* characteristic.

very rich in harmonics. The amplifier is saturated while the range *AB* is traversed, and the decay is determined by the time-constant of *C* in series with a resistor R_s defined by the mean slope of this part of the characteristic; to a good approximation the characteristic may be replaced by three straight lines, and the duration of *AB*, which is half the period, is $R_s C \ln (i_2/i_1)$. If i_2/i_1 is not much over unity the period may be substantially less than the time-constant, $R_s C$.

So long as *L* is present in the circuit, the phase trajectory is a continuous curve, as in fig. 37(*a*), and the *i–V* characteristic is traversed backwards and forwards during each cycle. As *L* is reduced the time taken for the negative-resistance range becomes steadily less, until when *L* is zero the transitions from *D* to *A* and from *B* to *C* are instantaneous, as represented by the broken lines. As *AB* is traversed the current discharges the capacitor and *V* decreases, but beyond *B* any attempt by the representative point to stray onto the negative-resistance range is foiled by the fact that the positive current must continue to reduce *V*. An inductor can allow this to occur, taking up the imbalance between the p.d. on the capacitor and that required to keep the point on the characteristic, though if *L* is small a very rapid change of current will be needed. Without L, the point has no option but to jump to the other side of the characteristic where the values of V and i are also consistent with the needs of the circuit elements. This argument has been spelt out in full to show that the jumping of the representative point is an idealized representation of what in practice is a continuous, but very rapid, process. We shall find similar behaviour in the next circuit to be analysed, the true multivibrator.

Multivibrator[13]

This is one of the classic circuits, invented early in the history of thermionic valves, and still used in updated versions as a generator of very angular waveforms. It also has the capability, as will be discussed in the following chapter, of being locked in subharmonic synchronism with an oscillator of much higher frequency, and therefore of acting as a frequency-divider. Apart from this, with small modifications it can be converted from a free-running oscillator into a bistable circuit with two equilibrium voltage configurations which can be switched over very rapidly; as such it has found application in logic circuits. The mode of operation is, rather naturally, treated in most textbooks in the semi-intuitive terms that electronic engineers come to adopt as their habitual descriptive language. In view of the complexity that would attend any direct mathematical analysis of the average practical circuit, especially if it depends for its operation on non-linear elements, one must respect the skill and economy with which a circuit designer picks out the relevant factors and sees how a given configuration will work. But the uninitiated reader often has difficulty in seeing how rigorous the argument is, and it is for his sake that we shall develop the theory of the multivibrator in some detail.

The multivibrator (fig. 39) is a two-stage RC-coupled amplifier connected back on itself so that if it were possible for the capacitors to be short-circuited without destroying the components there would be positive feedback at zero frequency. Two versions are shown in the diagram, the original with triode valves and a modern version with transistors. When the circuit is connected the resulting oscillations take the form of alternate stages switching fully on and off, the potentials at A, B, X and Y varying as shown in fig. 40. Almost all the time the amplifiers operate well outside their linear range of behaviour, and this means that their detailed characteristics are relatively unimportant, transistors and valves giving very similar results. We shall take the original circuit as the basis for discussion, simply because the connection of the resistors R to ground rather than to the positive line makes for less complexity while retaining the essential features. We shall also assume that the current passed by each valve is determined solely by the grid voltage, independent of the anode voltage. This is an oversimplification but not one that matters; we need know hardly more about the amplifier characteristics than that there is a range in which amplification occurs, bounded by cut-off when the input voltage is

Fig. 39. (*a*) The original multivibrator and (*b*) a modern transistorized version. The resistors *R* which in (*a*) are connected from grid to earth can with advantage be connected instead to the HT line, as in (*b*), but the analysis is slightly simpler for (*a*).

Fig. 40. Waveforms of the voltages on the grids (*A* and *B*) and anodes (*X* and *Y*) of the multivibrator, fig. 39(*a*).

strongly negative and by saturation when it is strongly positive. In the latter circumstance, however, electrons readily move to the grid to eliminate the positive bias, but this will not concern us for the moment. We shall start by deriving the differential equations governing the circuit and, having seen that they are of uncomfortably high order, shall then discuss certain special cases to elucidate the processes taking place.

Fig. 41. Schematic diagram of one amplifier stage in a multivibrator, to show notation. It is assumed that no current flows out at B.

The left-hand stage of the circuit is represented diagrammatically in fig. 41, in which it will be seen that a capacitor C_1 has been introduced, to take account of the capacitance between the anode and other electrodes. It does not matter that the model is not exact – what is important, as will become clear later, is not to forget the existence of inter-electrode capacitance. The current i_1, as remarked above, is assumed to be determined solely by the grid voltage V_1, independent of the anode voltage, V_X:

$$i_1 = i_1(V_1). \tag{27}$$

The equation of continuity for the current at X takes the form:

$$i_1 + C_1 \dot{V}_X + (V_X - V_0)/R_1 + V_2/R = 0, \tag{28}$$

while continuity of the current through R and C gives the equation:

$$C(\dot{V}_2 - \dot{V}_X) + V_2/R = 0. \tag{29}$$

Eliminating V_X from (28) and (29), and using (27), we have:

$$\tau\tau_1 \ddot{V}_2 + \tau'[\dot{V}_2 + \mu(V_1)\dot{V}_1] + V_2 = 0, \tag{30}$$

where $\tau = RC$, $\tau_1 = R_1 C_1$, $\tau' = \tau + \tau_1 + R_1 C$ and $\tau'\mu(V_1) = \tau R_1 di_1/dV_1$; μ is, as near as no matter, the differential amplification factor of the valve with R_1 and R in parallel in its anode circuit, i.e. the change in anode potential for unit change in grid potential. With pentode valves μ can be much larger than unity in the normal operating range, but falls to zero outside. A typical valve characteristic, and the derived form of μ, are shown in fig. 42. A similar equation applies to the right-hand stage:

$$\tau\tau_1 \ddot{V}_1 + \tau'[\dot{V}_1 + \mu(V_2)\dot{V}_2] + V_1 = 0. \tag{31}$$

Together, (30) and (31) yield a non-linear fourth-order equation for either V_1 or V_2, but fortunately one or two special cases tell us enough to make full solution unnecessary. First let us examine the behaviour near the

364

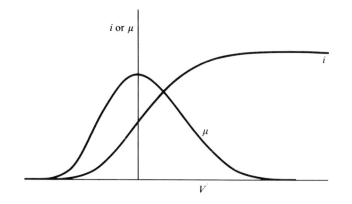

Fig. 42. Anode current, i, as a function of grid voltage V; and the derivative $\mu \propto di/dV$.

equilibrium point, $V_1 = V_2 = 0$, on the assumption that μ is constant.† Then (30) and (31) are linear equations, inspection of which shows that solutions can be found in which $V_2 = \pm V_1$. Since four independent solutions are generated this way, the general solution is a sum of all four in arbitrary proportions. Each has the exponential form e^{pt}, where p obeys the equation

$$\tau_1 \tau p^2 + \tau'(1 \pm \mu)p + 1 = 0, \tag{32}$$

the positive sign referring to the case $V_2 = V_1$ and the negative to the case $V_2 = -V_1$. The four values of p are therefore as follows:

$$p = \frac{1}{2\tau_1 \tau}\{-\tau' \mp \mu\tau' \pm [(1 \pm \mu)^2 \tau'^2 + 4\tau_1\tau_2]^{\frac{1}{2}}\}. \tag{33}$$

So long as μ is large, there is always a positive real value of p which will guarantee the instability of the equilibrium state; any displacement grows exponentially with a time-constant approximately equal to τ_1/μ. This is obtained by taking the positive sign in (32), and corresponds to V_2 and V_1 diverging in opposite senses. The time-constant τ_1 is short enough in its own right, since C_1 is a very small capacitance, and the presence of μ makes the divergence even more rapid. If we had started by neglecting C_1 altogether we should not have deduced a discontinuous jump from equilibrium equivalent to taking τ_1 as zero, but should have missed this solution altogether, and hit on something entirely incorrect. For if $\tau_1 = 0$ in (32) the dominant positive value of p is approximately $1/\mu\tau'$, giving a characteristic time-constant μ times as long as the already substantial τ'. Having recognized this, however, we know enough to keep τ_1 in as long as it matters and thenceforth to neglect it if we are so minded.

The virtually abrupt breakdown of equilibrium involves a jump to quite different values of V_1 and V_2, well outside the linear regime of the

† Strictly we should have taken note that a small current flows to the grid when its potential is zero, and the equilibrium point has slightly negative values of V_1 and V_2. This is not important for the analysis.

amplifiers. To see what happens let us consider a special case, not too far from physical reality, in which $\mu(V)$ is a symmetrical function, $\mu(-V) = \mu(V)$. This allows V_1 and V_2 to stay equal and opposite at all times, and reduces the problem to easily manageable proportions. For (31) now takes the form, applying to either V_1 or V_2:

$$\tau_1 \ddot{V} + \tau'[1 - \mu(V)]\dot{V}/\tau + V/\tau = 0, \tag{34}$$

the equation of a harmonic vibrator for which the viscous damping term depends on position (i.e. V), being strongly negative for small displacements and not so strongly positive for large. One can immediately visualize how such a system will behave. When displaced from zero, it will be impelled by a force of the order of $\mu\dot{V}$ and will continue to accelerate away from the origin until μ has fallen nearly to unity. The motion will continue while the kinetic energy is either dissipated or stored as potential energy. On the return journey the speed will be limited by the longest time-constant in the solution of (34), i.e. $\tau'[1 - \mu(V)]$ if τ_1 is very short. For much of the time this will be almost equal to τ', since $\mu \sim 0$, but as the displacement approaches the value where μ begins to rise once more the motion will speed up until suddenly, as μ passes through unity, a very rapid reversal of displacement will occur, the motion being again dominated by the shortest time-constant. Something like the same process will now occur with reversed signs, and it will not be long before a steady state is reached, with V_1 and V_2 alternating in their positive and negative excursions. If we now neglect τ_1 the pattern is clear – everything happens outside the range in which $\mu > 1$; every time the system reaches the edge of this range it jumps instantaneously to a far point on the opposite side, from which it relaxes back until it reaches a critical point once more. The essential feature of multivibrator action are displayed by this model, and we need take the argument only a little further.

In the light of this discussion, being assured that almost everything of importance takes place with one amplifier cut off and the other saturated, let us ignore τ_1 but let us no longer assume a symmetrical form for $\mu(V)$, so that V_1 and V_2 need not be equal and opposite. The fact that this model demands instantaneous jumps of V_1 and V_2 tells us that there must be different pairs of values of (V_1, V_2) consistent with the same charges on the capacitors C; for the charges cannot be altered abruptly. Let q_1 be the charge on the left-hand capacitor, as shown in fig. 41, and q_2 the charge on the right-hand. The condition for q_1 to take a given value is obtained by writing $V_X = V_2 + Cq_1$ in (28), which then assumes the form:

$$V_2 + \tau R_1 i_1(V_1)/\tau' = F(q_1). \tag{35}$$

This equation defines a family of curves of constant q_1 on a (V_1, V_2)-plane, each identical in shape with the amplifier characteristic of fig. 42, except for a sign reversal. A similar set of curves of constant q_2 is obtained by interchanging q_1 and q_2. Two of each set are shown in fig. 43, the others being obtained by vertical displacement of any one constant-q_1 curve,

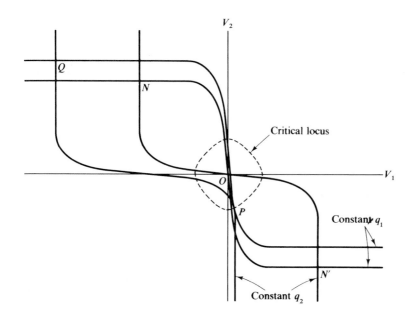

Fig. 43. Typical curves of constant q_1 and q_2; when the curves touch, as at P, P lies on the critical locus defined by the equation $\mu(V_1)\mu(V_2) = 1$.

horizontal displacement of any one constant-q_2 curve. The two curves that intersect at the origin and also at N and N' illustrate the two possibilities immediately after switching on; the system is unstable at O and may switch to N or N', depending on chance. From then on the oscillation is under way, as already described; and the detailed process is made clear by considering the other two curves, drawn so as to touch at P and intersect at Q. The condition for touching follows from (35), which shows that the slope dV_2/dV_1 of the constant-q_1 curve is $-\mu(V_1)$; similarly dV_1/dV_2 for the constant-q_2 curve is $-\mu(V_2)$. For the curves to have the same slope at a given point (V_1, V_2),

$$\mu(V_1)\mu(V_2) = 1. \tag{36}$$

This is the generalized form of the critical condition, $\mu = 1$, which we derived for the symmetrical amplifier characteristic, and the points which satisfy (36) define the critical locus for the problem, in the following sense. It is easy to show from (30) and (31), with τ_1 neglected, that trajectories on the plane all tend towards this locus. Moreover V_1 and V_2 can move along the common tangent to the constant-q curves at P without incurring any changes in q_1 and q_2; hence the representative point arrives at the critical locus with infinite speed (finite if C_1 is included in the calculation), but can proceed no further. It therefore jumps immediately to Q. This is exactly analogous to the process illustrated in fig. 38.

At this point we must take account of a hitherto neglected factor. At Q, V_2 is highly positive, and the grid of amplifier 2 will quickly collect enough electrons to reduce V_2 (V_B in figs. 39 and 40) to a low value. As this happens the anode current falls slightly from its saturation value and the potential at Y rises a little. These effects can be traced in the waveforms of fig. 40, and are schematically represented in fig. 44. Starting at Q, the first stage QR is the rapid elimination of the positive grid potential V_2, while V_1 changes only slightly, being limited by the time-constant τ' which controls the discharge of the right-hand capacitor through R and R_1. From R to S this slow discharge continues and only as it comes very near to S, which lies on the critical locus, does it speed up. From S the point jumps to T, the mirror of Q in a 45° line. The elimination of the positive grid potential V_1 and the subsequent slow discharge of the other capacitor account for TU and UP, and finally the jump from P to Q completes the cycle. The period is essentially defined by the time taken on the paths RS and UP.

Fig. 44. Schematic representation of multivibrator cycle. The broken lines indicate virtually instantaneous jumps, from P to Q and from S to T.

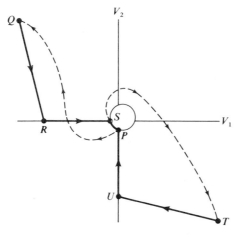

It is now clear that the alternating saturation of each amplifier is automatic, depending in no way on fluctuations (as seems to be implied by some accounts), and that the whole process takes place outside the regime in which the amplifiers combine to give a gain greater than unity. There is a very close resemblance in both respects to the behaviour of the 'multivibrator' circuit of fig. 38, and we may feel justified in dropping the inverted commas and allowing the same name to be attached both to the pioneer circuit and to its simpler successor.

Relaxation oscillators

The multivibrator, whether in its original version or the simplified version of fig. 38, turns out to present a character remarkably similar to the linear line-base circuit of fig. 2.27, in which a gas-filled tube serves as an automatic switch, causing the RC combination, like Sisyphus, to strive ever vainly towards a state of equilibrium. The multivibrator is more symmetrical in its operation, having two states of opposite polarity from each

41

of which it relaxes exponentially (like an *RC* combination), only to be denied the attainment of equilibrium by a sudden switch to the alternate state. This behaviour may be taken as typical of the rather loosely defined *relaxation oscillator*, of which other examples were given in chapter 2, including the stick–slip phenomenon discussed earlier in the chapter. In this case the system has a natural frequency and perhaps cannot strictly be said to be a relaxation oscillator if by relaxation we mean exponential or quasi-exponential return to the equilibrium. More significant, however, is the fact that the driving force for the maintenance of oscillation comes from sudden changes in the physical parameters, and that the changes are triggered by the system itself, not by an external agency. This we take to be the fundamental feature, if any can be found, that defines this class of oscillators.

337

Relaxation oscillators are very common outside the physical laboratory, a very fine type being the geysers of Iceland, New Zealand, and the Yellowstone Park.[14] We shall not go into details here, especially as there seems to be no general agreement about how they work. Indeed, different geysers probably have different mechanisms, but one simple model that should apply to some builds on a feature common to all – an underground water-filled chamber connected to the surface by a pipe, and steadily heated by volcanic action. The water in the chamber can reach a temperature well above 100 °C without boiling (20 m below the surface the boiling point is 134 °C) and before it gets hot enough to boil an unstable situation develops some way up the pipe, where the temperature may be only slightly less than in the chamber but the pressure considerably less. Boiling therefore starts in the pipe, and blows out the water above, so that the hydrostatic head over the chamber is reduced. Now the water there boils catastrophically, blowing large volumes of water and steam to a great height (especially if the pipe enters the chamber well below its roof, and allows much of the water therein to be expelled before the pressure is released). At the end of the eruption, surface or underground water flows back into the chamber and cools it so that the steam condenses and the chamber is refilled almost completely. The heating process starts again and in due course another eruption takes place.

As described, this is clearly a relaxation process like the time-base circuit, but it is normally much more irregular. Among the large geysers, only Old Faithful in Yellowstone Park can be relied on to erupt at the predicted time, and this may be more legend than fact; most are irregular in their periods and violence. This need not be thought of as remarkable, for the process is far more complicated than a simple electrical circuit, and the exact sequence followed must depend on just where the first ebullition occurred, and how much water it managed to eject; after that there is scope for variation in every subsequent stage of the action. In really complex situations irregularity is the rule rather than the exception, for there is no resonant circuit to control the frequency, which is at the mercy of every stray influence. This is particularly true of oscillatory processes in biological systems, many of which (heart beat, menstrual cycle) can be seen to follow the general pattern of relaxation oscillations. Take, for example, the egg-laying of the domestic fowl; the alternating processes of build-up and discharge are clearly exhibited, and so is the variability in period between eggs and in their size, occasioned by too extensive a range of possible causes to be catalogued here. It is worth noting in this connection a study of the egg-weight in ducks which shows how a conflict between the natural period of egg laying and the daylight rhythm leads to systematic variations of weight and periodic non-laying days.[15] This is a process than can be

405   modelled by a relaxation oscillator under the influence of a periodic signal at a frequency other than the natural frequency of the oscillator. Not every such oscillator is susceptible to being forced out of its natural period, and even those that can be locked to the controlling frequency can only be persuaded to shift to a limited extent. The duck that continues to lay eggs of more or less the same weight day after day is one that has responded to the locking signal; the duck that lays successively lighter eggs for a few days, and then after a short gap begins the cycle again, is one whose natural period is significantly greater than a day and cannot be brought under control.

This has brought us to the threshhold of a major new topic, the mutual influences of oscillatory systems on one another, to which we shall turn in the next chapter. There is, however, one aspect of the disturbance of an oscillator by extraneous signals that is conveniently dealt with here, and that is the effect of noise voltages on the phase and amplitude of a class C oscillator, the simplest example to demonstrate the characteristic behaviour.

The effect of noise on a class C oscillator

When Johnson noise or other random disturbances are neglected, a class C oscillator settles down to regular oscillation at the equilibrium amplitude, measured, for example, by q_0, the peak charge on the capacitor. An impulsive e.m.f., of magnitude P, injected into the circuit sets up an additional current P/L which, one-quarter of a cycle later, has been exchanged for an additional charge q, of magnitude $P/\omega_0 L$, on the capacitor. If this impulse is injected at a moment when the oscillatory current is at a peak, it alters the amplitude of oscillation without affecting its phase. Subsequently the amplitude relaxes back to its equilibrium value with a characteristic time-constant $2Q/\omega_0$, as illustrated in fig. 10(a). If, on the other hand, the impulse is injected at a moment when the oscillatory current is zero, it adds a small quadrature component to the oscillation, changing its phase but not its amplitude. There is no recovery from this disturbance, the oscillation proceeding thenceforward as though nothing had occurred. Noise, therefore, considered as a random sequence of impulses arriving at all phases of the oscillation, gives rise to fluctuations of the amplitude about its mean, and to an irregular drift of phase, so that the frequency spectrum is no longer perfectly sharp. As we shall see, under ideal conditions the effect is extremely small and cannot be held to blame for the frequency fluctuations exhibited by a typical oscillator in practice. These are caused by variations in the values of circuit parameters as a result of temperature changes or various aging processes. Nevertheless it is worth analysing the noise problem, if only to satisfy oneself how small the effect is.

A vibration diagram would show the vector representing q_0 rotating uniformly at angular velocity ω_0 in the absence of noise. A voltage impulse P would add a vector of length $P/\omega_0 L$, parallel to the imaginary axis. From the point of view of a frame rotating with q_0, the new vector may point in any direction according to the phase at the moment of the impulse. A random sequence of impulses therefore contributes a randomly oriented set of vectors, whose resultant has a Gaussian probability distribution, as expressed by (4.37). If for the moment we forget the tendency of the **83** amplitude to relax to its equilibrium value, we picture the end-point of the vector q_0 as the source of a cloud of points which diffuse out in a spreading circular patch, depicting the way in which the probability distribution of the vibration amplitude evolves with time. Introducing formally a diffusion constant D for this process (similar to, but not identical with, the D in (4.37)), we have the normal diffusion law for ρ, the density of points on the plane,

$$D\nabla^2\rho = \dot{\rho}, \tag{37}$$

with solution after time t,

$$\rho \propto t^{-1} e^{-r^2/4Dt}, \tag{38}$$

r being measured from the end of the initial vector q_0. The mean-square displacement follows from (38):

$$\overline{r^2} = \int_0^\infty r^3 \rho(r)\, dr \int_0^\infty r\rho(r)\, dr = 4Dt. \tag{39}$$

But we know that if there are N impulses per unit time, the resultant of Nt random vectors of length $P/\omega_0 L$ is given by

$$\overline{r^2} = NtP^2/(\omega_0 L)^2,$$

from which we identify the diffusion constant,

$$D = \tfrac{1}{4}NP^2/(\omega_0 L)^2 = \tfrac{1}{4}\pi I_\omega/(\omega_0 L)^2 \quad \text{from (4.45),} \tag{40} \quad \textbf{92}$$

if I_ω is the spectral intensity of the white voltage noise.

Having determined D, we may now tackle the real problem, taking the amplitude fluctuations first. Because the amplitude tends to relax back to q_0, after a long time the cloud of points representing possible resultants will lie close to a circle of radius q_0, phase information having been lost. The radial distribution may be determined by the same argument as is applied to sedimentation in Brownian theory – a point at radius $q_0 + y$ is not merely wandering as a result of noise impulses, it also moves purposefully back to the radius q_0. For exponential relaxation the radial velocity must be proportional to y, so that $\dot{y} = -\frac{1}{2}\omega_0 y/Q$; this ensures a time-constant $2Q/\omega_0$ for the noise-free relaxation process shown in fig. 10. If, then, the cloud of points has a density $\rho(y)$, the flux across unit length from this cause is $-\frac{1}{2}\omega_0 y\rho/Q$. This, however, is opposed by the diffusive flux, $-D\,d\rho/dy$, and the equilibrium distribution results from the balance of the two fluxes:

$$D\,d\rho/dy = -\tfrac{1}{2}\omega_0\rho y/Q.$$

Hence

$$\rho \propto e^{-\omega_0 y^2/4QD}.$$

The distribution of amplitudes is Gaussian around the mean q_0, with

$$\overline{(\Delta q_0)^2} = \overline{y^2} = 2QD/\omega_0 = \tfrac{1}{2}\pi QI_\omega/\omega_0^3 L^2 \quad \text{from (40).} \tag{41}$$

This may conveniently be expressed in terms of the energy, $\bar{\varepsilon}$, which would be possessed by an oscillation of this amplitude:

$$\bar{\varepsilon} = \tfrac{1}{2}\overline{(\Delta q_0)^2}/C = \tfrac{1}{4}\pi QI_\omega/\omega_0^3 L^2 C = \tfrac{1}{4}\pi I_\omega/R, \tag{42}$$

since $\omega_0^2 = 1/LC$ and $Q = \omega_0 L/R$. To appreciate the magnitude of $\bar{\varepsilon}$, suppose the noise to be due entirely to the resistor R in the resonant circuit, giving a Johnson noise voltage whose intensity I_ω is $2Rk_B T/\pi$, according to (6.49). Then

158

$$\bar{\varepsilon} = \tfrac{1}{2}k_B T. \tag{43}$$

Johnson noise does no more than add to the existing oscillation a random element of about the same magnitude as would exist in the unexcited circuit anyhow. In reality there will be more than this, for the negative-resistance device will also contribute, probably several times as much but still not necessarily enough to raise $\bar{\varepsilon}$ to anything perceptible. For an oscillating circuit with a voltage amplitude of 5 V and a capacitor of $0.1\ \mu$F has oscillatory energy of about $1\ \mu$W, more than 10^{14} times $\bar{\varepsilon}$ as given by (43).

In the light of this analysis it is clearly a good approximation to take the amplitude q_0 as constant when discussing the phase drift due to noise. The diffusion of the cloud of points round the circumference of a circle of radius q_0 is a one-dimensional process governed by the same diffusion constant D, and if $s(b)$ is the drift distance after time t, measured round the circle,

$$\overline{s^2} = 2Dt = \tfrac{1}{2}\pi I_\omega t/(\omega_0 L)^2.$$

The phase drift, θ, corresponding to s is s/q_0 and therefore

$$\overline{\theta^2} = \tfrac{1}{2}\pi I_\omega t/(\omega_0 L q_0)^2. \tag{44}$$

To put this result in perspective, let us use (41) to express I_ω in terms of $\overline{(\Delta q_0)^2}$:

$$\overline{\theta^2} = \omega_0 t\overline{(\Delta q_0/q_0)^2}/Q. \tag{45}$$

If we wait long enough for the mean-square phase drift to be 1 radian, the number of radians of oscillation that must elapse is the value of $\omega_0 t$ when $\overline{\theta^2} = 1$, i.e. $Q/(\Delta q_0/q_0)^2$, perhaps 10^{16} with the figures quoted above and with Q equal to 100. No real oscillator begins to approach this frequency stability – long-term stability of one part in 10^{12} is very good indeed – and it seems fair to conclude that, even with much less energetic oscillators, thermal noise is unlikely to be the limiting factor in precision timekeeping.

It may therefore be thought rather academic to pursue the matter further, but it is worth restating the last argument, if only as an illustration of the use of the autocorrelation function. The oscillator output takes the form $q_0 \exp[-i(\omega_0 t - \theta)]$, θ being a phase angle that wanders randomly. In a time τ the change in θ, which we write as $\theta(\tau)$, has a Gaussian distribution about zero, with mean-square value $\overline{\theta^2(\tau)}$ proportional to τ, as in (44). Hence the autocorrelation function, $\Gamma(\tau)$, according to (4.47) takes the form $\overline{\exp i\theta(\tau)} \exp(-i\omega_0\tau)$, i.e. $\overline{\cos \theta(\tau)} \exp(-i\omega_0\tau)$ since positive and negative values of $\theta(\tau)$ are equally likely. If the probability of $\theta(\tau)$ taking a certain value is proportional to $\exp[-\frac{1}{2}\theta^2/\overline{\theta^2(\tau)}]$ the mean value of $\cos \theta(\tau)$ is readily evaluated* as $\exp[-\frac{1}{2}\overline{\theta^2(\tau)}]$.†

Hence

$$\Gamma(\tau) = e^{-|\tau|/\tau_c - i\omega_0\tau}, \tag{46}$$

where, from (45),

$$\tau_c = 2Q/\omega_0\overline{(\Delta q_0/q_0)^2}$$

The Fourier transform of $\Gamma(\tau)$ gives the spectral intensity of the oscillator output, being a Lorentzian line shape of the form

$$I_\omega \propto 1/[1 + (\omega - \omega_0)^2\tau_c^2].$$

The effect of noise is to spread the output from a perfectly sharp frequency into a band of width about $2/\tau_c$, or $\omega_0(\Delta q_0/q_0)^2/Q$. This is an alternative to (45) as a specification of the frequency stability, but of course it is equivalent.

Exercise. An alternative approach to some of the examples in this chapter is revealing. Apply the Nyquist diagram to the circuits of fig. 14, breaking the circuit at the positive amplifier input to determine the loop gain, $G(\omega)$; remember that $A(\omega)$ for the amplifier falls off as ω rises, and that the Kramers–Kronig relations demand that A be complex. The series and parallel arrangements, fig. 14(a) and (b), give significantly different Nyquist diagrams. Try this approach also on the tunnel-diode oscillator and the multivibrator. In the latter case, neglect C_1 in (28) and instead allow inter-electrode capacitance etc. to play their part in reducing μ at higher frequencies, so rendering it complex.

94

313, 327

110

333
355

232

† The reader may like to derive the form of $\Gamma(\tau)$ directly from the diffusion equation, using the same argument (but for diffusion on a circle rather than on a sphere) as was used to arrive at (8.29).

12 Coupled vibrators

A resonant system acted upon by an oscillatory force presents a straight-forward enough problem if it is passive and linear, especially if the force is applied by some prime mover that is uninfluenced by the response it excites. Such problems are the subject matter of chapter 6, while non-linear passive systems, as discussed in chapter 9, are more complicated to handle. If the prime mover is influenced by the response, additional complexities enter, and this chapter treats of some of these. As is to be expected, linear systems present the least difficulty, and we shall begin with the behaviour of two coupled resonant systems, each of which may be thought of as driving the other and being perturbed by the reaction of the other back on it. Examples have already appeared earlier, as for instance the coupled pendulums discussed in chapter 2, and the coupled resonant lines in chapter 7. In both cases we noted a very general characteristic of such systems, that even if they are tuned to the same frequency before being coupled, they do not vibrate at this frequency when coupled, but have resonances which move progressively further from the original frequency as the coupling is strengthened. It would perhaps be logical, having considered this problem, to proceed to coupled, passive, non-linear vibrators; but these are so difficult that we shall leave them alone. It is not quite so hard to derive useful results for the behaviour of self-maintained oscillators when perturbed either by the injection of a steady signal or by being coupled to a similar oscillator. These are significantly different from passive resonant systems in that they exhibit the Huygens phenomenon, locking in frequency to the input signal or to one another, with characteristic critical behaviour at that strength of coupling which initiates locking. There are also special features worth investigating when a self-maintained oscillator is coupled to a passive resonant system, and these are particularly interesting when the oscillator is a multivibrator or something similar with a highly non-sinusoidal waveform. Some aspects of the behaviour here are strongly reminiscent of the way a Josephson junction between superconducting materials can be brought into oscillation and locked in frequency to a passive resonant circuit, and we shall conclude by giving a brief account of this phenomenon.

Coupled passive vibrators

It is convenient to develop a general picture of the mode-splitting phenomenon in coupled vibrators through electrical circuits, and then show the analogy with mechanical and other systems afterwards. In the circuit of fig. 1(*a*), the two series resonant circuits are coupled by means of a four-terminal element, of which a few realizations are given in figs. 1(*b*)–(*f*).† If it is a linear coupler, the e.m.f.s across its terminals in circuits 1 and 2 are expressible in the form:

$$V_1 = \mathfrak{z}_{11}i_1 + \mathfrak{z}_{12}i_2, \qquad V_2 = \mathfrak{z}_{21}i_1 + \mathfrak{z}_{22}i_2, \tag{1}$$

all $\mathfrak{z}$s being functions of frequency, though those shown in the diagram vary only slowly with ω. Except for the gyrator in fig. 1(*f*) the coupler is symmetrical, having $\mathfrak{z}_{21} = \mathfrak{z}_{12}$, but the gyrator has $\mathfrak{z}_{21} = -\mathfrak{z}_{12}$. We consider the gyrator separately and concentrate first on symmetrical couplers. The impedances $\mathfrak{z}_{11}$ and $\mathfrak{z}_{22}$ may be incorporated in the individual circuits along with the other elements, so that the natural resonant frequency of circuit 1, ω_1, is taken as that for which $j\omega_1 L_1 + R_1 + 1/j\omega_1 C_1 + \mathfrak{z}_{11}(\omega_1) = 0$; ω_1 in general is complex. If the circuit has a high Q, as we shall assume throughout this discussion, its impedance at frequency Ω, close to resonance, is very nearly $2jL_1(\Omega - \omega_1)$. The equations for the coupled circuits then take the form

135

$$2jL_1(\Omega - \omega_1)i_1 + \mathfrak{z}_{12}i_2 = 0 \tag{2}$$

and
$$2jL_2(\Omega - \omega_2)i_2 + \mathfrak{z}_{21}i_1 = 0, \tag{3}$$

from which the possible resonant frequencies, $\Omega_{1,2}$, are seen to be solutions of the equation

$$(\Omega - \omega_1)(\Omega - \omega_2) = \kappa^2, \tag{4}$$

where $\kappa^2 \equiv -\mathfrak{z}_{12}\mathfrak{z}_{21}/4L_1L_2$. The coupling constant κ is real if $\mathfrak{z}_{12}$ and $\mathfrak{z}_{21}$ are equal and purely reactive (figs. 1(*b*)–(*d*)), imaginary if they are purely resistive (fig. 1(*e*)). There are two separate solutions (*normal modes*) to (4), and the system may, if properly excited, be set into free oscillation at either frequency. In general, however, both solutions coexist independently and the observed variations in frequency and amplitude of the currents in the two circuits are the result of the linear addition of the two modes.

---

† The circuit shown in fig. 1(*g*), in which two parallel resonant circuits are coupled by, for example, a capacitor C_3, is very similar in general behaviour to that in fig. 1(*a*) but a little more complicated. The difference can be pinpointed by observing that if $\mathfrak{z}_c$ is a capacitor, as in 1(*b*), the energy of oscillation in the circuit 1(*a*) can be expressed entirely in terms of q_1 and q_2, the charges on C_1 and C_2, and their first derivatives – the coupling introduces no more coordinates than were already needed for the uncoupled systems. The same is not true of the circuit 1(*g*), which has an additional degree of freedom in that there are three loops to the network instead of two.

Fig. 1. (*a*) Two resonant
circuits coupled by $\mathscr{z}_c$;
(*b*)–(*e*) forms of $\mathscr{z}_c$, having
reciprocal properties;
(*f*) non-reciprocal
(gyratory) coupling; (*g*) an
alternative form of
coupling.

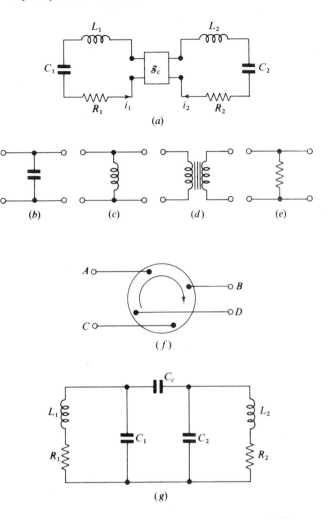

The currents in the two circuits are in general different when the system
is excited in a pure normal mode. From (2) and (3) it is seen that

$$L_1^{\frac{1}{2}}i_1/L_2^{\frac{1}{2}}i_2 = \kappa/(\Omega - \omega_1) = (\Omega - \omega_2)/\kappa = (\Omega - \omega_2)^{\frac{1}{2}}/(\Omega - \omega_1)^{\frac{1}{2}}. \quad (5)$$

The form of (5) shows that it is natural to measure the response of a circuit
not by the currents but by coordinates $x_{1,2}$, defined so that

$$L_1 i_1^2 = x_1^2, \qquad L_2 i_2^2 = x_2^2. \quad (6)$$

In other words, something proportional to the square root of the energy of
excitation is taken instead of current as a suitable coordinate, and then the
different magnitudes of L_1 and L_2 play no further part in the equations,
except in so far as they help determine ω_1 and ω_2. Rewriting (2), (3), (4)
and (5), with κ_{12} defined as $\frac{1}{2}j\mathscr{z}_{12}/(L_1 L_2)^{\frac{1}{2}}$, we have that

$$(\Omega - \omega_1)x_1 - \kappa_{12}x_2 = 0, \quad (7)$$

$$(\Omega - \omega_2)x_2 - \kappa_{21}x_1 = 0, \tag{8}$$

$$(\Omega - \omega_1)(\Omega - \omega_2) = \kappa_{12}\kappa_{21} = \kappa^2, \tag{9}$$

and
$$x_1^2/x_2^2 = (\Omega - \omega_2)/(\Omega - \omega_1). \tag{10}$$

These equations form the basis for the following rather extended discussion, which we begin by considering loss-free systems, coupled reactively; ω_1 and ω_2 are therefore real, and so are all solutions of (9) provided κ^2 is positive – the normal modes are undamped harmonic vibrations. The variation of normal mode frequencies as the two circuits are tuned is best represented by rewriting (9) in terms of $\bar{\omega}$, defined as $\frac{1}{2}(\omega_1 + \omega_2)$, and $\Delta\omega$, defined as $\frac{1}{2}(\omega_1 - \omega_2)$. Then

$$(\Omega - \bar{\omega})^2 = \kappa^2 + (\Delta\omega)^2. \tag{11}$$

This reveals the character of the hyperbolic dependence (fig. 2) of Ω on $\Delta\omega$. If the tuning of the two circuits is arranged so that their mean frequency, $\bar{\omega}$, stays constant the straight asymptotes of the hyperbola represent the natural frequencies of the uncoupled circuits, crossing over at the centre, while the hyperbola itself shows how the two normal-mode frequencies are split, the effect being greatest when the circuits are perfectly tuned. Then $\Delta\omega = 0$ and $\Omega = \bar{\omega} \pm \kappa$. When $\Delta\omega \gg \kappa$, however, the normal modes have frequencies almost the same as the uncoupled circuits; an oscillation in one is so far off the resonance of the other than it excites little response and is itself correspondingly little perturbed. This point is made more precisely by noting the graphical interpretation of (10). At a certain value of $\Delta\omega$, P and Q represent the normal modes, Ω_1 and Ω_2, while R and S are the frequencies of the uncoupled circuits, ω_1 and ω_2. The ratio $PS/PR = (\Omega_1 - \omega_2)/(\Omega_1 - \omega_1) = x_1^2/x_2^2$, from (10), and this is the ratio of oscillatory energy in the two circuits for the mode at Ω_1. The greater energy is always in that circuit whose natural frequency ω is nearer Ω. As one moves along the upper branch of the hyperbola from negative to positive $\Delta\omega$, at first $\omega_2 > \omega_1$ and the upper mode has almost all its energy in circuit 2; when $\Delta\omega = 0$ both circuits are equally excited, and ultimately as $\Delta\omega$ goes positive

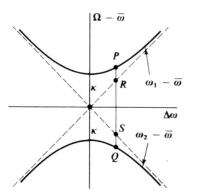

Fig. 2. Splitting of normal-mode frequencies.

the energy is concentrated in circuit 1, which now has the higher natural frequency. The opposite, of course, occurs with the lower mode, for which the energy partition is in the ratio QR/QS, the same as PS/PR.

The relative phase of the currents is given by (5); if κ is positive as with capacitive coupling, the currents are in phase in the upper mode ($\Omega > \omega_1$) and in antiphase in the lower mode. This is obvious from the circuit – if i_1 and i_2 oscillate in phase, the coupling capacitor, C_c of fig. 1(b), is charged and discharged periodically and introduces extra capacitance in series with C_1 in circuit 1, or C_2 in circuit 2. The combined capacitance being smaller, the natural frequency is increased; the effect is less marked, or even reversed, when the currents are in antiphase.† With inductive coupling ($\kappa < 0$), however, as with fig. 1(c), the effective inductance is increased and the frequency reduced when the currents are in phase. There is a sound energetic argument for the currents to be in phase or in antiphase in a pure normal mode – only thus can there be no mean power flowing from one circuit to the other, the current in one circuit being in phase quadrature with the e.m.f. due to the other. With capacitive coupling, for example, any mean power flow is the result of i_1 flowing through C_c in the presence of the e.m.f. $i_2/j\Omega C_c$, and is zero.

The result just derived, that x_1/x_2 in one mode is the same as $-x_2/x_1$ in the other, makes possible an elegant and valuable graphical representation of the normal modes, for the lines $x_1 = cx_2$ and $x_2 = -cx_1$, c being a constant, are orthogonal when x_1 and x_2 are chosen as cartesian coordinates. On the (x_1, x_2)-plane of fig. 3, therefore, the normal modes appear

Fig. 3. Normal mode coordinates, ξ_1 and ξ_2, shown as a linear combination of x_1 and x_2 equivalent to a rotation of axes.

† If the currents are so phased that the effect of one circuit on the other is equivalent to introducing an extra capacitance, so also does the first circuit acquire extra capacitance. This reciprocity holds in general for reactive coupling when the currents are either in phase or in antiphase. Thus both natural frequencies are shifted in the same direction, and this explains why the normal modes lie outside the range between ω_1 and ω_2. The circuit whose frequency is less affected is obviously that which is more strongly excited, for the extra e.m.f. due to the other is relatively weak.

as harmonic vibrations along two orthogonal lines, ξ_1 and ξ_2. To find the orientation of the lines, put $\kappa/\Delta\omega = \tan 2\theta$; then from (11) the frequencies of the normal modes are given by

$$\Omega = \omega \pm \Delta\omega \sec 2\theta.$$

Taking the positive sign for the upper mode frequency, Ω_1, we evaluate $(\Omega_1 - \omega_2)/(\Omega_1 - \omega_1)$ and substitute in (10), to find that $x_1/x_2 = \cot \theta$, so that θ is the angle shown in fig. 3. This is for the case of capacitive coupling, when x_1 and x_2 have the same sign in the upper mode, represented by ξ_1; as $\Delta\omega$ runs from large negative to large positive values, θ swings round from 90° to 0, being 45° when $\Delta\omega = 0$. With inductive coupling the process is reversed in sign, θ swinging round from 0 to 90°; ξ_2 is now the upper mode, with the currents in antiphase.

A simple experiment to demonstrate these points is shown in fig. 4. When the two strips are joined at right angles their vibrations are independent; the natural frequency of the upper strip is unaffected by whatever length of lower strip may be protruding. The variations of normal mode frequencies are shown in fig. 4(b), and ideally even at the point where the curves cross there should be no coupling and no splitting of the modes. On the other hand, when the strips are not at right angles the bending of one exerts a bending moment on the other, and the vibrations are coupled. Consequently the modes are split in the characteristic 'hyperbolic' fashion, with the unperturbed modes as asymptotes far from the cross-over point. With a calibrated stroboscopic flash one can measure both the frequencies and the vibration directions of the normal modes. Only in two orthogonal directions can one set the system in motion and have it stay vibrating along the same line.† When deflected in some other direction the two modes are simultaneously excited and, being slightly different in frequency, beat with one another so that the end of the strip describes an evolving Lissajous pattern, alternating linear and elliptical motions. The indicators at various points along the curves in fig. 4(b) show how the directions of the normal modes rotate systematically. When the lower strip is short the vibration direction of the upper mode is that of the lower strip. As one lengthens the lower strip the mode takes on a proportion of the upper strip vibration, and eventually the upper mode is entirely composed of vibration of the upper strip.

The Lissajous figure just remarked on provides the explanation for the characteristic energy exchange between two pendulums, described in chapter 2. The effect is most marked when the two pendulums are exactly tuned, but something similar occurs even when they are not. With the parameters

† It is not strictly true that the normal modes have the end of the upper strip vibrating in orthogonal directions, but the discrepancy is hardly noticeable. The reader may care to work out for himself how the model does not quite fit the theoretical framework and is therefore enabled to violate, ever so slightly, this prediction of the theory.

Fig. 4. (*a*) Simple
experiment on coupled
vibrations;
(*b*) experimental results;
the lines without points
show the behaviour when
the strips are nearly at
right angles, those with
points when the strips
make an angle of 80°. The
lines in circles indicate the
vibration direction at
various points along the
curves.

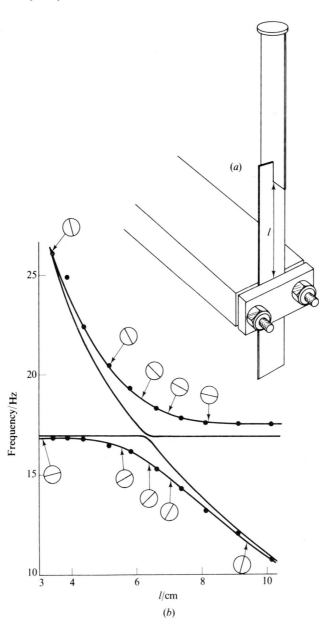

such as to give normal mode orientations as shown in fig. 3, for example,
displacing oscillator 1 to x_1 and releasing it, while $x_2 = 0$ at the moment of
release, means starting the system at a point A, at which both normal
modes are excited, ξ_1 with amplitude $x_1 \cos \theta$ and ξ_2 with amplitude
$x_1 \sin \theta$. Subsequently, as each mode continues to vibrate independently of
the other, the representative point executes a Lissajous figure contained
within the rectangle $ABA'B'$, the time for a complete beat-cycle being

$2\pi/(\Omega_1-\Omega_2)$. The furthest excursion of the vibration from the line AA' occurs half a beat-cycle after the start, when the point is vibrating along BB'; the amplitude of x_1 has dropped from OA to OC, and that of x_2 has risen from zero to OD.† Only when the two oscillations are perfectly in tune before coupling do ξ_1 and ξ_2 lie at 45°, so that the rectangle becomes a square with its corners on the x_1- and x_2-axes. There is then the possibility of complete transfer of energy from one oscillator to the other and back again. It will be observed that this does not depend on the two oscillators being identical, only on their being well-tuned. If they are weakly coupled a very small degree of detuning is enough to cause ξ_1 and ξ_2 to lie close to the x_1- and x_2-axes. The response of the initially stationary oscillator then rises and falls periodically, but to so small an amplitude that the other continues oscillating at almost constant amplitude. This is a close approximation to the case of a vibrator excited by an oscillatory force of constant amplitude.

146

Energy and normal modes

In addition to the examples of reactive coupling shown in fig. 1 we may use an inductor and capacitor in series, as in fig. 5; L_c is much smaller than either L_1 or L_2, C_c much larger than either C_1 or C_2, so that their combined impedance is always small and the coupling weak. It does not matter in what follows where the resonant frequency, $(L_cC_c)^{-\frac{1}{2}}$, of the

Fig. 5. As in fig. 1(a), with z_c in the form of L_c and C_c in series.

coupling impedance lies in relation to the resonant frequencies of the separate circuits. If the charges on C_1 and C_2 are q_1 and q_2, as shown, the charge on C_c is q_1+q_2 and the currents through all inductors are determined in terms of $\dot{q}_1$ and $\dot{q}_2$. The energy of the circuit at any instant may then be written in the form:

$$E = \Phi + T,$$

† The reader may care to verify that when two normal modes are present together, the e.m.f. due to one circuit and the current in the other upon which it acts, at the coupling impedance, are not in phase quadrature. There is thus a steady power flow, as of course there must be when the beating phenomenon takes place. As the beat-cycle progresses, the phase relationship alters steadily, and energy flows to and fro through z_c.

where the potential energy,

$$\Phi = \tfrac{1}{2}q_1^2/C_1 + \tfrac{1}{2}q_2^2/C_2 + \tfrac{1}{2}(q_1+q_2)^2/C_c$$

$$= \tfrac{1}{2}q_1^2/\bar{C}_1 + \tfrac{1}{2}q_2^2/\bar{C}_2 + q_1q_2/C_c, \tag{12}$$

27

and the kinetic energy,

$$T = \tfrac{1}{2}L_1\dot{q}_1^2 + \tfrac{1}{2}L_2\dot{q}_2^2 + \tfrac{1}{2}L_c(\dot{q}_1+\dot{q}_2)^2$$

$$= \tfrac{1}{2}\bar{L}_1\dot{q}_1^2 + \tfrac{1}{2}\bar{L}_2\dot{q}_2^2 + L_c\dot{q}_1\dot{q}_2. \tag{13}$$

In these expressions $1/\bar{C}_{1,2}$ is written for $1/C_{1,2}+1/C_c$, $\bar{L}_{1,2}$ for $L_{1,2}+L_c$. The energy is a positive definite quadratic form in the variables q_1, q_2, $\dot{q}_1$ and $\dot{q}_2$. If contours of Φ are plotted on a (q_1, q_2)-plane they form a family of similar ellipses whose axial ratios are all the same and which are all oriented alike with respect to the coordinate axes; any one may therefore be taken as representative. Similarly for the contours of T on a $(\dot{q}_1, \dot{q}_2)$-plane. If the two planes are superposed, with $\dot{q}_1 \parallel q_1$, the orientations of the ellipses need bear no special relationship to one another. We now enquire what determines the normal modes – in which directions can one draw a line representing the vibrations of q_1 and q_2 in phase and unchanging in amplitude?

To answer this question, note first that

$$\partial\Phi/\partial q_1 = q_1/C_1 + (q_1+q_2)/C_c, \quad \text{from (12)}, \tag{14}$$

and $\quad\partial T/\partial\dot{q}_1 = L_1\dot{q}_1 + L_c(\dot{q}_1+\dot{q}_2), \quad \text{from (13)}.$

Hence $\quad\dfrac{d}{dt}(\partial T/\partial\dot{q}_1) = L_1\ddot{q}_1 + L_c(\ddot{q}_1+\ddot{q}_2). \tag{15}$

Now (14) is the sum of the potential differences across the capacitors in circuit 1, and (15) the sum of e.m.f.s across the inductors. Hence

$$\frac{\partial\Phi}{\partial q_1} + \frac{d}{dt}\left(\frac{\partial T}{\partial\dot{q}_1}\right) = 0, \tag{16}$$

and similarly, for circuit 2,

$$\frac{\partial\Phi}{\partial q_2} + \frac{d}{dt}\left(\frac{\partial T}{\partial\dot{q}_2}\right) = 0. \tag{17}$$

These equations could have been derived by Lagrangian methods, which provide the natural formalism in which to analyse more complicated problems in coupled oscillatory systems. The derivation from first principles is so easy here, however, that we have preferred to do this. Let us now introduce two vectors,

$$\text{grad } \Phi \equiv \left(\frac{\partial\Phi}{\partial q_1}, \frac{\partial\Phi}{\partial q_2}\right) \quad \text{and} \quad \text{grad } T \equiv \left(\frac{\partial T}{\partial\dot{q}_1}, \frac{\partial T}{\partial\dot{q}_2}\right).$$

Then (16) and (17) may be combined into the equation†

$$\operatorname{grad} \Phi + \frac{\mathrm{d}}{\mathrm{d}t}(\operatorname{grad} T) = 0. \qquad (18)$$

Since Φ is constant on any Φ-contour, grad Φ at any point has no component along the direction of the tangent to the contour passing through that point; it is therefore directed normal to the contour. Similarly grad T is everywhere normal to the T-contour. If the representative point is to oscillate on a straight line through the origin, and still obey (18), the similarity of the T-ellipses implies that grad T remains constant in direction all along the line, and therefore that $(\mathrm{d}/\mathrm{d}t)(\operatorname{grad} T)$ is parallel to grad T. It is only possible to satisfy (18) along such lines as cut the Φ and T ellipses at the same angle, so that grad Φ and grad T are parallel to one another, though not in general parallel to the line representing the trajectory. An example of the possible normal modes is shown in fig. 6.

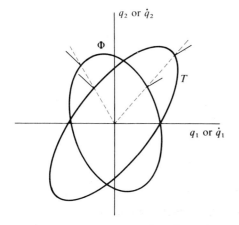

Fig. 6. Normal modes for the general case when Φ-contours are ellipses on the (q_1, q_2)-plane and T-contours different ellipses on the $(\dot{q}_1, \dot{q}_2)$-plane. The broken lines, on which the normals to the ellipses are parallel, define the ratio q_2/q_1 in each normal mode.

† The significance of this equation is perhaps more easily appreciated in the corresponding mechanical system. A mass moving on a line has kinetic energy $T = \frac{1}{2}m\dot{x}^2$, and $\partial T/\partial \dot{x} = m\dot{x}$, the momentum. Two masses, m_1 and m_2, with coordinates x_1 and x_2, have $T = \frac{1}{2}m_1\dot{x}_1^2 + \frac{1}{2}m_2\dot{x}_2^2$, represented by elliptical contours of constant T on an $(\dot{x}_1, \dot{x}_2)$-plane. The velocity vector $(\dot{x}_1, \dot{x}_2)$ and momentum vector $(m_1\dot{x}_1, m_2\dot{x}_2)$ do not point in the same direction; in fact grad T is the momentum vector, $\boldsymbol{P}$, normal to the contours of T and only parallel to the velocity on an axis of the ellipse. A body having tensorial mass, with principal values m_1 and m_2, is governed by the same equation. As for grad Φ, this is just $-\boldsymbol{F}$, the vector representing the forces on each mass. The general problem here is therefore to be thought of, if one wishes, as that of a sphere with tensorial mass rolling in an elliptical cup, and (18) is the equation of motion $\dot{\boldsymbol{P}} = \boldsymbol{F}$. The directions in which it can oscillate to and fro on a straight path are not in the principal planes of the cup, but oriented so that the restoring force, itself not radial, acts on the tensorial mass to accelerate it, not parallel to $\boldsymbol{F}$, but in a radial direction.

When q_1 and q_2 are chosen as variables the normal modes are not in general orthogonal. Unless one of them lies along an axis, however, it is always possible to scale the axes so as to make them orthogonal. The rule is simple when either Φ or T is free of cross-terms (q_1q_2 or $\dot{q}_1\dot{q}_2$). For one of the ellipses then has its axes along the coordinate axes and may be scaled into a perfect circle; the other ellipse, similarly scaled, remains elliptical and its axes are the orthogonal normal mode directions, for along these axes the normals to both ellipse and circle are radial. With capacitive coupling it is the T-ellipse which is to be scaled to a circle, and since the axis lying along $\dot{q}_1$ is $(L_2/L_1)^{\frac{1}{2}}$ times the length of the other axis, lying along $\dot{q}_2$, the q_2 and $\dot{q}_2$ values must be scaled up by $(L_2/L_1)^{\frac{1}{2}}$, leaving q_1 and $\dot{q}_1$ unchanged, to convert it into a circle. This is the same rule as yielded x_1 and x_2 as suitable coordinates. With inductive coupling the corresponding scaling factor is $(C_1/C_2)^{\frac{1}{2}}$, to convert the Φ-ellipse into a circle. In general, when both Φ and T have cross-terms, the factor is $[\bar{L}_2(\omega_2^2 - \omega_c^2)/\bar{L}_1(\omega_1^2 - \omega_c^2)]^{\frac{1}{2}}$, in which $1/\omega_{1,2}^2 = \bar{L}_{1,2}\bar{C}_{1,2}$ and $1/\omega_c^2 = L_cC_c$. The proof of this is left to the reader.

A very important property of normal modes is that they contribute independently to the energy of the vibrating system. To show this, we express an arbitrary state of oscillation as the superposition of the two normal modes, each vibrating at its own frequency, i.e. $\xi_1 = A\cos(\Omega_1 t + \phi_1)$ and $\xi_2 = B\cos(\Omega_2 t + \phi_2)$. Since the coordinates $q_1, \dot{q}_1$ etc. are linear combinations of $\xi_1, \dot{\xi}_1$ etc., and the total energy at any instant is a quadratic function of the qs and $\dot{q}$s, it is also a quadratic function of the ξs and $\dot{\xi}$s:

$$E = \tfrac{1}{2}a\xi_1^2 + b\xi_1\xi_2 + \tfrac{1}{2}c\xi_2^2 + \tfrac{1}{2}d\dot{\xi}_1^2 + e\dot{\xi}_1\dot{\xi}_2 + \tfrac{1}{2}f\dot{\xi}_2^2, \tag{19}$$

As written, the cross-terms $b\xi_1\xi_2$ and $e\dot{\xi}_1\dot{\xi}_2$ imply that the energy is not a simple additive function of contributions due to each mode separately; but the requirement that E shall be independent of time is enough to show that b and e vanish. The phase-constants ϕ_1 and ϕ_2 can be omitted, so that when the oscillatory forms of ξ_1 and ξ_2 are substituted, (19) takes the form:

$$E = \tfrac{1}{2}A^2(a\cos^2\Omega_1 t + d\Omega_1^2\sin^2\Omega_1 t) + \tfrac{1}{2}B^2(c\cos^2\Omega_2 t + f\Omega_2^2\sin^2\Omega_2 t)$$
$$+ \tfrac{1}{2}AB[(b-e)\cos(\Omega_1 + \Omega_2)t + (b+e)\cos(\Omega_1 - \Omega_2)t].$$

For E to be independent of time the first two brackets, containing oscillatory terms at frequencies $2\Omega_1$ and $2\Omega_2$, must be constrained by choice of Ω_1 and Ω_2 to be constant; and the third bracket must vanish. Hence

$$a = d\Omega_1^2, \quad c = f\Omega_2^2 \quad \text{and} \quad b = e = 0.$$

The normal mode frequencies (which of course are the same as found by circuit analysis) are thus determined in terms of the energy coefficients, and the cross-terms eliminated. In the end (19) may be rewritten in terms of two coefficients only:

$$E = \tfrac{1}{2}d(\Omega_1^2\xi_1^2 + \dot{\xi}_1^2) + \tfrac{1}{2}f(\Omega_2^2\xi_2^2 + \dot{\xi}_2^2), \tag{20}$$

in which d and f play the roles of mass, inductance etc. associated with each independent normal mode, considered as a simple vibrator.

Examples of coupled vibrators

The general features of the behaviour of the following examples are very similar, and there is no need to analyse each in detail. The first three categories are distinguished by the expressions for their energies and the forms taken by Φ and T as defined in (12) and (13); the remaining two are outside this classification.

(1) *Potential coupling:* Φ contains a cross-term, but T does not. The circuit of fig. 1(a), with capacitive coupling as in fig. 1(b), is typical; so are the two pendulums on a slack string, shown in fig. 2.14, and the strip vibrator in fig. 4.   **24**

(2) *Kinetic coupling:* T contains a cross-term, but Φ does not. The inductive couplers, fig. 1(c) and (d), provide typical examples. So also does the circuit of fig. 7, a parallel resonant circuit with a series resonant circuit connected directly across it. If q_1 and q_2 are the charges on the capacitors,

$$\Phi = \tfrac{1}{2}q_1^2/C_1 + \tfrac{1}{2}q_2^2/C_2$$

and

$$T = \tfrac{1}{2}L_1(\dot{q}_1 + \dot{q}_2)^2 + \tfrac{1}{2}L_2\dot{q}_2^2$$
$$= \tfrac{1}{2}L_1\dot{q}_1^2 + \tfrac{1}{2}(L_1 + L_2)\dot{q}_2^2 + L_1\dot{q}_1\dot{q}_2.$$

Fig. 7. Coupled series and parallel resonant circuits.

The circuit is no different from the standard form, fig. 1(a), with L_1 acting as coupling inductor, and C_1 having no inductor in its own arm. Then $\omega_1^2 = 1/\bar{L}_1C_1 = 1/L_1C_1$, $\omega_2^2 = 1/(L_1 + L_2)C_2$ and $\kappa^2 = \tfrac{1}{4}L_1/L_2$. It is necessary that $L_2 \gg L_1$ for weak coupling. The acoustic resonator illustrated in fig. 7.11 is of this type, the side arm having the characteristics of a series   **178** resonant circuit, while the lengths of main pipe on either side are represented by L_1 and C_1 in parallel. As the coupling point is brought near a pressure node, L_1 and $1/C_1$ get smaller and the coupling is reduced. The splitting of the resonances has already been discussed in connection with this experiment, and is clearly typical of coupled resonant systems.

In a circular waveguide large enough to propagate both modes shown in fig. 7.16, it is possible to filter out the TE_{11} mode and let through the TM_{01} **184** mode by mounting a metal ring as shown in fig. 8. When the perimeter of the ring is about one wavelength it is excited by the TE_{11} mode, which is reflected. The other mode does not excite currents in the ring and is virtually unaffected by its presence. The ring is highly resonant and only works over a narrow frequency range. If one attempts to enhance the filtering by mounting a similar ring close behind it, the resonance is found to be split by magnetic (i.e. kinetic) coupling, and a measurable fraction of the microwave power is transmitted at the natural frequency of each ring alone, which now lies between the two peaks of reflection.[1]

Fig. 8. Electric lines in a circular waveguide exciting resonant current oscillations in a metal ring (drawn heavily). The perimeter of the ring is about one free-space wavelength.

(3) *Potential and kinetic coupling:* both Φ and T contain cross-terms. A triatomic linear molecule modelled by the arrangement of masses and springs in fig. 9, and vibrating along its axis, falls into this category. If the displacements of the atoms are x_1, x_2 and x_3,

$$\Phi = \tfrac{1}{2}\mu_1(x_1-x_2)^2 + \tfrac{1}{2}\mu_2(x_2-x_3)^2,$$

and

$$T = \tfrac{1}{2}m_1\dot{x}_1^2 + \tfrac{1}{2}m_2\dot{x}_2^2 + \tfrac{1}{2}m_3\dot{x}_3^2.$$

Fig. 9. Mechanical spring-and-mass model of a linear triatomic molecule.

$$m_1 \quad\quad \mu_1, \quad\quad m_2 \quad\quad \mu_2 \quad\quad m_3$$

Now one must remember that the centroid of the three atoms does not vibrate, so that $m_1x_1 + m_2x_2 + m_3x_3 = 0$. Using this to eliminate x_2, we have

$$\Phi = \tfrac{1}{2}\left[\mu_1\left(1+\frac{m_1}{m_2}\right)^2 + \mu_2\left(\frac{m_1}{m_2}\right)^2\right]x_1^2 + \tfrac{1}{2}\left[\mu_2\left(1+\frac{m_3}{m_2}\right)^2 + \mu_1\left(\frac{m_3}{m_2}\right)^2\right]x_3^2$$

$$+ \left[\mu_1\left(1+\frac{m_1}{m_2}\right)\frac{m_3}{m_2} + \mu_2\left(1+\frac{m_3}{m_2}\right)\frac{m_1}{m_2}\right]x_1x_3,$$

and

$$T = \tfrac{1}{2}m_1\left(1+\frac{m_1}{m_2}\right)\dot{x}_1^2 + \tfrac{1}{2}m_3\left(1+\frac{m_3}{m_2}\right)\dot{x}_3^2 + \frac{m_1m_3}{m_2}\dot{x}_1\dot{x}_3.$$

These expressions provide the basis for a complete analysis, but we shall not pursue the matter in that generality. In the simpler case of a symmetrical molecule, like CO_2, with two oxygen atoms of mass M flanking a carbon of mass m, and the same spring constants for each bond, the two modes can be seen by inspection. The symmetric mode has both bonds stretching or contracting in phase, and the carbon atom at rest, while the antisymmetric has both oxygens moving in the same direction at the same speed while the carbon moves in the opposite direction at a speed $2M/m$ times as great. For a given displacement of an oxygen atom the restoring force in the latter case is greater than in the former by a factor $1+2M/m$, and the frequencies of these two modes should therefore be in the ratio $(1+2M/m)^{\frac{1}{2}}$, i.e. 1.91. Spectroscopic measurements give 1.75 and the discrepancy indicates the limitations of a simple spring model of the valence bond.[2] The fault does not lie so much in the assumption of a linear force law as in the assumption that the same spring constant applies to symmetrical and antisymmetrical vibration.† Shifts in the electron distribution, however, when an antisymmetrical displacement occurs, allow the potential energy to take a lower value than otherwise, and so reduce the spring constant and the resonance frequency.

(4) *Gyrator coupling:* because of the non-reciprocal character of a gyrator, for which $\mathcal{z}_{21} = -\mathcal{z}_{12}$, this example does not follow the pattern of those discussed previously. To consider only an idealized loss-free model, represented schematically in fig. 1(*f*), we assume that unit current flowing from A to C generates no p.d. across AC but does generate a p.d. of magnitude G across BD, B being positive; while unit current flowing from B to D generates no p.d. across BD, but generates G across AC, C being positive. The tendency of a gyrator to redirect current in a given sense (clockwise in this case) is only achievable at the expense of time-reversal symmetry, and requires a magnetic field or a rotating frame. The Hall effect in semiconductors offers the possibility of realizing something very near ideal gyratory properties in a circuit element. The circuit equations (2) and (3) keep the same form, and since $\mathcal{z}_{12} = -\mathcal{z}_{21} = G$, which is real, (4) also is unchanged and $\kappa^2 (=G^2/L_1L_2)$ is positive, just as for reciprocal reactive coupling. The splitting of the normal modes follows the same pattern, as in fig. 2, but their character is different; for (2) and (3) show that i_1 and i_2 are now in quadrature. In the representation used in fig. 3 the two modes are similar ellipses with major axes along x_1 and x_2 respectively, and with opposite senses of rotation.

239

† The force law does, of course, contain anharmonic terms which happen to play a significant role in determining some of the energy levels of CO_2. This arises because the bending vibration of the molecule accidentally has a natural frequency almost exactly half that of the symmetrical stretching vibration. Anharmonicity in the bending process causes the spacing of the atoms to vibrate at twice the frequency of the bending vibration, and to couple this vibration with the symmetrical stretching mode. The phenomenon of *Fermi Resonance* is clearly a quantum-mechanical analogue of parametric excitation.[3]

286

210   The Foucault pendulum illustrates this point, the Coriolis force providing gyrator coupling between the pendulum vibrations in two orthogonal planes. A perfectly constructed pendulum is isotropic, no direction of swing being different from any other; $\omega_1 = \omega_2$ and the normal modes are circular with slightly different frequencies for the two senses of rotation. It is the combination of these two in equal proportion that yields the usual manifestation of the effect, the slow precession of the plane of vibration. This is the analogue in a gyratory system of the energy exchange in a reciprocally coupled system. An imperfect Foucault pendulum, having different effective length for different directions of swing, behaves like two circuits with different ω_1 and ω_2; the normal modes are now elliptical, and when the pendulum is swung in a plane the beating of the modes causes the trajectory to open into an ellipse, whose axes may precess at a quite different rate from that expected with a perfect pendulum.[4] This is why it requires great care to construct a reliable Foucault pendulum.

(5) *Resistive coupling:* if $\mathscr{z}_c$ is a resistor, as in fig. 1(e), $\mathscr{z}_{12} = \mathscr{z}_{21} = R_c$, and (4) holds but with κ^2 now negative. This implies that $\Omega_{1,2}$ lie between ω_1 and ω_2, the modes being attracted by the coupling rather than repelled. It must be remembered, however, that R_c is included in the circuits when ω_1 and ω_2 are derived, so that in general (not unexpectedly) ω_1, ω_2, Ω_1 and Ω_2 are complex. There is no point in taking separate note of resistively coupled loss-free circuits, which are a special case of the general problem of coupled lossy vibrators.

Coupled lossy vibrators

The determination of the normal modes of coupled vibrators whose Q-values are high, but not infinite, involves no extension of the foregoing theory beyond finding the solutions of (4) when ω_1 and ω_2 are complex. Proceeding as before to (11), we note that for reactive, resistive or gyratory coupling (though not for arbitrary mixtures) κ^2 is real; $(\Delta\omega)^2$ is in general complex but constant in an experiment in which only the coupling is changed, ω_1 and ω_2 remaining fixed. If then we write $\Omega - \bar{\omega}$ as $u + iv$, we have that

$$u^2 - v^2 = \kappa^2 + \text{Re}\,[(\Delta\omega)^2] \qquad (21)$$

and
$$2uv = \text{Im}\,[(\Delta\omega)^2]. \qquad (22)$$

Since the right-hand side of (22) is constant, $\Omega - \bar{\omega}$ lies on a rectangular hyperbola, having the u- and v-axes as asymptotes, as shown in fig. 10, and passing through ω_1 and ω_2, which are the solutions when $\kappa^2 = 0$. Stronger coupling moves the solutions further from these points. For any real choice of κ^2, ω_1 and ω_2 together with the normal mode frequencies Ω_1 and Ω_2 form a parallelogram. The vectors $\Omega_1 - \omega_1$ and $\Omega_1 - \omega_2$, as shown in the diagram, have the real axis as bisector and their product is therefore real and positive; according to (4) Ω_1 and Ω_2 are solutions for reactive coupling

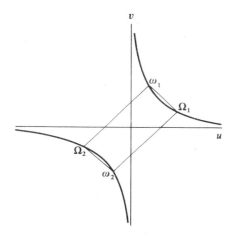

Fig. 10. Relationship of normal modes (Ω_1 and Ω_2) and natural frequencies of uncoupled lossy vibrators (ω_1 and ω_2), when represented on the complex frequency plane.

($\kappa^2 > 0$). Any point Ω on the hyperbola above ω_1 or below ω_2, on the other hand, has $\Omega - \omega_1$ and $\Omega - \omega_2$ with the imaginary axis as bisector, and corresponds to resistive coupling ($\kappa^2 < 0$). Reactive coupling repels the real parts of the normal mode frequencies, and attracts the imaginary parts; resistive coupling does the reverse. In the particular case when both circuits have the same Q, ω_1 and ω_2 lie at the same distance above the real axis and the hyperbola degenerates into two lines, vertical and horizontal, intersecting at $\bar{\omega}$. With reactive coupling the normal modes lie on the horizontal line and have the same Q at all strengths of coupling. The same degeneracy occurs when ω_1 and ω_2 have the same real part and lie on a vertical line. As κ^2 is increased the normal modes move towards one another on the vertical line and coalesce at $\bar{\omega}$, having the same frequency but different decrements until this point. Thereafter their frequencies diverge while their decrements remain fixed at the mean value for the uncoupled circuits.

If ω_1 and ω_2 do not have the same imaginary parts $\Omega - \omega_1$ does not approximate to a real quantity unless Ω is a great distance from ω_1 along the hyperbola. It follows from (5) that, even with reactive coupling, i_1 and i_2 may depart considerably from the in-phase or antiphase relationship characteristic of loss-free circuits. The requirement for different rates of dissipation in the two circuits makes a significant phase shift necessary, especially when the coupling is weak, to allow a steady transfer of energy from the less dissipative to the more dissipative circuit. Various expressions for the ratio of the currents in the two circuits follow from (5). If we write $x_1 = cx_2$ in the upper mode, of frequency Ω_1,

$$c = \kappa/(\Omega_1 - \omega_1) = -\kappa/(\Omega_2 - \omega_2) = (\Omega_1 - \omega_2)/\kappa = -(\Omega_2 - \omega_1)/\kappa, \quad (23)$$

the complex numbers $(\Omega_1 - \omega_1)$ etc. referring to the sides of the parallelogram in fig. 10; κ is positive for capacitive, negative for inductive, coupling. Further,

$$cc^* = |\Omega_1 - \omega_2|/|\Omega_1 - \omega_1| = |\Omega_2 - \omega_1|/|\Omega_2 - \omega_2|, \quad (24)$$

the ratio of the lengths of the sides. In the lower mode $x_1 = c'x_2$, where

$$c' = -1/c. \tag{25}$$

Only for loss-free circuits can one expect c to be real.

Impedances and impulse response functions for coupled circuits

Let us begin the discussion of the effect of an injected e.m.f. by calculating the input impedance of the coupled resonant circuits when the left-hand circuit in fig. 1(a) is opened and an e.m.f. is applied at frequency ω. The equations for the current are like (2) and (3), except that a source V is added to the right-hand side of (2). Since the input impedance of the circuit, $Z_{11} = V/i_1$, we may instead add $-Z_{11}$ to the coefficient of i_1 on the left-hand side of (2), and rewrite (4) in the form

$$[(\omega - \omega_1) - Z_{11}/2jL_1](\omega - \omega_2) = \kappa^2,$$

or
$$Z_{11} = \frac{2jL_1}{\omega - \omega_2}[(\omega - \omega_1)(\omega - \omega_2) - \kappa^2]$$

$$= \frac{2jL_1}{\omega - \omega_2}(\omega - \Omega_1)(\omega - \Omega_2), \tag{26}$$

since Ω_1 and Ω_2 are the solutions of (4). Similarly the input impedance of the right-hand circuit takes the form

$$Z_{22} = \frac{2jL_2}{\omega - \omega_1}(\omega - \Omega_1)(\omega - \Omega_2). \tag{27}$$

A third coefficient of the system is the transfer impedance, Z_{12} or Z_{21}, the e.m.f. needed in one circuit to set up unit current in the other. This can be derived from (26) by noting that Z_{11} is the e.m.f. needed to set up unit current in the same circuit, and that (3) gives the ratios of the currents. Hence

$$Z_{12} = -2jL_2(\omega - \omega_2)Z_{11}/\mathcal{z}_{21} = -\frac{\mathcal{z}_{12}}{\kappa^2}(\omega - \Omega_1)(\omega - \Omega_2). \tag{28}$$

168

This expression is symmetrical with respect to the two circuits if $\mathcal{z}_{12} = \mathcal{z}_{21}$, and provides another illustration of the Rayleigh–Carson reciprocity theorem.

The transfer impedance is low only in the vicinity of the resonances, and this makes the coupled circuit a useful feed-through arrangement, acting as a simple band-pass filter. For this purpose one is usually more interested in the magnitude of Z_{12} than in the phase relation between input e.m.f. and secondary current. Even so, (28) in its general form, with Ω_1 and Ω_2 different in both real and imaginary parts, develops into a cumbersome expression when one writes out $|Z_{12}|$ explicitly. A diagram is far more revealing, and fig. 11(a) shows $(\omega - \Omega_1)$ and $(\omega - \Omega_2)$ as vectors on a complex plane. Since Z_{12} is proportional to the product of the lengths of these vectors, one may easily sketch how this varies as ω moves along the real axis. Clearly if the imaginary parts of Ω_1 and Ω_2 are small compared with their separation, there will be minima of $|Z_{12}|$ at two points where ω lies almost as close as it can get to either Ω_1 or Ω_2. This is equivalent to the

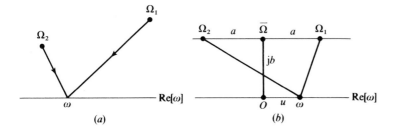

Fig. 11. To illustrate transfer impedance of coupled circuits.

statement that if the modes are separated by rather more than the widths of their resonances they will be resolved in the response function. On the other hand, if Ω_1 and Ω_2 are closer to each other than to the real axis, there is only a single minimum. The former case is referred to as over-coupled, the latter as under-coupled, and in between is the state of critical coupling when $|Z_{12}|$ has a flat minimum. For two circuits having the same Q, so that Ω_1 and Ω_2 are equidistant from the real axis, as in fig. 11(b), critical coupling occurs when $\mathrm{Im}\,[\Omega_{1,2}] = 2(\Omega_1 - \Omega_2)$. The current response $|A_{12}|$ ($\equiv 1/|Z_{12}|$) to an e.m.f. of constant magnitude but variable frequency, is shown in fig. 12 for this case and for typical under-coupled and over-coupled cases. To derive the form of these curves, take $\mathrm{Re}\,[\bar{\Omega}]$ as the origin

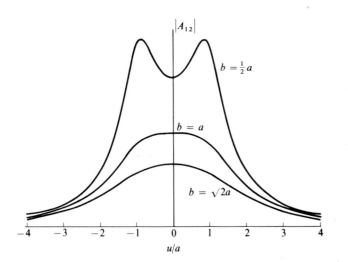

Fig. 12. Transfer admittance of coupled identical circuits, undercoupled ($b = \sqrt{2}a$), critically coupled ($b = a$) and overcoupled ($b = \frac{1}{2}a$).

on the complex plane, so that $\Omega_{1,2} = \pm a + jb$, and write ω as $\mathrm{Re}\,[\bar{\Omega}] + u$. Then

$$|(\omega - \Omega_1)(\omega - \Omega_2)|^2 = u^4 + 2(b^2 - a^2)u^2 + (a^2 + b^2)^2$$

and

$$|A_{12}| = \frac{\kappa^2}{|\mathscr{z}_{12}|}[u^4 + 2(b^2 - a^2)u^2 + (a^2 + b^2)^2]^{-\frac{1}{2}}. \tag{29}$$

Critical coupling occurs when the coefficient of u^2 vanishes, i.e. when $b = a$. If the normal modes arise from the coupling of two identical resonant circuits, $\omega_2 = \omega_1$ in (4) and $\Omega_1 - \Omega_2 = 2\kappa$; the condition for critical coupling is then that $\kappa = \frac{1}{2}\bar{\Omega}/Q$.

The transfer admittance A_{12} may be written, from (28), in the form

$$A_{12} = -\frac{\kappa^2}{\mathscr{z}_{12}(\Omega_1 - \Omega_2)}[1/(\omega - \Omega_1) - 1/(\omega - \Omega_2)]. \tag{30}$$

135   The current is made up of contributions due to each normal mode responding independently to the applied e.m.f. The character of each contribution is just the same as for a simple sharply resonant oscillator, as given in (6.17). In the secondary circuit the amplitudes are equal and opposite, for a reason which becomes obvious when one thinks of the impulse response function. An impulsive e.m.f. sets up an instantaneous current in the primary circuit, to which it is applied, but there is either no immediate response by the secondary circuit (when coupling is capacitive) or a much weaker response (when coupling is inductive). In the weak-coupling limit, which is implied in (30) by the approximate form of the resonant response, one may take the secondary current as initially zero, whatever the mode of coupling, and that means that the two damped normal-mode oscillations initiated by the impulse must be such that in the secondary circuit they start equal in amplitude and in antiphase.

To put the argument quantitatively, let an impulsive e.m.f. of unit strength be applied to circuit 1, so that immediately afterwards a current $1/L$ flows in circuit 1, but none in circuit 2. The impulse response function for x_1 has the form of decaying vibrations due to each mode, with as yet undetermined initial amplitudes a_1 and a_2:

$$h_{11}(t) = a_1\,e^{j\Omega_1 t} + a_2\,e^{j\Omega_2 t},$$

from which and from (25) it follows that in circuit 2, due to injection in circuit 1, the corresponding response has the form:

$$h_{12}(t) = \frac{a_1}{c}\,e^{j\Omega_1 t} - a_2 c\,e^{j\Omega_2 t}.$$

Now not only does h_{12} vanish at the outset, but its derivative does also, since until the coupling capacitor begins to be charged there is no e.m.f. to drive current through the inductor of the second circuit. Both the real and imaginary parts of $h_{12}(0)$ must vanish, so that

$$a_2 = a_1/c^2,$$

$$\left. \begin{array}{l} h_{11}(t) = a_1 \, e^{j\Omega_1 t} + a_1/c^2 \, e^{j\Omega_2 t} \\ h_{12}(t) = a_1(e^{j\Omega_1 t} - e^{j\Omega_2 t})/c. \end{array} \right\} \tag{31}$$

and

For a short while, until the difference between Ω_1 and Ω_2 asserts itself, $h_{11}(t)$ behaves like $a_1(1 + 1/c^2) \, e^{j\Omega_1 t}$. This must be matched to the response of circuit 1 which, just like a simple series resonant circuit of high Q, starts with the current taking its maximum value of $1/L$, and x its maximum value of $1/L^{\frac{1}{2}}$. Hence $a_1(1 + 1/c^2)$ must be equated to $1/L_1^{\frac{1}{2}}$; and

$$a_1 = c^2/L_1^{\frac{1}{2}}(1 + c^2), \qquad a_2 = 1/L_1^{\frac{1}{2}}(1 + c^2).$$

Translated into currents by writing $i_{11}(t) = h_{11}(t)/L_1^{\frac{1}{2}}$, $i_{12}(t) = h_{12}(t)/L_2^{\frac{1}{2}}$, (31) gives for the current impulse response functions:

$$\left. \begin{array}{l} i_{11}(t) = \dfrac{c}{L_1(1 + c^2)} \left(c \, e^{j\Omega_1 t} + \dfrac{1}{c} \, e^{j\Omega_2 t} \right) \\[3mm] i_{12}(t) = \dfrac{c}{(L_1 L_2)^{\frac{1}{2}}(1 + c^2)} (e^{j\Omega_1 t} - e^{j\Omega_2 t}). \end{array} \right\} \tag{32}$$

and

Since the first and fourth forms of (23) show that $c + 1/c = (\Omega_1 - \Omega_2)/\kappa$, i_{12} may be recast in the form:

$$i_{12}(t) = \frac{\kappa}{(L_1 L_2)^{\frac{1}{2}}(\Omega_1 - \Omega_2)} (e^{j\Omega_1 t} - e^{j\Omega_2 t}),$$

from which, by application of (5.4) to each term separately, $A_{12}(\omega)$ may be derived: **106**

$$A_{12} = \frac{-\frac{1}{2}j\kappa}{(L_1 L_2)^{\frac{1}{2}}(\Omega_1 - \Omega_2)} [1/(\omega - \Omega_1) - 1/(\omega - \Omega_2)],$$

which is the same as (30) since $\kappa = \frac{1}{2}j\dot{s}_{12}/(L_1 L_2)^{\frac{1}{2}}$. Similarly A_{11} and A_{22} may be derived from i_{11} and i_{22}, and shown after a little manipulation to be reciprocals of Z_{11} and Z_{22} as given by (26) and (27).

This argument has been presented in some detail for two reasons: first, to show that one must exercise care in using the normal modes of lossy coupled resonators, even if the losses are quite small, since the complex nature of c introduces significant phase shifts into the impulse responses given by (32). Secondly, the ideas expressed and some of the results derived here will find immediate application to the next example of coupled systems, where the description in terms of normal modes in non-linear circuits proves to be both feasible and helpful, but where the physical principles must be kept firmly in mind if one is not to fall into plausible errors.

Coupled active and passive vibrators[5]

There is nothing in the foregoing analysis to prevent it being applied to coupled circuits one of which has negative resistance, so that ω_1 (say) has negative imaginary part. The parallelogram of fig. 10 then has ω_1 lying below the real axis, but if it is not too far below it may be possible to find a strong enough coupling constant for both normal modes to lie above the real axis; the circuit is then stable, behaving like any of the examples already discussed. If, however, a normal mode lies below the axis, it will grow exponentially until saturation of the amplifier leads to a steady state of maintained oscillation. With strong enough amplification, both normal modes will initially lie below the axis and be capable of growing, and it may be that the mode which happens to be more strongly excited when saturation sets in will survive in the steady state. On the other hand it may be that only one of the two is capable of survival. This is the problem which we shall resolve in this section, taking the model of a class C oscillator that was developed in chapter 11; a resonant circuit incorporates a resistor which is negative at low currents and positive at high, the steady state being characterized by such a current that the resistance is zero. It must be remembered that the circuit itself is of such high Q that it oscillates sinusoidally and does not respond appreciably to harmonics of its own frequency. We are interested therefore only in the average response of the amplifier and feedback over one cycle, and it is this that we express in the model by a resistance which differs from that of a passive resistor only in the fact that it is a function of the amplitude of the current oscillation. Once the oscillation has settled in its steady state the resistance in the active circuit stays constant, and one may therefore assign two frequencies, ω_1 and ω_2, to the uncoupled circuits, exactly as was done on setting up (2) and (3) (subscript 1 refers to the active circuit, 2 to the passive); The only difference is that we do not know *a priori* the imaginary part of ω_1; we only know that ω_1 moves on a vertical line to such a point that one or other of the normal modes, Ω_1 or Ω_2, lies on the real axis, representing steady oscillation. An example of the modified form of fig. 10 is given in fig. 13. For a given ω_2 and with ω_1 lying on a given vertical line there are, in this case, three separate solutions for the same real value of the coupling constant, κ^2. Each parallelogram is drawn so that the horizontal axis bisects the angle through which it passes; the product of the two sides, considered as complex numbers, is therefore real and is, moreover, the same for each. Since the negative resistance depends on the current, each of these solutions can only exist at a certain level of excitation. For example, steady oscillation at Ω_{2c} demands a large negative resistance, and therefore a small current, to make ω_{1c}'' large enough. This arrangement, however, is unlikely to be stable since the same circuit parameters allow an alternative mode, Ω_{1c}, which can grow and swamp Ω_{2c}. If this were a passive circuit, with current-independent resistances, Ω_{2c} would certainly be unstable for this reason, but we shall find the matter less certain when the negative resistance varies with current.

318

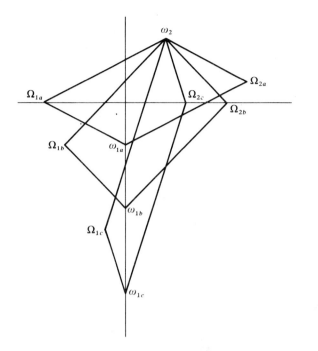

Fig. 13. Three possible normal-mode diagrams (after fig. 10) for a passive resonant circuit of natural frequency ω_2 coupled to an active circuit for which ω_1 lies on a given vertical line. In each parallelogram one corner lies on the real axis and is bisected by the axis; also the product of the sides is the same for each.

Nevertheless, if it is true in this case that the mode at Ω_{2c} is swamped by the growth of Ω_{1c}, the rise of current will bring ω_1 up towards the real axis, and the process may continue so long as any growing mode is available. If it continues to the point where the circuit oscillates at Ω_{1a} it is undoubtedly stable, for the alternative, Ω_{2a}, is a decaying mode. We shall find on proceeding with the analysis that in an active circuit stability may be possible even though an alternative mode lies below the real axis, and our task is to determine under what conditions this is so. The essence of the explanation lies in the fact that any attempt on the part of the alternative mode to grow in amplitude changes the negative resistance, and the ensuing changes in the properties of the circuit may suppress this growth.

The first stage in a quantitative discussion is to enumerate the possible steady-state solutions, whether stable or not, but being steady having the resistances constant. Writing (4) with the real and imaginary parts of ω_1 and ω_2 displayed, we have that a real solution Ω obeys the equation:

$$\Omega - \omega_1' - j\omega_1'' = \kappa^2/(\Omega - \omega_2' - j\omega_2'') = \kappa^2(\Omega - \omega_2' + j\omega_2'')/[(\Omega - \omega_2')^2 + \omega_2''^2].$$

Since it is only ω_1'' that adjusts itself to circumstances, the real part of this equation is free from uncertainty:

$$(\Omega - \omega_1')/\kappa^2 = (\Omega - \omega_2')/[(\Omega - \omega_2')^2 + \omega_2''^2]. \tag{33}$$

The solutions, one or three in number, are exhibited in fig. 14 as the intersection(s) of the straight line L representing the left-hand side with the

Fig. 14. The solutions $\Omega_{a,b,c}$ of (33) exhibited as the intersections of the resonance curve R with the straight line L. In curve C the solutions are replotted with $\Omega_{a,b,c} - \omega'_1$ as ordinates above ω'_1.

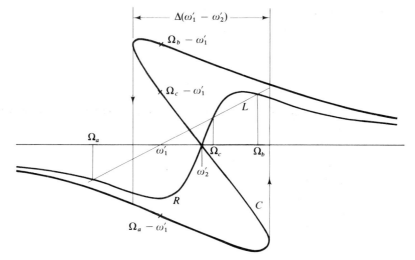

resonance curve R representing the right. For three solutions to be possible, as here, ω'_1 and ω'_2 must be close enough, or κ^2 large enough. The diagram can be replotted as curve C to show how the possible real values of Ω vary with $\omega'_2 - \omega'_1$, κ^2 being kept constant as one of the circuits is tuned. For any choice of ω'_1 relative to ω'_2 draw a line L through ω'_1 at a gradient $1/\kappa^2$; drop perpendiculars from the intersections to find the real solutions, here shown as Ω_a, Ω_b and Ω_c; replot the distances $\Omega_a - \omega'_1$ etc. as ordinates above ω'_1. The resulting curve is C, the required variation of $\Omega - \omega'_1$ with ω'_1. If all these solutions were stable one might expect, on tuning the passive circuit so that ω'_2 fell from a value well above ω'_1 to one well below, to find the system oscillating at a frequency initially slightly below ω'_1, as indicated by the left-hand side of the curve, falling steadily as the lower heavy curve was followed, and switching modes sharply at a certain point when $\omega'_2 < \omega'_1$, so that thenceforward Ω was greater than ω'_1 but by a steadily decreasing amount. On reversing the procedure the upper heavy curve would be followed, and there would be hysteresis. That something of this sort happens is clear from the experimental curves in fig. 15. The coupling was the same in each case, but extra resistances were inserted in the circuits to obtain curves b and c. The resistance in case c is high enough to preclude triple intersections, the maximum slope of the resonance curve R in fig. 14 being just less than $1/\kappa^2$. In curve b there is a pronounced mode switch with a narrow range of hysteresis. In curve a the hysteresis is much more marked, but clearly the whole range indicated in C of fig. 14 is not achieved, since the switch occurs before Ω turns over towards ω'_1. The very fact that any hysteresis occurs shows that linear theory is inadequate, since under no conditions is there more than one solution where the alternative mode is dissipative (e.g. Ω_{2a} in fig. 13).

Since graphical representations will prove convenient in what follows, we may start by recasting the above argument in geometrical form. Some

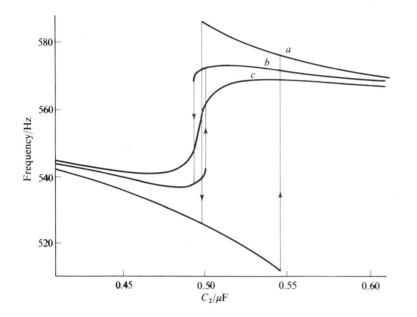

Fig. 15. Experimental measurements of oscillation frequency of coupled active and passive circuits; the natural frequency of the latter was varied by varying its capacitance C_2. See text for further details.

of the assertions made will be left unproved; in each case it is fairly easy, by coordinate geometry, to verify the statements. For convenience, the passive circuit is assumed to remain unchanged, with natural frequency ω_2, while the active circuit is tuned with respect to it. Thus ω_1' is variable, but κ^2 is taken to be constant. The possible solutions to (33), including now the imaginary parts, are represented as in fig. 16, by a family of parallelograms of which one corner (ω_2) is fixed, a second (Ω_1) lies on the real axis, a third (Ω_2) lies on the circle whose diameter is κ^2/ω_2'', and the fourth is ω_1. All these parallelograms have the required properties: $\omega_2\Omega_1$ and $\omega_1\Omega_1$ are bisected by the real axis, and the product of the lengths of the sides is constant ($=\kappa^2$). As Ω_2 runs round the circle, ω_1 describes a cubic trajectory;[†] the range labelled $\Delta(\omega_1' - \omega_2')$ in fig. 14 and fig. 16 is the tuning range which permits three separate solutions. The diagram is drawn for a value of ω_1' at the extreme edge of this range, which can be shown to possess the convenient property that the diagonal $\Omega_1\Omega_2$ makes a right angle with the side $\omega_1\Omega_2$.

Let us now suppose the system to be oscillating steadily in a pure normal mode at Ω_1, and enquire into the effect of adding a little of the normal mode at Ω_2. If the current due to the principal mode is $i^{(1)}$ in the active circuit, it is $i^{(1)}/c$ in the passive circuit,[‡] and $i^{(1)}$ takes such a value that the

---

[†] Note that the line joining Ω_1 and ω_1 always goes through ω_2^*. This makes it easy to construct the trajectory of ω_1.

[‡] From now on we take the two circuits to be virtually identical, with $L_1 = L_2 = L$.

Fig. 16. Alternative to fig. 13, with ω_2 fixed, showing trajectory of ω_1 that is consistent with steady oscillation. Q is the point $\omega_2 - j\kappa^2/\omega_2''$.

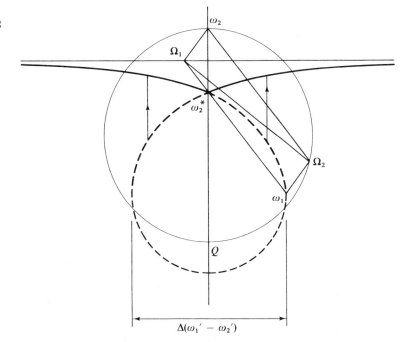

negative resistance R_a just balances dissipation by the positive resistances R in each circuit;

$$(R + R_a)|i^{(1)}|^2 + R|i^{(1)}|^2/cc^* = 0,$$

i.e. $$R_a = -R(1 + 1/cc^*). \qquad (34)$$

A small extra current $i^{(2)}$ in the active circuit, accompanied by $-ci^{(2)}$ in the passive circuit, and oscillating at the subsidiary mode frequency Ω_2, would develop independently of $i^{(1)}$ if R_a remained unchanged. But if the magnitude of the total current i_1 in the active circuit is changed, by δi_1 say, with the appearance of the subsidiary mode, so too is R_a changed, by an amount $-sR_a \delta i_1/i_1$ if s is defined as $-\mathrm{d}(\ln|R_a|)/\mathrm{d}(\ln i_1)$, having the value $+1$ for a fully excited class C oscillator. The effect of this change in resistance is the same as if a source had been inserted into the active circuit, with an e.m.f. of $sR_a\delta i_1$. By this device we are enabled to continue treating the coupled circuits as a linear network, with the normal modes developing independently; but the new e.m.f. generates increments of the two modes, so that the changes of amplitude and phase they suffer are different from what they would be in a truly linear system. Since the e.m.f. generates each mode in roughly equal amplitude the relative effect on $i^{(2)}$ is much greater than on $i^{(1)}$ and we may neglect the latter while concentrating on the development of the alternative mode. We shall now take $i^{(1)}$ as real, and express $i^{(2)}$ at any instant by the complex number $\xi + j\eta$.

It is important to recognize that the feedback circuit behaves like an adjustable resistor, generating an e.m.f. in phase with the current through it. Thus if the increment δi_1 is due to a small current $i^{(2)}$ at a phase angle ϕ with respect to $i^{(1)}$, $\delta i_1 = i^{(2)} \cos \phi = \xi$, and the extra e.m.f., $sR_a\xi$, is in phase with $i^{(1)}$ (if $\eta \ll i^{(1)}$ its effect on the phase of the total current is negligible). The amplitude and phase of the extra contribution to $i^{(2)}$ follow from the impulse response function. According to (32), that part of $i_{11}(t)$ attributable to the mode at Ω_2 is $\exp{(j\Omega_2 t)}/L(1+c^2)$, and (6.32) shows that the e.m.f. $sR_a\xi$ acting for time δt will generate an amplitude of $i^{(2)}$ in this mode in accordance with the equation:

146

$$\delta i^{(2)} = \frac{sR_a\xi}{2L(1+c^2)}\,\delta t.$$

Hence, due to this cause,

$$\dot{\xi} + j\dot{\eta} = sR_a\xi/2L(1+c^2) \equiv (p+jq)\xi. \tag{35}$$

This is the additional effect superposed on the linear behaviour of $i^{(2)}$ as it rotates round $i^{(1)}$ with $\dot{\phi}$ equal to $\Omega_2'-\Omega_1$ and changes in amplitude exponentially under the influence of Ω_2''. The complete equations for the variation of $i^{(2)}$ therefore take the form:

$$\dot{\xi} = -\Omega_2''\xi - (\Omega_2'-\Omega_1)\eta + p\xi,$$
$$\dot{\eta} = -\Omega_2''\eta + (\Omega_2'-\Omega_1)\xi + q\xi;$$

that is to say, in the notation of chapter 9,

255

$$\dot{\xi} = \alpha_1\xi + \alpha_2\eta, \qquad \dot{\eta} = \alpha_3\xi + \alpha_4\eta,$$

where

$$\left.\begin{aligned}\alpha_1 &= -\Omega_2'' + p,\\\alpha_2 &= -\Delta,\\\alpha_3 &= \Delta + q,\\\alpha_4 &= -\Omega_2'';\end{aligned}\right\} \tag{36}$$

$\Delta = \Omega_2'-\Omega_1$ and $p+jq = sR_a/2L(1+c^2) = -s\omega_2''(1+1/cc^*)/(1+c^2)$, by use of (34) and the fact that $\omega_2'' = \frac{1}{2}R/L$. Then $T = \alpha_1+\alpha_4 = -2\Omega_2''+p$, and $D = \alpha_1\alpha_4 - \alpha_2\alpha_3 = \Omega_2''^2 + \Delta^2 - p\Omega'' + q\Delta$.

So long as $T<0$ and $D>0$ the alternative mode cannot grow spontaneously and the oscillation at Ω_1 is stable. In the absence of p and q instability would occur as Ω_2'' moved down through the real axis, and T became positive. In fig. 16, ω_1 would follow that part of its cubic trajectory drawn as a full line, the cusp marking the point at which $\omega_2' = \omega_1'$ and a reversible mode switch would occur. The presence of p (which is negative), as a consequence of non-linearity, delays the instability and allows ω_1 to proceed beyond the cusp in fig. 16 into the region indicated by the heavy broken curve; T may become positive before the extreme limit of ω_1' is reached, in which case ω_1 makes a vertical mode switch as shown in the diagram. This is what happens in curve a of fig. 15. In curve b, on the other hand, there is no instability along the heavy broken curve, and the mode switch occurs as ω_1' reaches the edge of the three-valued region. It will be appreciated that fig. 16 shows the mode-switching in terms of the behaviour of ω_1, which is not directly measured. The corresponding behaviour of $\Omega_1 - \omega_1'$ is shown as C in fig. 14. This indicates how far the passive circuit is able to pull the frequency of the active circuit.

In the last paragraph it was taken for granted that the critical condition for instability was the vanishing of T rather than of D. The latter is intrinsically positive when p and q are absent, and usually, if not always, remains positive when they are present until T reaches zero. We shall not carry through the rather tedious reduction of these quantities in terms of the parameters of the system; the geometrical interpretation of ω_2'', κ^2, Ω_2'', Δ, p and q is given in fig. 17, and it is very quick in any given case to measure off the diagram and check the stability criteria that way. On the assumption that mode-switching occurs when $T = 0$, i.e. $p = 2\Omega_2''$, the information in the diagram leads by elementary geometry to the criterion:

$$RU/RS = 2/s. \tag{37}$$

For a class C oscillator this means that mode-switching occurs when S bisects RU, unless this condition does not operate before ω_1 reaches the end of the heavy broken lines in fig. 16. So long as the coupling is weak enough for $\kappa^2 < \frac{9}{2}\omega_2''^2$ the full range of hysteresis is observed (as in curve (b) of fig. 15), while for stronger coupling the criterion (37) applies (as in curve (a)). Finally we note that the transition is smooth (curve (c)) when $\kappa^2 < \omega_2''^2$.

Fig. 17. Geometrical interpretation of the parameters in (36); $\omega_2'' = BN$, $\kappa^2 = BN \cdot BQ$, $\Omega_2'' = -RC$, $\Delta = AR$, $p = -s \cdot BN \cdot MC/AP$, $q = s \cdot BN \cdot BM/AP$.

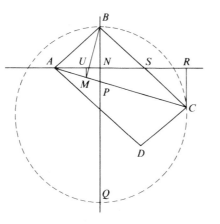

The Huygens phenomenon (entrainment)

Writing to his father in 1665, Christian Huygens, the inventor of the pendulum escapement, reported that two clocks mounted on the same yielding panel or bracket could pull each other into synchronism. The effect has frequently been commented on since in connection with many different self-excited oscillators, including electronic circuits. A not dissimilar biological example was remarked on in chapter 11. A critical degree of coupling is required; with less the two oscillators carry on at their own natural frequencies, slightly perturbed in frequency perhaps and exchanging a fraction of their energy at the beat frequency. The transition to the frequency-locked state is abrupt and is sometimes accompanied by hysteresis – before they can resume their natural oscillations the coupling must be reduced to a value significantly below that which initiated locking. Entrainment can be a nuisance in precise measurements. The ring laser, in which light is made to circulate by means of mirrors and is maintained by the amplifying action of the excited low-pressure gas through which it travels, is potentially a valuable navigating instrument because of its sensitivity to rotation.[6] So long as the frame supporting mirrors and gas tubes is at rest, the natural frequency is the same for both senses of circulation of the light; but rotation of the frame causes one mode to increase and the other to decrease in frequency, and the beat frequency which can be detected should, in ideal conditions, measure the angular velocity with very great precision and sensitivity. Unfortunately, however, scattering of light from one mode into the other, and other effects of minute imperfections, lock the modes together when their frequencies are not too different, and the instrument fails at low angular velocities, just where it is most needed as a substitute for gyroscopes. On the other hand, entrainment can be put to practical use in, for example, stabilizing the frequency of a multivibrator and using it to generate harmonics and subharmonics of the output from a primary clock, which might be a quartz or caesium frequency standard. We shall consider this application later in the chapter, but as a preliminary we analyse the Huygens phenomenon in class C oscillators.

In principle the method applied in the last section could be extended to two active oscillators, but description of the state in terms of normal modes is not convenient when the imaginary parts of both ω_1 and ω_2 have to be adjusted to be compatible with the current in each circuit. We shall therefore abandon normal modes in favour of the currents themselves as variables describing the state at any instant. To illustrate the method, and because it will turn out that the problem is complicated enough to discourage a complete solution, we shall begin with something simpler that can be fully worked through without too great an effort. This simpler problem is one of interest in its own right – the locking of a class C oscillator to an injected signal of fixed frequency and amplitude.

As usual, let us represent the oscillator as a series resonant circuit in which the resistance has a constant positive term, R, and a negative term whose magnitude is inversely proportional to the amplitude of current oscillations. The total resistance can be written as $R(1 - i_0/i)$, i_0 being the amplitude at which the oscillation at the natural frequency ω_0 establishes itself if left unperturbed. Into the circuit we now inject an e.m.f. $v\,e^{j\omega t}$, v being taken as real, which in accordance with (6.32) causes an oscillatory current, in phase with v, to build up at a rate $\frac{1}{2}v/L$. If at any instant the oscillation is represented in magnitude and phase by $i\,e^{j(\omega t+\phi)}$, i and ϕ change in accordance with the equations†

$$\frac{\mathrm{d}i}{\mathrm{d}t} = -\frac{1}{2}R(i-i_0)/L + \frac{1}{2}v\cos\phi/L, \tag{38}$$

and

$$\frac{\mathrm{d}\phi}{\mathrm{d}t} = (\omega_0-\omega) - \frac{1}{2}v\sin\phi/Li. \tag{39}$$

In these two equations the first terms describe the behaviour in the absence of injected signal, the amplitude of oscillation relaxing towards its equilibrium, i_0, with time-constant $2L/R$ just like a passive circuit, while the phase difference evolves at the difference frequency. The second terms describe the effect of the injected oscillation on the amplitude and phase of what is already present. These are the basic equations from which t may be eliminated by division:

$$\frac{\mathrm{d}y}{\mathrm{d}\phi} = \frac{-y(y-y_0-\cos\phi)}{\varepsilon y - \sin\phi} \tag{40}$$

in which $y \equiv Ri/v$, $y_0 \equiv Ri_0/v$ and $\varepsilon \equiv 2L(\omega_0-\omega)/R$.

To sketch the solutions of (40) we construct on a y–ϕ diagram the loci on which the numerator and denominator vanish; on the former, labelled 0, the trajectories are horizontal while on the latter, labelled ∞, they are vertical. The two examples in fig. 18 differ in the choice of ε, only the ∞-loci being affected; they refer to poor tuning (ε large) and better tuning (ε small), the amplitude v of the injected signal being kept constant. Only if ε is small enough does the ∞-locus rise high enough to intersect the 0-locus and provide the possibility of a stable solution in which $\dot{\phi}=0$ and mode-locking has occurred. Where there is no intersection, as in fig. 18(a), there is a certain trajectory, labelled P, which enters and leaves the diagram at the same ordinate, y, and which is clearly periodic. As the oscillator slips in phase relative to the injected signal, the amplitude fluctuates periodically, and the appearance is of a beating between the free oscillation and the signal. When y_0 is small (strong perturbing signal) the beat modulation can be very pronounced, the oscillation amplitude dropping very nearly to zero at the beat minima. An expression for the beat period, T_b, can be derived from (39) which, rearranged and integrated, takes the form

$$T_b = \frac{2L}{R\varepsilon}\int_0^{2\pi}\mathrm{d}\phi/(1-y_1/y), \tag{41}$$

† An alternative derivation, not involving the impulse response function, treats the injected e.m.f. as if it were due to an extra impedance, of magnitude $-v\,e^{-j\phi}/i$, inserted in the circuit. The real part, $-v\cos\phi/i$, gives rise to additional decrement according to the standard formula, $\mathrm{d}i/\mathrm{d}t = -\frac{1}{2}Ri/L$, and this produces the second term in (38). The imaginary part, $v\sin\phi/i$, acts like an additional inductance and lowers the frequency, as expressed by the second term in (39).

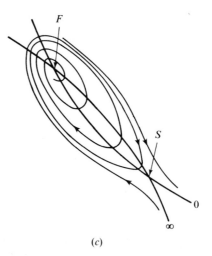

Fig. 18. 0-loci, ∞-loci and sketched trajectories for solutions of (40); (a) $y_0 = 2$, $\varepsilon = 1$, (b) $y_0 = 2$, $\varepsilon = 0.287$, (c) illustrating form of trajectories when F and S are about to merge.

in which y_1 is written for $\sin \phi/\varepsilon$, the ordinate of the ∞-locus. The speed of the traverse, $\dot{\phi}$, is not uniform, but is much diminished when the trajectory runs close to the ∞-locus, and the denominator in (41) becomes small. This is particularly noticeable just before locking takes place, the oscillator feeling, as it were, a strong attraction towards the perturbing signal but breaking away only to repeat the process one beat period later.

When the loci intersect, as in fig. 18(b), curve-drawing shows immediately that S is a saddle-point but that F has possibilities for being a stable locked state. To investigate this further, let the coordinates of F be (Φ, Y); then, since both numerator and denominator of (40) vanish at F,

$$Y - y_0 - \cos \Phi = 0 = \varepsilon Y - \sin \Phi, \tag{42}$$

from which it follows that

$$(Y - y_0)^2 + \varepsilon^2 Y^2 = 1, \tag{43}$$

i.e.
$$Y = \{y_0 \pm [1 + \varepsilon^2(1 - y_0^2)]^{\frac{1}{2}}\}/(1 + \varepsilon^2). \tag{44}$$

The solution with positive sign is F; that with negative sign is S. When $y_0 < 1$ and the 0-locus drops below the axis, intersection occurs at all values of ε, but at weaker signal inputs, when $y_0 > 1$, (44) has a solution only if $\varepsilon > \varepsilon_c$, where

$$\varepsilon_c = (y_0^2 - 1)^{-\frac{1}{2}}. \tag{45}$$

In the vicinity of F, write $\phi = \Phi + \xi$ and $y = Y + \eta$; then by differentiating numerator and denominator of (40) the linear approximation is derived in the form

$$\frac{d\eta}{d\xi} = -Y\frac{\eta + \xi \sin \Phi}{\varepsilon \eta - \xi \cos \Phi} = \frac{\alpha_3 \xi + \alpha_4 \eta}{\alpha_1 \xi + \alpha_2 \eta},$$

where the coefficients can be put in the following form by use of (42):

$$\alpha_1 = y_0 - Y, \qquad \alpha_2 = \varepsilon, \qquad \alpha_3 = -\varepsilon Y^2 \quad \text{and} \quad \alpha_4 = -Y.$$

255 For F to be a stable solution, $D > 0$ and $T < 0$. The first criterion demands that $Y(1 + \varepsilon^2) > y_0$, which is seen from (44) to be satisfied by F, if it exists; it fails at the same moment as the loci cease to intersect. Two intersecting loci such as we find here cannot generate the same type of stationary point at both intersections, and the presence of one saddle-point guarantees that the other is not a saddle-point.[†] This argument is illustrated in fig. 18(c). The stability of F can therefore be tested by the second criterion alone; this demands that $Y > \frac{1}{2}y_0$, and is obviously satisfied in fig. 18(b) but is not necessarily satisfied when $y_0 < 1$. For ε can then be made as small as one likes without losing the intersection, and there is no problem in arranging for Y to be less than $\frac{1}{2}y_0$, whereupon F becomes an unstable focus.

[†] The same point is made by noting that at the moment the loci cease to intersect they run parallel at the point of contact. Hence $\alpha_4/\alpha_3 = \alpha_2/\alpha_1$, or $\alpha_1\alpha_4 - \alpha_2\alpha_3 = 0$.

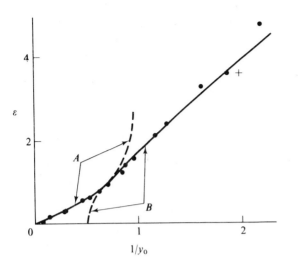

Fig. 19. Stability curve
derived from (40);
entrainment occurs
everywhere below the line.
The points are
experimental values for
the stability limit.

Fig. 19 illustrates the stability criteria as a plot of critical mistuning (ε) against strength of injected signal ($1/y_0$). Curve A is a plot of (45); when $1/y_0 < 1$ it is only for points below this curve that F exists and the possibility of locking need be considered. Curve B is the hyperbola $\varepsilon^2 = 4/y_0^2 - 1$ on which, as follows from (43), $Y = \frac{1}{2}y_0$. On the upper part of B, beyond the point ($1/\sqrt{2}$, 1) at which it touches A, the saddle-point S either cannot have $Y = \frac{1}{2}y_0$ (in the range $1/\sqrt{2} < 1/y_0 < 1$) or does not exist ($1/y_0 > 1$); there is no question therefore but that this part of B refers to F alone, and that only points lying below B can show stable locking. The lower part of B, shown broken, refers to S, not F, and is irrelevant to this discussion. The full curve is therefore the limit of stability of the locked mode. The experimental points were taken by fixing the level of the injected signal and detuning its frequency away from ω_0 (both upwards and downwards) until locking failed – a very obvious moment since strong beats suddenly appear on the waveform of the oscillation. The value of ε has been taken as proportional to the difference between the upper and lower critical frequencies, the constant of proportionality being adjusted to get a good fit. No arbitrary adjustment was needed for $1/y_0$, for (44) shows that when the signal is tuned to ω_0, making ε vanish, the oscillation amplitude Y rises to $1 + y_0$; far from the locking condition, however, $y = y_0$ and from observation of the amplitude peak y_0 is unambiguously derived. The experimental points confirm the theory very adequately. Further confirmation is provided by the observation that when the injected signal is strong, but not when it is weak, the amplitude has dropped to half its unperturbed value, i.e. $Y = \frac{1}{2}y_0$ as demanded by the stability analysis, at the moment of unlocking.

Finally it is worth noting that when the injected signal is strong hysteresis appears, entrainment occurring, as ε is reduced, at a smaller value than that at which it fails when ε is increased again. This means that under certain conditions a periodic unlocked trajectory and a stable focus can be found on the same diagram. A set of computed trajectories confirms this; fig. 20 is drawn for the case $\varepsilon = 3.65$, $y_0 = \frac{1}{2}$ which lies, as shown in fig. 19, just below the critical curve B. A slowly converging trajectory surrounds the focus F which is clearly stable. The linear approximation begins to fail as one moves away from F; convergence becomes slower, and L is an unstable closed trajectory; trajectories within L converge, and those outside it diverge, breaking away after some circuits of F and gradually approaching the stationary unlocked trajectory, which lies about midway between T_1 and T_2. It is clear that between the limiting forms of behaviour – on the one hand a stable focus into which all trajectories converge, and on the other an unstable focus and a stable unlocked trajectory – one must normally expect a transition regime in which both stabilities coexist, and that hysteresis will accompany the switch from one to the other. That it is not observed at lower signal strengths is probably because the width of the hysteretic regime is narrow. There is no very obvious reason, even in the situation covered by curve A of fig. 19, when locking takes place as soon as the 0-loci and ∞-loci intersect, why the unlocked trajectory should simultaneously become unstable. We shall meet the same behaviour in the next example, and experimental data will be presented in fig. 23 to show the marked differences in detail between hysteretic and non-hysteretic locking. Observations of the system just analysed show it to have very similar characteristics, and there is no need to give an account of any further measurements in this case.

Fig. 20. Computed trajectories for (40) with $y_0 = \frac{1}{2}$, $\varepsilon = 3.65$, corresponding to cross in fig. 19.

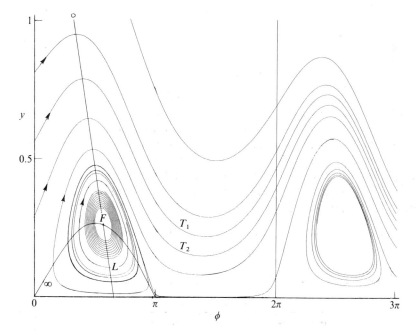

Let us proceed, therefore, to the Huygens phenomenon proper, extending the argument to cover two class C oscillators coupled by a mutual inductance M, so that each injects a perturbing signal into the other, tending to lock it but inevitably being itself perturbed in the process. Let the oscillators be nearly identical, except for oscillator 2 having a slightly higher natural frequency than oscillator 1. At any instant the currents in the two circuits may be written as $i_1 e^{j\phi_1}$ and $i_2 e^{j\phi_2}$, i_1 and i_2 being real, and the time variation of ϕ_1 and ϕ_2 including the natural frequency as well as any perturbing effects. Then by the same argument as led to (38) and (39), we write the equations for the currents and phases:

$$\frac{di_1}{dt} = -\tfrac{1}{2}R(i_1 - i_0)/L - \mathrm{Re}\,[\tfrac{1}{2}j\omega M i_2\, e^{j(\phi_2 - \phi_1)}/L], \tag{46}$$

$$\frac{di_2}{dt} = -\tfrac{1}{2}R(i_2 - i_0)/L - \mathrm{Re}\,[\tfrac{1}{2}j\omega M i_1\, e^{j(\phi_1 - \phi_2)}/L], \tag{47}$$

$$\frac{d\phi_1}{dt} = \omega_1 - \tfrac{1}{2}\,\mathrm{Im}\,[j\omega M i_2\, e^{j(\phi_2 - \phi_1)}/L i_1], \tag{48}$$

$$\frac{d\phi_2}{dt} = \omega_2 - \tfrac{1}{2}\,\mathrm{Im}\,[j\omega M i_1\, e^{j(\phi_1 - \phi_2)}/L i_2], \tag{49}$$

in which ω_1 and ω_2 are the frequencies of the two oscillators when decoupled. Write $\omega_2 - \omega_1 = \Delta$, $i_2/i_1 = \rho$, $\phi_2 - \phi_1 = \phi$, $2L\Delta/\omega M = \varepsilon$, $\omega M/R = \mu$; as before, ε is a measure of the mistuning of the oscillators, while μ is a measure of the coupling. These equations may now be rearranged with the following outcome:

$$d\rho/dt = \frac{\Delta\rho}{\mu\varepsilon}[(1 - \rho)i_0/i_2 - \mu(\rho + 1/\rho)\sin\phi], \tag{50}$$

and
$$d\phi/dt = \Delta[1 + (\rho - 1/\rho)/\varepsilon \cos\phi]. \tag{51}$$

But for the presence of i_0/i_2 in (50), the variables in the problem would have been reduced to two, as in (40), with ρ playing the part previously played by y. It is worth forcing the problem into this mould, while recognizing the dubious legitimacy of the process, a matter to which we shall return afterwards. A conceivably valid assumption which achieves this end is that the energy content of the system as a whole does not vary during the beating cycle when the oscillators are not locked, so that the total dissipation can be equated to zero:

$$R(1 - i_0/i_1)i_1^2 + R(1 - i_0/i_2)i_2^2 = 0,$$

i.e.
$$i_0/i_2 = (1 + \rho^2)/(1 + \rho).$$

When this is substituted in (50), it and (51) together yield the equation:

$$\frac{d\rho}{d\phi} = \frac{1 + \rho^2}{\mu}\frac{(1 - \rho)/(1 + \rho) - \mu\sin\phi}{\varepsilon + (\rho - 1/\rho)\cos\phi}. \tag{52}$$

In like manner to fig. 18 we plot the loci of zero and infinite gradient, which intersect either four times or not at all; fig. 21 is an example where intersecting loci give the possibility of locking at either F_1 or F_2, but not at S_1 or S_2 which are saddle-points. The two modes of locking, F_1 and F_2, are closely related in that they occur at values of ϕ that are π apart and at values of ρ that are reciprocals; substituting $1/\rho$ for ρ and $\phi + \pi$ for ϕ in (52) leaves the equations unchanged.† It is only necessary, therefore, to consider the conditions for the occurrence of one intersection, F_1, and its stability, and we shall assume from now on that $\rho < 1$. We proceed as before; if F_1 is (Φ, P), then

$$\mu \sin \Phi = (1-P)/(1+P) \quad \text{and} \quad \cos \Phi = -\varepsilon/(P-1/P). \qquad (53)$$

Fig. 21. 0-loci, ∞-loci and sketched trajectories for solutions of (52).

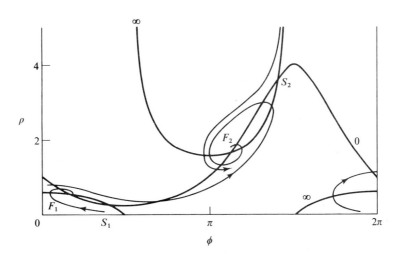

By squaring and adding, to eliminate Φ, we find after rearrangement that

$$(P-\rho_1)^2(P-1/\rho_1)^2 = \mu^2 P^2[4\mu^2 - \varepsilon^2(1-\mu^2)]/(1-\mu^2)^2, \qquad (54)$$

where $\rho_1 + 1/\rho_1 = 2/(1-\mu^2)$. Now every term except that in square brackets on the right is essentially positive. The condition for P to exist is then seen to be that the term in square brackets must be positive. If $\mu \geq 1$ this is automatically satisfied, but if $\mu < 1$ it is necessary that

$$\varepsilon^2 \leq 4\mu^2/(1-\mu^2). \qquad 55)$$

At the moment this criterion fails to be satisfied, each focus $F_{1,2}$ coincides with its accompanying saddle-point $S_{1,2}$ at values of ρ equal to ρ_1 and $1/\rho_1$.

† At F_1, $i_2 < i_1$ and the circuit of higher natural frequency, being less excited than the other, suffers the greater shift of frequency; the two are therefore locked at a frequency below the natural frequency of either. At F_2, $i_2 > i_1$, and the frequency of locking is higher than either natural frequency.

To investigate the stability of F_1, put $\phi = \Phi + \xi$, $\rho = P + \eta$ to derive, with the help of (53), the linearized versions of the numerator and denominator of (52):

$$\frac{\mathrm{d}\eta}{\mathrm{d}\xi} = \frac{\alpha_3\xi + \alpha_4\eta}{\alpha_1\xi + \alpha_2\eta},$$

where $\alpha_1 = (1-P)^2/P$, $\alpha_2 = \mu\varepsilon(1+P^2)/P(1-P^2)$, $\alpha_3 = -\mu\varepsilon P(1+P^2)/(1-P^2)$ and $\alpha_4 = -2(1+P^2)/(1+P)^2$. As in the previous example, the merging of F and S at the moment when the loci cease to intersect is also the moment of failure of the stability criterion $D > 0$. This criterion is therefore the same as (55), and it is to the second criterion, $T_1 < 0$, that attention must be directed. It may be written in the form

$$(P + 1/P)^2 - 2(P + 1/P) - 4 < 0, \tag{56}$$

so that $P + 1/P$ must lie between 2 (since $P > 0$) and $1 + \sqrt{5}$. Now we have seen that intersection fails when $P = \rho_1$, i.e. when $P + 1/P = 2/(1-\mu^2)$. So long as the coupling is weak enough for μ^2 to be less than $\frac{1}{2}(3 - \sqrt{5})$ locking fails when the possibility of a stationary solution disappears, not because a stationary solution has become unstable; for larger μ, however, there is still an intersection F as the circuits are detuned, but it changes from a stable to an unstable focus.

The condition for locking is presented graphically in fig. 22, which is based on an alternative version of (54):

$$\varepsilon^2 = (P - 1/P)^2 - (P^{\frac{1}{2}} - 1/P^{\frac{1}{2}})^4/\mu^2. \tag{57}$$

Each choice of P (or $1/P$) defines a straight line on a graph with $1/\mu^2$ as abscissae and ε^2 as ordinates, and the envelope of these lines is the hyperbola $\varepsilon^2 = 4\mu^2/(1-\mu^2)$ which, according to (55), defines the limiting

255

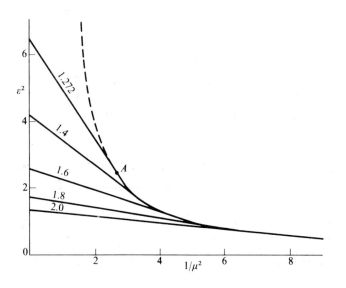

Fig. 22. Stability curve derived from (52), consisting of the straight line labelled 1.272 and the curve to the right of A. The numbers on the lines are values of δ/Δ, δ being the separation of the two frequencies at which the entrained system can oscillate, for a given separation, Δ, of the frequencies when the circuits are uncoupled.

condition for intersection of the loci. Not all points below the envelope allow stable locking, however, but only those that also satisfy (56) by lying below the line for $P+1/P$ equal to $1+\sqrt{5}$, i.e. $P=0.346$. Every point below the envelope lies on two straight lines defined by different values of P. This is because for a given choice of μ and ε, F is always accompanied by S. Since the value of P at the unstable intersection S is always less than at F, the diagram may be limited to F by drawing the lines only on the left-hand side of the point where they touch their envelope, and this has been done in fig. 22. The diagram serves also to indicate the frequency ω at which the oscillators lock, which is the value of $d\phi_1/dt$ in (48) or $d\phi_2/dt$ in (49), both being the same. Hence

$$\omega = \tfrac{1}{2}(d\phi_1/dt + d\phi_2/dt) = \bar{\omega} \pm \tfrac{1}{2}\delta,$$

where $\bar{\omega} = \tfrac{1}{2}(\omega_1 + \omega_2)$ and

$$\delta = \frac{1}{2}\frac{\omega M}{L}(P+1/P)\cos\Phi = \Delta(1+P^2)/(1-P^2) \quad \text{from (53)}.$$

As with reactively coupled passive resonant circuits, $\delta > \Delta$ and the frequency at which steady oscillation can occur lies outside the range defined by the natural frequencies, and for the same reason, that the reactive coupling either depresses both or elevates both.

Some results of an experimental test of this analysis are shown in fig. 23. Two class C oscillators, with inductors of rather large radius (10 cm) were arranged so that the spacing between the inductors, and thus the coupling,

320

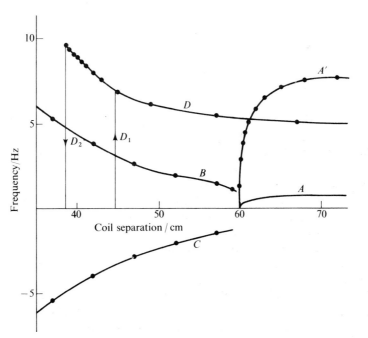

Fig. 23. The Huygens phenomenon with two class C oscillators. The circuit parameters were held constant while the coupling was changed by changing the coil separation. For curves A, B and C the frequencies of the uncoupled oscillators were 556.0 and 556.7 Hz. Curve A is the beat frequency (multiplied by 10 in A') when the coupling is too weak for entrainment; B and C are the two modes of entrainment, the frequencies here being measured from the mean, 556.4 Hz. For curve D the frequencies of the uncoupled oscillators were 555.9 and 560.8 Hz.

could be varied. When the frequency difference was small weak coupling sufficed to produce entrainment, and the approach to the critical point had the character shown in curve A; as the spacing is reduced the beat frequency drops, ultimately with a vertical tangent at the moment entrainment begins. From then on, further increase of coupling causes a progressive frequency shift away from the mean, along either curve B or curve C according as entrainment has taken place in the upper or lower mode. It is worth remarking that the parameters (e.g. amplifier gain) of the two oscillators must be carefully matched in order to obtain either B or C with equal likelihood. If one amplifier is stronger it will cause the system to lock in the mode whose frequency is nearer to its own. With carefully matched oscillators, however, it is possible to select either mode by pushing the two coils together at the right phase in the beat cycle. As with the previous example of entrainment, there is no sign of hysteresis when the frequency difference is small and the critical point is that at which 0-loci and ∞-loci first intersect. With a larger frequency difference stronger coupling is needed and hysteresis then appears, as in curve D. Failure of entrainment (D_1) on reducing the coupling takes place as T changes sign and the focus F becomes unstable; the onset of entrainment (D_2) on increasing the coupling takes place when the periodic unlocked trajectory becomes unstable. This is a problem which we have not analysed – it is distinctly more difficult than anything tackled here. Attention may be drawn to the steady increase in the beat frequency along D as the coupling is increased. A rough comparison with the difference between B and C is enough to show that the unlocked oscillators are beating at much the same frequency as if they were passive coupled circuits. There is no sign of the attraction that is so characteristic a feature of the weakly coupled system (curve A).

Let us now consider whether it was justifiable to assume, in the argument leading to (52), that the energy content of the circuits was constant. It is obvious that when the oscillators are locked they must settle into a state in which the total energy remains constant, and the description of the amplitudes and phase relation given here by P and Φ is correct. The same is not necessarily true for any unlocked trajectory, and therefore the analysis of stability which assumed that a transition could be made only to an energy-conserving trajectory is not necessarily valid. For example, when detuning causes the stable focus F to become unstable and the trajectory begins to spiral slowly outwards, the energy may oscillate with the beat frequency – the frequency of the trajectory round F – and this fact is almost certain to make some difference to the precise condition for instability to set in. It is on the left-hand side of fig. 22 that this criticism becomes serious, as the following argument makes clear. The characteristic time for a class C oscillator to settle into its stationary state is $2L/R$, just like a passive resonant circuit, and we must expect oscillations of energy if the trajectory has a period of this order or shorter. In the present problem this occurs when $\varepsilon\mu \gtrsim 1$, which defines the region to the left of a 45° line through the origin in fig. 22. On the other hand, at the breakdown of locking on the right-hand side, when D goes to zero as the loci cease to intersect, the trajectories around where F has just disappeared are traversed exceedingly slowly. There is every expectation, therefore, that the assumption of zero net energy production applies here both to the stable and to the just-unstable regime, and that the condition for stability (i.e. the existence of F) is correctly applied. We may hazard a guess that the boundary of the stable region in fig. 22 is the whole of the hyperbolic envelope up to A where it joins the limiting straight line, and that as we proceed along the straight line it becomes progressively less correct, the true limit falling below as other possibilities for instability make themselves felt. To go into the matter in more detail involves considerably more effort. Alternative approaches are (a) to define the system by means of three parameters, i_1, i_2 and ϕ, rather than two, and consider the conditions of stability of a linearized three-variable system, with three criteria to be satisfied; or (b) to extend the normal mode treatment used for the coupled active and passive oscillators. Both approaches should lead to precise criteria rather than the imprecise ones developed here but, as we have remarked before, there must be some doubt as to the value of the effort in terms of improved understanding of the mechanisms involved. We propose to remain satisfied with a treatment that reveals the underlying physics, even if the answer is not correct in all details.

An altogether simpler example of the Huygens phenomenon is provided by the *resistive* coupling of two active oscillators, such as the arrangement of fig. 1(e) but with both circuits maintained as class C oscillators. The equations have the same form as (46)–(49), except that $j\omega M$ is replaced by R_c. Then if ε is redefined as $2L\Delta/R_c$, and μ as R_c/R,

$$di_1/dt = \frac{\Delta}{\varepsilon}[(i_0 - i_1)/\mu - i_2 \cos \phi], \tag{58}$$

$$di_2/dt = \frac{\Delta}{\varepsilon}[(i_0 - i_2)/\mu - i_1 \cos \phi] \tag{59}$$

and

$$d\phi/dt = \frac{\Delta}{\varepsilon}[\varepsilon + (\rho + 1/\rho) \sin \phi]. \tag{60}$$

From (58) and (59) we have that

$$\frac{d}{dt}(i_2 - i_1) = \frac{\Delta}{\varepsilon}[-(i_2 - i_1)/\mu + (i_2 - i_1) \cos \phi]. \tag{61}$$

If entrainment occurs, $di_1/dt = di_2/dt = 0$, and (58) and (59) show that $i_1 = i_2$. As we shall see in a moment, only states in which $\cos \phi$ is negative can be stable; it is clear from (61) that in the neighbourhood of such a state any departure of $i_2 - i_1$ from zero must eliminate itself automatically. In assuming $i_1 = i_2$, then, when determining the conditions for locking we are

on much safer ground than we were in the previous calculation. Writing
$i_1 = i_2 = i$, we have from (58) and (60)

$$\frac{di}{d\phi} = \frac{1}{\mu} \frac{i_0 - (1 + \mu \cos \phi)i}{\varepsilon + 2 \sin \phi}. \tag{62}$$

Typical loci of zero and infinite gradient are shown in fig. 24 for $\mu < 1$, and
it needs no further analysis to see that S is a saddle-point and N a stable
node, which always exists provided $\varepsilon < 2$ so that the denominator of (62)
can vanish. Thus the oscillators lock when $\Delta < R_c/L$, and there is no
possibility of hysteretic behaviour since non-locking trajectories crossing
the diagram can only be present when the ∞-loci are absent.

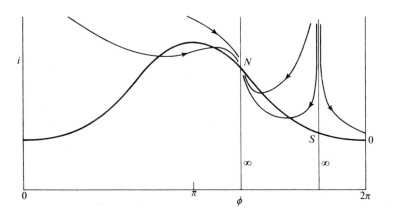

Fig. 24. 0-loci, ∞-loci and
sketched trajectories for
solutions of (62).

This analysis is verified by experiment. When the coupling is weak the
currents oscillate in amplitude with a beat-like pattern, and the amplitude
oscillations in the two circuits are in phase. This contrasts with the react-
ively-coupled system where they oscillate in antiphase. As the coupling is
increased the beat period gets longer without limit, and as it goes to infinity
locking sets in, only to be broken once more without any sign of hysteresis
when the coupling is reduced. When the beat period is long it exhibits clear
signs of non-linearity, the envelope departing markedly from sinusoidal
form, as the record in fig. 25 shows. This is not an isolated example – the
other systems showing the Huygens phenomenon behave similarly. In this
particular case the oscillators, when uncoupled, differed in frequency by 1
part in 41, but the beat period when coupled is not 41 but 67 cycles. This
shows, as expected for resistive coupling, and in contrast to curve D of fig.
23, the attraction of the two frequencies towards the mean.

To conclude, let us return to Huygens himself and his observation of
entrainment in pendulum clocks. It is quite easy to demonstrate the effect
by extracting the mechanisms from two metronomes and mounting them
on a light frame, which is suspended on threads so that its swinging can
couple the two vibrations. With the adjustable weights removed the
vibration frequency is about 100 cycles (200 clicks) a minute and strong

Fig. 25. Non-sinusoidal beat envelope for resistively-coupled oscillators not quite entraining.

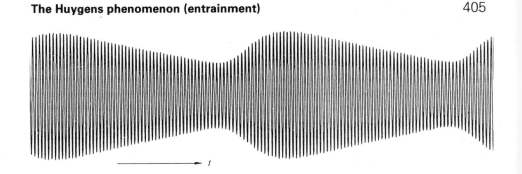

coupling is observed. The records shown in fig. 26 are redrawn from a chart recording of the clicks picked up by a microphone. The records in (*a*) and (*b*) show how the metronomes vibrate independently, with a frequency difference of 2.5%, when the frame is clamped, but lock into a steady phase relationship when it is free to swing. The graph shows that the approach to the steady state is oscillatory, as expected for reactive (i.e. loss-free) coupling. An extra complication here, apart from the escapement mechanism, is the large amplitude of swing, up to ±45° during the oscillatory approach to the steady state; changes in frequency with amplitude must be a significant factor in determining the details of the behaviour.

353   Frequency-locking of a multivibrator

The most striking difference between a class C oscillator and a relaxation oscillator, as regards their response to an injected sinusoidal disturbance, is that the former responds only to a nearly synchronous signal and is drawn into synchronism, while the latter can be controlled by a signal at a harmonic of its own oscillation. Perhaps the distinction is slightly exaggerated by this statement, in that a class C oscillator may indeed by locked by a powerful enough signal at the third or fifth harmonic, but this is a very mild affair compared with the way a multivibrator may be diverted, if only by a small amount, from its natural frequency into an exact submultiple of an injected frequency hundreds of times higher. It is this property that has given the multivibrator its place in the repertoire of electronic devices, for with two or three stages of subdivision one can convert the frequency of a standard source (e.g. a quartz crystal oscillator) from several megahertz down to a suitable frequency for running a clock.

The multivibrator is not only sensitive to harmonics; it can also be locked, though rather weakly, at a frequency which is a rational fraction of the injected frequency. An example is shown in fig. 27, where it can be seen that the frequency of the multivibrator is pulled in an extremely complicated way as the injected frequency is varied. The details depend critically on the amplitude of the perturbation, but the fundamental response is always the most marked. The odd harmonics pull the multivibrator more strongly than the even, and certain intermediate lockings are obvious. What is not obvious in the diagram is that every point marked on

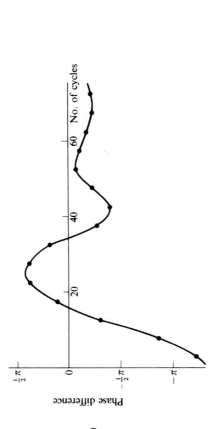

Fig. 26. The Huygens
phenomenon with two
metronome movements;
(*a*) and (*b*) are transferred
from a record of the clicks
of the two metronomes
(shown different in height
here). In (*a*) the beam on
which they were mounted
was clamped, and they
vibrated independently at
different frequencies; in
(*b*) the beam was swinging
freely and entrainment
occurred, as plotted in
(*c*) which shows that the
phase difference between
the clicks settled down to
a constant value. The
oscillatory approach
indicates a case of focal
stability.

Fig. 27. Multivibrator as in fig. 11.38, entrained by a sinusoidal input of variable frequency.

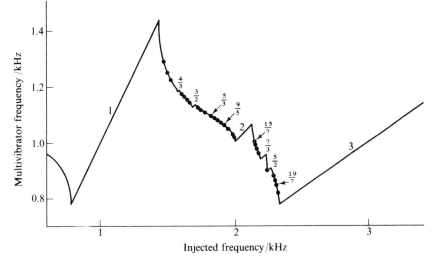

the curve represents a locking, if only over a very short frequency interval (1 Hz or less), but nevertheless discernible; doubtless many more could be found by patient searching. Below the fundamental, however, it is hard to convince oneself that any locking occurs, though vestiges can be discovered, for example at the third subharmonic, by turning up the amplitude of the injected signal. We shall provide a rather superficial discussion of these observations, concentrating on generalities and not attempting a detailed theory of why some steps are long and others short; and one notable failure will be the absence of any explanation for the weakness of the sub-fundamental steps. This is not to imply that the model analysed is at fault, but only that elementary mathematics has failed to cope with a remarkably elusive phenomenon.

Let us begin with the contrast, already noted, between the class C oscillator and the multivibrator in their sensitivity to other than synchronous locking. Qualitatively, the insensitivity of the one and the sensitivity of the other to a harmonic disturbance is easily understood. The model we have adopted for a class C oscillator is essentially that of a free-running high-Q circuit kept going by a very weak impulse during each cycle. The injection of a third-harmonic signal can produce only the weakest perturbation compared with the effect of a signal near the resonance frequency. By contrast, the multivibrator is non-resonant, spending almost all the time relaxing, like an RC circuit, towards an unattainable point of rest ($\pm V_0$ in fig. 28(a)), only to be regularly tripped into a different state at B or D. The circuit can be idealized as in fig. 28(b), the device labelled N being the feedback network and a series resistor (10 Ω in fig. 11.39), which together have the characteristic shown in (a); a locking e.m.f. v may be injected as shown. If $V_0 \gg \Delta V$, i and V move along AB or CD at a very nearly constant rate, $\dot{V} = \pm V_0/\tau$, τ being the time-constant $R_s C$. This is what happens when $v = 0$, but when a signal is

354

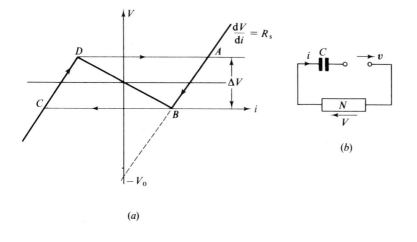

Fig. 28. (*a*) Assumed negative-resistance characteristic of the device *N* in (*b*) which represents the multivibrator circuit with sinusoidal input *v*.

injected the immediate response to a sudden change Δv is that $-\Delta v$ is developed across *N*, since the p.d. across *C* cannot change suddenly. Only over a time of the order of τ does *C* respond to the stimulus, and if v oscillates as $v_0 \cos \omega t$, with $\omega \tau \gg 1$, we may ignore everything but the initial effect; on the steady change of *V* is superposed an oscillation $-v_0 \cos \omega t$:

$$V = \text{const} \pm V_0 t/\tau - v_0 \cos \omega t. \qquad (63)$$

Here, and in what follows, the upper sign refers to motion along *CD*, the lower to motion along *AB*; the constant changes its value at every switch from one side to the other, but it will soon become clear that it is of no importance to the argument.

It is convenient to introduce V' as a variable that increases steadily at the same rate as $|V|$, rather than oscillating to and fro; $\dot{V}' = |\dot{V}|$, and

$$V' = A + V_0 t/\tau \mp v_0 \cos \omega t. \qquad (64)$$

Every time V' reaches a value that is a multiple of ΔV, as defined in the diagram, the sign changes, and so does the value of *A* to keep V' continuous. The switches occur at successive times, t_0, t_1 etc. such that

$$\Delta V = (V_0 t_1/\tau - v_0 \cos \omega t_1) - (V_0 t_0/\tau - v_0 \cos \omega t_0),$$

$$\Delta V = (V_0 t_2/\tau + v_0 \cos \omega t_2) - (V_0 t_1/\tau + v_0 \cos \omega t_1),$$

and so on with alternating signs. The process is illustrated in fig. 29(*a*), and the actual variation of V' in (*b*) where each element of vertical height ΔV has been shifted by appropriate choices of *A* so as to give a continuous curve. This is irrelevant, however, since all the interest lies in the placing of the successive segments on the sinusoids, to see whether it shows any sign of that regularity which must accompany entrainment.

In this example there is no obvious relationship between the mean switching frequency and ω the injected frequency though possibly, by continuing the diagram for long enough, a periodic pattern would be

Fig. 29. (*a*) Each successive heavily-drawn segment has vertical height ΔV, and at the end of each the point is transferred to the other sinusoid to start the next segment; in (*b*) the successive segments are joined to show the variation of V' with time, as given by the solution of (64). (*c*) is a simplified form of the sinusoids in (*a*), used in constructing fig. 31.

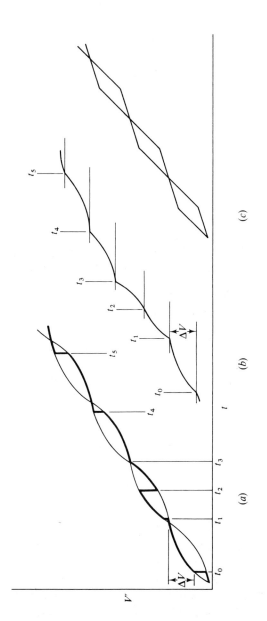

revealed. It is much easier to see how locking arises in the specially simple case when switching is synchronized with the fundamental of the injected signal. If this has been achieved, the switch occurs at the same phase in each half-cycle, as fig. 30 shows for several different cases. For a given choice of ω, however, the change in V' during half a cycle depends on ϕ, defined as the phase of the injected signal at the moment of switching from D to A; from (64), the value of ω to ensure that V' changes by ΔV is given by the equation:

$$\Delta V = \pi V_0/\omega\tau - 2v_0 \cos \phi,$$

or
$$\omega = \pi V_0/\tau(\Delta V + 2v_0 \cos \phi) \sim \omega_1\left(1 - \frac{2v_0}{\Delta V}\cos \phi\right), \qquad (65)$$

if $v_0 \ll \Delta V$; here ω_1 is written for $\pi V_0/\tau \Delta V$, the unperturbed frequency of the multivibrator. There is a range of ω over which (65) may be satisfied by appropriate choice of ϕ, as shown in fig. 30(b). Not all values of ϕ, however, represent stable locking. Suppose, for example, the system were started at A when the frequency and phase of the injected signal were given by the point X. Since X lies above the curve, ω is too low for synchronous switching, and B will be reached before half a cycle has been executed; the next switch occurs at a smaller value of ϕ, and successive positions of X migrate to the left until they reach the curve, in this case at X', which represents stable locking, as do all points on the full curve $(0 < \phi < \pi)$. Consequently locking occurs, as (65) shows, if ω lies within the range

$$(1 + 2v_0/\Delta V)^{-1} < \omega/\omega_1 < (1 - 2v_0/\Delta V)^{-1}.$$

Discussion of locking at other than the fundamental frequency is considerably more difficult, and we shall do no more than indicate basic principles, leaving many problems unsolved. It makes matters easier, without seriously altering the character of the argument, to replace the sinusoidal input by a sawtooth, as in fig. 29(c). The trajectory of fig. 29(b) then takes the form of a succession of lines with alternating gradient, a switch from one to the other occurring every time V' reaches a multiple of ΔV, and every time t reaches a multiple of π/ω. These trajectories may be drawn without difficulty by dividing the (V', t)-plane into rectangular cells in which the gradients take alternate values, shown as $+$ and $-$ in fig. 31. Two examples of the construction are shown here. On the left the gradients have been chosen so that a trajectory starting at O passes through an equivalent point, P, two cycles of t and three of V' later – the multivibrator is oscillating at $\frac{3}{2}$ the input frequency. The matching of one to a rational multiple of the other does not automatically imply, however, that they are locked; in this particular case, indeed, they are not – they are merely adjusted to coincidence, as the following argument makes clear. Between the identical trajectories OP and O_1P_1, one cycle apart in time, there are five others with the same periodicity, obtained by starting the construction

Fig. 30. Synchronization
(entrainment) of a
multivibrator at the
fundamental, the curves in
(a) showing how different
values of ϕ demand a
different value of ΔV for a
given injected frequency,
ω. The heavy line in
(b) indicates the range of
ϕ and ω over which
synchronization occurs,
and illustrates the process
by the two phase
migrations XX' and YY'.

Fig. 31. Illustrating
entrainment at other than
the fundamental
frequency.

at O', O'' etc.; all those shown in the diagram have exactly the same shape but are displaced by the diagonal of the cell. One can in this case draw identical trajectories through every lattice intersection. It now follows that intermediate trajectories also have the same periodicity; a single drawn example elucidates the reasoning. The broken line starts at the bottom of the diagram, halfway between its neighbours and most of its path lies between them as they run parallel. The only exceptions occur when one neighbour passes through a lattice intersection, but this does not stop the broken line lying halfway between. It would be different if the two neighbours could pass on either side of an intersection, but this is precluded, every intersection already lying on a trajectory. It follows that for every starting point the same frequency serves to make the trajectory periodic, and that the curve corresponding to fig. 30(b) is simply a horizontal line with no locking properties.

This argument holds equally well for any frequency ratio n/m (n multi-vibrator cycles to m input cycles), if either n or m is even. In these circumstances one can draw a periodic trajectory that passes through the two different types of intersection (e.g. through O and P the gradient is low, while through Q it is high), and this allows every intersection to lie on an identical trajectory. By contrast, if both n and m are odd there are two different trajectories, as illustrated in fig. 31(b); that through R and S is periodic while that through T is not, nor are intermediate trajectories.

Consequently, to force periodicity with a given choice of gradients, ω must depend on ϕ, providing, as in fig. 30(b), the right conditions for locking. It is apparent in fig. 27 that locking does in practice occur even if n or m is even, but the weakness of the effect when $n/m = 2$, compared with what happens at 1 and 3, suggests that small asymmetries in the multivibrator waveform may be invoked as an explanation.

The argument given above, although illustrated by means of a sawtooth waveform, does not depend on this assumption, but can be applied to any waveform that lacks even harmonics. In particular a sinusoidal input should behave very similarly, with no locking for frequency ratios n/m if either n or m is even, and locking in principle for every other rational fraction, whether greater or less than unity. It is worrying, therefore, that signs of subharmonic locking should be so elusive, while such a wealth of examples show up when the input frequency is greater than that of the multivibrator. One would expect in such a case to find a mathematical identity guaranteeing for subharmonic frequencies that property which we have noted for n or m even. But this has not been found, for the good reason that it is not true. Careful computation of the frequencies needed to give coincidence at the third subharmonic (i.e. $n/m = 3$) starting from different values of ϕ, reveals very small variations. In principle subharmonic locking can occur, but in practice it seems to be too weak to be discerned.

Locking of multivibrator to a resonant circuit

The problem presented by a passive resonant circuit coupled to a multivibrator is similar to that of the coupled class C oscillator and resonant circuit, though in one respect it is simpler. The class C oscillator acts on the passive circuit and itself suffers a reaction that can change both the amplitude and frequency of its oscillation; the multivibrator is less sensitive – its frequency can be altered, but the waveform and amplitude are determined by the amplifier characteristic and do not change. If the coupling is reactive, by means of a mutual inductance, say, the square waveform of the multivibrator current induces impulsive e.m.f.s, of alternating sign, in the passive circuit, which will be strongly excited if its natural frequency is near an odd harmonic of the multivibrator. If it is not so well tuned it will respond less strongly, but always in the end will come into synchronism with the nearest odd harmonic (we shall suppose the Q of the resonant circuit to be high enough, and tuning close enough, for one harmonic to dominate). In this sense the multivibrator is always entrained by the resonant circuit, and the problem is to determine at what frequency the combination will oscillate. We shall consider only locking at the fundamental, the natural frequencies of the multivibrator, ω_1, and of the resonant circuit, ω_2, being close together.

According to (65) and the arguments leading to it, the multivibrator frequency is affected only by that component of injected e.m.f. which is in phase quadrature ($\phi = 0$ or π) with the fundamental of i_1; if $\phi = 0$ the frequency is reduced. In this particular phase relationship the e.m.f. leads the current by $\pi/2$, as if an extra inductance had been added. The response of the multivibrator is thus qualitatively similar to that of a class C oscillator, in that extra inductance lowers the frequency, while extra resistance can be compensated for and is of no consequence. There are, of course, differences in detail which are readily supplied. Thus in determining the possible frequencies for steady oscillation one must remember that if the multivibrator current oscillates, as we shall assume, between $\pm i_1$ with square waveform, its fundamental Fourier component has magnitude $4i_1/\pi$. Since the impedance at frequency ω presented by the passive circuit, through the mutual inductance, is $-\mathrm{j}\omega^2 M^2/2L(\omega - \omega_2)$, ω_2 being complex, the reactive component of the e.m.f., which must be substituted for $v_0 \cos \phi$ in (65), is $4i_1/\pi$ times the imaginary part. Hence, from the approximate form of (65), ω is a possible frequency for the combination if

$$\omega - \omega_1 = \kappa^2 \frac{\omega - \omega_2'}{(\omega - \omega_2')^2 + \omega_2''^2}, \tag{66}$$

in which $\kappa^2 = 4\omega^2\omega_1 M^2 i_1/\pi L\Delta V$. This is exactly the same as (33), only the meaning of κ^2 being changed. The general pattern of locking must clearly be similar to that shown in fig. 14, but it still remains to determine the stability range of the various branches. The treatment in terms of normal modes is inapplicable here, on account of the highly non-linear behaviour of the multivibrator, and we shall proceed in a manner more closely analogous to what was used for the Huygens phenomenon.

We shall suppose the oscillation in the passive circuit to be not perfectly in synchronism with the multivibrator and to have a magnitude slightly different from its equilibrium value; we shall then be able to determine whether it tends towards or away from equilibrium. At any instant let the (real) frequencies of multivibrator and passive circuit be Ω_1 and Ω_2, with the phase difference ϕ defined as before, so that $\dot\phi = \Omega_2 - \Omega_1$. When the amplitude of current oscillation in the passive circuit is i_2, the induced e.m.f. in the multivibrator circuit is $\Omega_2 M i_2$, and (65) takes the form:

$$\Omega_1 = \omega_1 \left(1 - \frac{2\Omega_2 M i_2}{\Delta V} \cos \phi \right). \tag{67}$$

The passive circuit at the same time receives an induced e.m.f. of magnitude $4\Omega_1 M i_1/\pi$, leading the current i_2 by $\frac{1}{2}\pi - \phi$. It therefore responds as to an additional impedance δZ, where

$$\delta Z = 4\mathrm{j}\Omega_1 M i_1 \, \mathrm{e}^{-\mathrm{j}\phi}/\pi i_2.$$

The imaginary part of δZ alters the natural frequency of the circuit, and the real part its decrement, so that

$$\Omega_2 = \omega_2 \left(1 - \frac{2M i_1}{\pi L i_2} \cos \phi \right),$$

and
$$di_2/dt = -i_2(R + 4\Omega_1 M i_1/\pi i_2 \sin\phi)/2L.$$

With the substitutions $\Delta = \omega_2 - \omega_1$, $\lambda = \frac{1}{2}R/L$, $x = (\pi\Omega_2 L)^{\frac{1}{2}} i_2/(\Delta V i_1)^{\frac{1}{2}}$ and κ as already defined, these equations, together with (67), take the form:

$$\dot\phi = \Delta + \kappa(x - 1/x)\cos\phi, \tag{68}$$

and
$$\dot x = -\lambda x - \kappa \sin\phi. \tag{69}$$

Hence
$$\frac{dx}{d\phi} = \frac{-\lambda x - \kappa \sin\phi}{\Delta + \kappa(x - 1/x)\cos\phi}, \tag{70}$$

from which trajectories of the same type as those in figs. 18 and 21 may be constructed. Examples are given in fig. 32. There is always one stable focus with x less than unity, shown as F_1 in the diagrams. As Δ is raised from a negative value to a positive, F_1 moves up to cross the line $x = 1$ and eventually merges with the saddle point S and disappears, leaving F_2 as the only stable solution. This behaviour provides an immediate association with curve C of fig. 14 – the upper and lower heavily drawn curves represent F_1 and F_2, while the lightly drawn curve is S. If F_1 and F_2 are stable up to the point of merger with S, the full range of hysteresis is attained, but as with the class C oscillator it may prove that F_1 and F_2 become unstable before this point is reached.

Fig. 32. 0-loci, ∞-loci and sketched trajectories for solutions of (70). As Δ goes from negative to positive, the ∞-loci change progressively from the form shown by full lines to that shown by broken lines; F_1 moves to $\bar F_1$, F_2 to $\bar F_2$, S to $\bar S$.

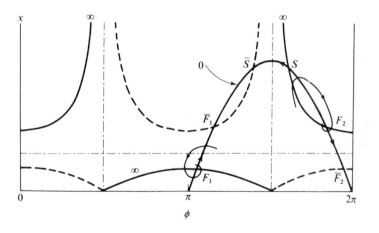

To investigate this we proceed exactly as before. An intersection of the loci occurs when $x = X$ and $\phi = \Phi$, such as to cause (68) and (69) to vanish together. Hence

$$\Delta = -\kappa(X - 1/X)\cos\Phi \quad \text{and} \quad \lambda X = -\kappa \sin\Phi. \tag{71}$$

Writing $\phi = \Phi + \xi$, $x = X + \eta$ in the neighbourhood of such a point, we have that

$$\frac{d\eta}{d\xi} = \frac{\alpha_3\xi + \alpha_4\eta}{\alpha_1\xi + \alpha_2\eta},$$

where
$$\alpha_1 = \lambda (X^2 - 1), \qquad \alpha_2 = -\frac{\Delta}{X}\frac{X^2 + 1}{X^2 - 1}, \qquad \alpha_3 = \frac{\Delta X}{X^2 - 1}, \qquad \alpha_4 = -\lambda.$$
$$(72)$$

The disappearance of the intersection above the line $x = 1$ is marked by the vanishing of the determinant D; what we have to discover is whether, as $|\Delta|$ is increased, this occurs before or after T passes from negative to positive and causes F to turn into an unstable focus. That is to say, we need to find Δ and κ such that D and T vanish together. The value of κ is then the critical coupling at which the full range of hysteresis ceases to be observable.

From (72), $T = 0$ when $X^2 = 2$, and $D = 0$ when $\lambda^2(X^2 - 1)^3 = \Delta^2(X^2 + 1)$. This is satisfied when $X^2 = 2$ if $\Delta^2 = \frac{1}{3}\lambda^2$. All that remains is to find the value of κ such that when $\Delta^2 = \frac{1}{3}\lambda^2$ an intersection occurs at $X^2 = 2$. By eliminating Φ from (71) an equation is obtained for the values of X^2 at the intersections:

$$\lambda^2 X^6 - (2\lambda^2 + \kappa^2)X^4 + (\lambda^2 + \Delta^2 + 2\kappa^2)X^2 - \kappa^2 = 0. \qquad (73)$$

Substituting $\Delta^2 = \frac{1}{3}\lambda^2$ we find:

$$\lambda^2 X^6 - (2\lambda^2 + \kappa^2)X^4 + (\tfrac{4}{3}\lambda^2 + 2\kappa^2)X^2 - \kappa^2 = 0, \quad \text{if } \Delta^2 = \tfrac{1}{3}\lambda^2. \qquad (74)$$

This can only have $X^2 = 2$ as a solution if $\kappa^2 = 8\lambda^2/3$; when this value of κ^2 is substituted, (74) factorizes in the required manner, with the expected double root at 2, indicating the merger of F and S:

$$(X^2 - 2)^2(3X^2 - 2) = 0.$$

From (69) it is seen that when κ is given a larger value the 0-locus in fig. 32 rises higher, and correspondingly the value of X at which F and S coalesce is higher; instability therefore sets in before this point is reached. Like the class C oscillator coupled to a resonant circuit, the multivibrator only shows the full range of hysteresis, as in fig. 15(b), when the coupling is relatively weak; when $\kappa^2 > 8\lambda^2/3$ something more like fig. 15(a) should be observed. On the other hand, if $\kappa < \lambda$ the 0-locus does not reach to the line $x = 1$ and only one solution is ever possible; this corresponds to fig. 15(c). Apart from numerical details there is a very close similarity between the two systems. Examples of the behaviour are given in fig. 33, the approach to a vertical tangent at the point of instability being clearly marked in the case of weaker coupling, and entirely absent when it is stronger.

Superconducting weak links (Josephson junctions)[7]

It is a far cry from multivibrators to the essentially quantum-mechanical Josephson effect in superconductors, but their behaviour shows enough common features to justify a short exposition. Let us admit at the outset that any account of the basic theory is precluded; we shall do little more than state the fundamental equation and proceed to develop mechanical

Fig. 33. Experimental observations of the locking of a multivibrator to a resonant circuit, the curve on the left being obtained with stronger coupling than for that on the right. The frequency of the multivibrator was altered by changing its capacitance, C.

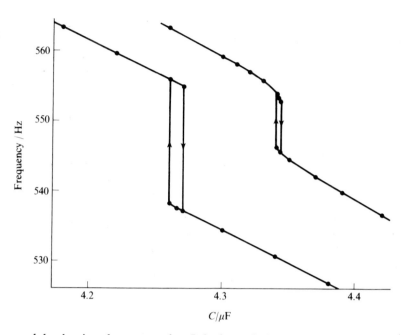

models obeying the same rules. It is through these models, and mechanical models of the multivibrator, that the similarity will be revealed.

The electrons responsible for the characteristic properties of a block of metal which has been cooled below the temperature (a few K at most) at which it becomes superconducting are collectively in a single isolated quantum state having a well-defined energy. The wave-function describing this state is something like the resonant mode of a wave in a cavity, though far more complicated – the analogy is, however, good enough to be helpful. When another superconducting block is brought so near to it that it becomes possible for electrons to pass through the intervening space (by quantum tunnelling as in the tunnel diode), the two are coupled together in a similar manner to the coupling of two resonant electromagnetic cavities which can exchange energy. The coupled cavities, which we suppose to have been originally tuned to the same frequency, can only oscillate steadily if they are in phase or in antiphase, these two phase relationships being characteristic of the normal modes, and in either normal mode there is no energy exchange between the cavities. So too with the superconductors; there are two normal modes in which the wave-functions oscillate either in phase or in antiphase. These are stationary states; when the system is in either there is no exchange of electrons between the blocks. But if the cavities are set with a phase difference ϕ that is not 0 or π, energy flows from one to the other at a rate proportional to $\sin \phi$. And if the superconducting blocks have a phase difference that is not 0 or π a current of electrons flows, the magnitude being proportional to $\sin \phi$.[8] In both cases the maximum current that can flow, when $\phi = \pm\frac{1}{2}\pi$, is determined by the strength of the coupling – the closer the superconducting blocks, the easier electrons find tunnelling, and the greater the magnitude of the Josephson current.

What is surprising about this phenomenon is that no potential difference is needed to cause the current to flow – it can pass through a thin layer of vacuum or dielectric as though there were a perfectly resistanceless link between the two blocks. Unless the blocks are extremely close, however, the maximum current that can flow without a potential difference is very small indeed. Typically an oxide layer four or five molecules thick may allow a few milliamperes across a junction 1 mm square. Alternatively the superconductors may be separated by thicker layers, even up to 10 μm or more, of normally resistive metal and still pass a Josephson current without resistance. The oxide junctions are made by evaporating a layer of super-conductor on a suitable (e.g. glass) substrate, allowing it to oxidize for a short while in air, and then depositing another superconducting layer on the oxide surface. The metal junctions can be made by evaporating, for example, lead on to both sides of a rolled copper foil. They behave rather similarly except in certain dynamical experiments where, as we shall see, the capacitance between the superconductors in the oxide junction, but not in the metal junction, may play an important part.

The steady flux of energy from one cavity to the other, and of particles from one superconductor to the other, continues only so long as the phase difference is constant, and this is only possible in Josephson junctions when there is no potential difference between the two superconductors. There is a basic association of energy with frequency in quantum mechanics, and two superconducting blocks, identical except for their electrostatic potential, will differ also in the frequency of their wave-functions; ϕ therefore changes with time, and the current oscillates to and fro at the difference frequency. Something very similar happens with resonant cavities; we may imagine the analogous behaviour to be what takes place when two cavities differing in resonant frequency by $\delta\omega$ but excited equally (and therefore in a mixture of normal modes) are coupled weakly; as the phase difference shifts steadily, so the energy flux oscillates at the difference frequency. One may describe this phenomenon, somewhat artificially perhaps, in quantum terms as follows: when one quantum of electromagnetic energy, $\hbar\omega$, is transferred the energy of the whole system must be changed by reason of the different quantum energy in the two cavities, $\delta E = \hbar\delta\omega$, and the frequency to be assigned to the process is $\delta E/\hbar$, which is just $\delta\omega$. The argument may be applied to the superconducting case by noting that the unit in which matter is transferred, the analogue of the quantum $\hbar\omega$, is an electron-pair, since in the superconducting ground state all electrons are associated in pairs, and one alone cannot be taken out or inserted. If a potential difference V is applied between the superconductors, an electron-pair changes its energy by $2eV$ in crossing the gap, and the frequency of oscillation of current in the gap is $2eV/\hbar$. This leads immediately to Josephson's expression for the dependence of phase on voltage:

$$\dot{\phi} = 2eV/\hbar, \tag{75}$$

to which must be added the relation between phase and Josephson current, i_s,

$$i_s = i_J \sin \phi, \tag{76}$$

where i_J is the maximum Josephson current that the particular junction can support, and is determined by the nature and dimensions of the junction.

Let us apply these equations to two special cases, the SIS junction in which superconducting layers are separated by a thin insulating layer, and the SNS junction in which they are separated by a metal foil having normal resistive properties. The former, SIS, in addition to its superconducting properties is also a capacitor. If its capacitance is C, when a current i enters one superconducting layer and leaves the other, charge builds up on the capacitor plates at a rate $i - i_s$, so that

$$C\dot{V} = i - i_s,$$

and, from (75) and (76),

$$\frac{\hbar C}{2e}\ddot{\phi} + i_J \sin \phi = i. \tag{77}$$

The SNS junction, on the other hand, has no capacitance between the superconducting layers, since no charges can remain within the metal at the interface between superconductor and normal metal, but if a potential difference is established a normal ohmic current will flow as well as the Josephson current. For a junction having ohmic resistance R,

$$i = V/R + i_s,$$

or
$$\frac{\hbar}{2eR}\dot{\phi} + i_J \sin \phi = i. \tag{78}$$

These equations, (77) and (78), describe how ϕ is affected by the injection of current i. The laboratory observer cannot observe ϕ directly, but (75) shows that $\dot{\phi}$ is uniquely associated with a potential difference, and therefore that observations of $V(t)$ in the presence of $i(t)$ provide information that can be compared directly with the theoretical predictions.

The appearance of $\sin \phi$ in both equations renders them non-linear and therefore difficult to solve in general. For this reason it is helpful to devise mechanical models that obey the same equations, so that physical intuition may replace mathematics up to the moment when exact solutions are required, and computing must take over. Fig. 34 shows some models. The pendulum (a), to which a torque Fa can be applied, obeys an equation of the same form as (77):

$$ml^2\ddot{\phi} + mgl \sin \phi = Fa,$$

while if it is immersed in so viscous a fluid as to be overdamped to the point where the initial term $\ddot{\phi}$ is negligible in comparison with the damping term $\lambda\dot{\phi}$, its equation of motion takes the form of (78):

$$\lambda\dot{\phi} + mgl \sin \phi = Fa.$$

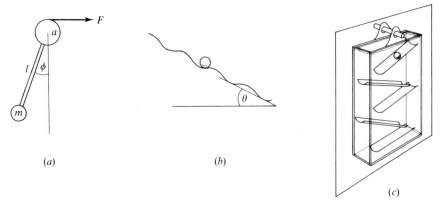

Fig. 34. Models of (a) and (b) Josephson junction, (c) multivibrator.

The same equations apply to the washboard model (b). If x represents distance measured down the slope, and the ripples have wavenumber k, the slope varies as $\theta - \alpha \sin kx$ if θ and α are both much less than unity. A ball rolling down in the absence of friction obeys the equation:

$$m\ddot{x} + g\alpha \sin kx = g\theta,$$

while immersing it in a viscous fluid, so that it always moves at its terminal velocity, causes it to obey the equation

$$\lambda\dot{x} + mg\alpha \sin kx = mg\theta.$$

In this model kx plays the part of ϕ, the maximum slope of the ripples ($k\alpha$) is related to i_J, and θ is related to i.

With the aid of these models the solutions of (77) and (78) under various conditions may be visualized. We shall devote more attention to (78), which is considerably easier to treat, though in the devices which employ Josephson effects it is the undamped SIS system, governed by (77), which is more important. If a steady current i is injected pure Josephson current, with no p.d., flows so long as $i < i_J$, for a solution $\phi = $ constant is possible.† In the models this corresponds to Fa being too weak to raise the pendulum beyond the horizontal position, or the slope θ being less than $\kappa\alpha$ so that the ball can sit at rest in any of the troughs. When $i > i_J$, however, the torque Fa pulls the pendulum over the top and it continues to rotate; with heavy damping the angular velocity is less on the way up than on the way down, so that the voltage, $\dot{\phi}$, has an oscillatory component superposed on a steady mean. In this case (78) is directly integrable:

$$t = \frac{\hbar}{2eR} \int d\phi/(i - i_J \sin \phi).$$

† But not unique. By injecting a current greater than i_J the junction is set in a mode analogous to the pendulum going over the top in a continuous, but irregular, rotation. When damping is weak the motion can be maintained by a force F much weaker than that required to initiate it. In contrast to an SNS junction, the SIS junction is hysteretic.

In particular T, the period for one cycle, is obtained by setting limits 0 and 2π to the integral, and is equal to $\pi\hbar/eR(i^2-i_J^2)^{\frac{1}{2}}$. The mean value of $\dot{\phi}$ is $2\pi/T$, and the mean voltage follows from (75):

$$\bar{V} = R(i^2-i_J^2)^{\frac{1}{2}}.$$

This hyperbolic $\bar{V}-i$ characteristic is shown in fig. 35(a).

Fig. 35. $V-i$ characteristics of SNS junction; (a) without a.c. injection, (b) with a.c. injection (unpublished measurements by Prof. J. Clarke). The frequency of the injected current was 260 kHz, and the numbers against each step are the ratio of the 'step frequency', $2e\bar{V}/\hbar$, to the injected frequency.

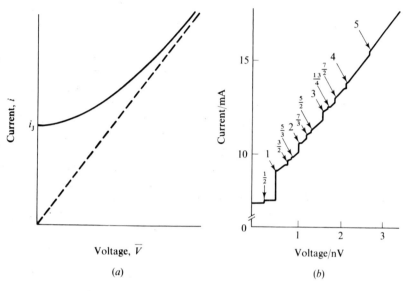

When an alternating current is added to the steady current something very interesting occurs; the $\bar{V}-i$ characteristic develops a series of steps, such as in fig. 35(b). The precise size of each step depends in a complicated way on the strength of the alternating current. The equation governing ϕ follows from (78):

$$\frac{\hbar}{2eR}\dot{\phi} + i_J\sin\phi = i_0 + i_1\sin\omega t$$

or
$$d\phi/dx = I_0 + I_1\sin x - I_J\sin\phi, \qquad (79)$$

where $x = \omega t$ and $I = 2eRi/\hbar\omega$. As soon as one thinks of sketching solutions of (79) on the (x, ϕ)-plane, the similarity to the entrainment problem of the multivibrator becomes obvious. For the slope of the trajectory of ϕ is periodic in x and ϕ, much like the periodicity shown in fig. 31. To emphasize this point, let us replace $\sin x$ and $\sin \phi$ by square waves of the same amplitude, as if the Josephson current were of a given magnitude, either positive or negative, and the alternating current had a square waveform. Then (79) may be written in shorthand notation

$$d\phi/dx = I_0 \pm I_1 \pm I_J,$$

and the square lattice into which the (x, ϕ)-plane is divided has four different types of cell, in each of which the gradient is constant; though a little more complicated than the simple alternating-gradient pattern in fig. 31, it is hardly different in principle.

The similarity of the two problems may be illustrated in a different way by comparing the washboard model of the SNS junction with a mechanical model simulating the multivibrator, fig. 34(*b*) and (*c*). In the latter model a ball rolls in a viscous fluid down a cascade of slopes, the horizontal displacement representing voltage in the multivibrator. If the whole arrangement is rocked the tilt speeds up or slows down the motion, and the reader will have no difficulty in appreciating that the equations governing a perturbed multivibrator apply here also. The extra complication of (*b*), when its sinusoidal ripple is replaced by a sawtooth, and the sinusoidal rocking of θ through a small angle is replaced by a square wave, lies in the alternation of four possible slopes; when (*c*) is given a square-wave rocking, only two alternatives are available. But in both cases the mean speed depends on the phase relationship between the position of the ball on a step and the perturbing motion; as a result, the system drifts towards a stable relationship in which *n* steps are traversed during *m* periods of the perturbation. So long as *n* and *m* remain the same, as they can do over a significant range of i_0 when *n* and *m* are small (just as the multivibrator can be entrained in the fundamental or a low harmonic mode over a wide range of input frequencies), the mean value of $\dot{\phi}$ remains constant at $n\omega/m$, and a step appears in the *V*–*i* characteristic at a voltage of $(n/m)(\frac{1}{2}\hbar/e)$. Entrainment at integral and fractional harmonics is to be expected, and with the additional complication of four gradients in the Josephson problem the symmetry property that eliminated even values of *n* and *m* is no longer present – all possible integral fractions should be candidates for entrainment. That they do not all appear in practice, but only a fairly rich selection, is probably a thermal effect, Johnson and other noise being adequate to disrupt the very weak phase linkage accompanying entrainment at a fractional harmonic ratio with large integers. As with the multivibrator, subharmonic entrainment is not observed.

When a steady current is injected into the SNS junction, the mean voltage is accompanied by an oscillatory voltage at a frequency $2e\bar{V}/\hbar$. A resonant circuit connected across the junction should therefore be excited when $\bar{V}$ takes something near the correct value, and its response should react back on the junction, changing the behaviour of $\dot{\phi}$. The locking of junction and resonator might be expected to introduce steps, possibly with hysteresis, into the *V*–*i* characteristic, analogous to the frequency-locking of a multivibrator to a resonant circuit. Unfortunately this experiment seems not to have been tried but there are many published results to show the variety of ways in which the behaviour of an SIS junction may be modified by coupling a resonator to it. These are all, however, much more difficult to analyse than anything involving an SNS junction. The second derivative in (77) makes discussion by graphical methods virtually useless. We shall therefore leave the topic at this point, having made clear, it is to be hoped, that the study of classical systems can still cast light on essentially quantum-mechanical problems, and that some of these modern problems present the same sort of difficulties as their classical counterparts, demanding of the physicist an outlook that does not reject old attitudes but welcomes and improves on them.

References

Chapter 2: The free vibrator

1 C. Huygens, *Horologium Oscillatorium* (1673).
2 F. Hope-Jones, *Electrical Timekeeping*, 2nd edition (London: N.A.G. Press, 1940).
3 H. L. F. Helmholtz (trans. A. J. Ellis), *The Sensations of Tone*, p. 68 (London: Longmans, Green, 1875).
4 P. C. Wraight, 1971, *Phil. Mag.* **23**, 1261.
5 D. G. Fincham and P. C. Wraight, 1972, *J. Phys.* A **5**, 248.
6 F. Homann, 1936, *Forsch. Geb. Ing.* **7**, 1.

Chapter 3: Applications of complex variables to linear systems

1 E. C. Titchmarsh, *Introduction to the Theory of Fourier Integrals*, p. 305 (Oxford: Clarendon Press, 1948).
2 P. M. Morse, *Vibration and Sound*, p. 114 (New York: McGraw-Hill, 1936).

Chapter 4: Fourier series and integral

1 K. F. Riley, *Mathematical Methods for the Physical Sciences*, chapter 8 (Cambridge University Press, 1974).
2 S. G. Lipson and H. Lipson, *Optical Physics*, p. 152 (Cambridge University Press, 1969).
3 K. G. Budden, *Radio Waves in the Ionosphere*, p. 256 (Cambridge University Press; 1961).
4 L. M. Milne-Thomson, *Theoretical Hydrodynamics*, 4th edition, chapter 14 (London: Macmillan, 1960).
5 J. L. Lawson and G. E. Uhlenbeck (editors), *Threshold Signals*, chapter 3 (New York: McGraw-Hill, 1950).
6 J. C. Dainty, The Statistics of Speckle Patterns, in *Progress in Optics* vol. 14, ed. E. Wolf (Amsterdam: North-Holland, 1977).
7 C. Kittel, *Thermal Physics* (New York: Wiley, 1969).
8 H. Feder, 1929, *Ann. Phys.* (Germany) **1**, 497.

Chapter 5: Spectrum analysis

1 G. A. Vanasse and H. Sakai, Fourier Spectroscopy, in *Progress in Optics* vol. 6, ed. E. Wolf (Amsterdam: North-Holland, 1967).
2 Ref. 4.2, pp. 210 and 225.
3 Ref. 4.1, chap. 16. *See also* E. G. Phillips, *Functions of a Complex Variable*, p. 96 (Edinburgh: Oliver and Boyd, 1940).
4 R. Kronig, 1926, *J. Opt. Soc. Amer.* **12**, 527.
   H. A. Kramers, 1927, *Atti congr. fis.* (Como) 545.

5   F. Abelès (editor) *Optical Properties and Electronic Structure of Metals and Alloys* (Amsterdam: North-Holland, 1966).

6   R. T. Schumacher and W. E. Vehse, 1963, *J. Phys. Chem. Solids* **24**, 297.

7   R. K. Potter, G. A. Kopp and H. C. Green, *Visible Speech* (New York: Van Nostrand, 1947).

8   A. H. Benade, 1973, *Scientific American* **229**, 24.

9   A. B. Pippard, 1947, *Proc. Roy. Soc.* A **191**, 370.

10 H. Bremmer, *Encylopaedia of Physics* **16**, 573 (Berlin: Springer, 1958).

Chapter 6: The driven harmonic vibrator

1   E. Kappler, 1931, *Ann. Phys.* (Germany) **11**, 233. *See also* R. B. Barnes and S. Silverman, 1934, *Rev. Mod. Phys.* **6**, 162.

2   R. V. Jones and C. W. McCombie, 1952, *Phil. Trans. Roy. Soc.* A **244**, 205.

3   J. B. Johnson, 1928, *Phys. Rev.* **32**, 97. *See also* ref. 4.5, chapter 4.

4   A. W. Hull and N. H. Williams, 1925, *Phys. Rev.* **25**, 148.

Chapter 7: Waves and resonators

1   B. I. Bleaney and B. Bleaney, *Electricity and Magnetism*, 3rd edition, chapter 9 (Oxford University Press, 1976).

2   Ref. 1, p. 72.

3   J. W. Strutt, Baron Rayleigh, *The Theory of Sound*, vol. 2, p. 196 (New York: Dover, 1945).

4   H. Lamb, *Hydrodynamics*, 6th edition, p. 74 (Cambridge University Press, 1924).

5   A. B. Pippard, Metallic Conduction at High Frequencies and Low Temperatures, in *Advances in Electronics and Electronic Physics*, vol. 6, ed. L. Marton (New York: Academic Press, 1954).

6   P. H. Ceperley, I. Ben-Zvi, H. F. Glavish and S. S. Hanna, 1975, *IEEE Trans. Nuc. Sci* **22**, 1153.

7   A. H. Cook, *Interference of Electromagnetic Waves*, chapter 6 (Oxford: Clarendon Press, 1971).

8   S. Tolansky, *Introduction to Interferometry* (London: Longmans, Green, 1955).

9   C. G. Montgomery (editor), *Technique of Microwave Measurements*, p. 319 (New York: McGraw-Hill, 1947).

10 L. I. Schiff, *Quantum Mechanics*, 3rd edition, chapter 5 (New York: McGraw-Hill, 1968).

11 J. D. Jackson, *Classical Electrodynamics*, 2nd edition, chapter 17 (New York: Wiley, 1975).

12 A. E. S. Green, *Nuclear Physics*, p. 454 (New York: McGraw-Hill, 1955).

13 D. J. Bennet, *The Elements of Nuclear Power*, p. 175 (London: Longmans, 1972).

14 A. B. Pippard, *Elements of Classical Thermodynamics*, p. 78 (Cambridge University Press, 1966).

15. J. W. Strutt, 1871, *Phil. Mag.* **41**, 107.

16. S. A. Korff and G. Breit, 1932, *Rev. Mod. Phys.* **4**, 471.

Chapter 8: Velocity-dependent forces

1   A. B. Pippard, *Forces and Particles*, pp. 33 and 223 (London: Macmillan, 1972).

2   Ref. 7.1, p. 750.

3   S. Timoshenko, *Vibration Problems in Engineering*, 2nd edition, p. 222 and chapter 5 (London: Constable, 1937).

4   Ref. 7.1, chapter 24.
5   E. Segré, *Nuclei and Particles*, p. 130 (Reading, Mass.: Benjamin, 1965).
6   Ref. 7.1, p. 729.
7   G. Dresselhaus, A. Kip and C. Kittel, 1955, *Phys. Rev.* **98**, 368.
8   N. W. Ashcroft and N. D. Mermin, *Solid State Physics*, p. 11 (New York: Holt, Rinehart and Winston, 1976).
9   Ref. 7.1, p. 236.
10   R. Bowers, C. Legendy and F. Rose, 1961, *Phys. Rev. Lett.* **7**, 339. *See also* ref. 4.3, chapters 5 and 6.
11   J. A. Delaney and A. B. Pippard, 1972, *Rep. Prog. Phys.* **35**, 677.

Chapter 9: The driven anharmonic vibrator; subharmonics; stability

1   N. Minorsky, *Non-linear oscillations* (Huntington, NY: Krieger, 1974).
     J. J. Stoker, *Non-linear vibrations* (New York: Interscience, 1950).
2   R. Thom, *Structural Stability and Morphogenesis* (Reading, Mass.:Benjamin. 1975).
3   A. E. H. Love, *Mathematical Theory of Elasticity*, 4th edition, p. 401 (Cambridge University Press, 1927).
4   R. A. Cowley, 1964, *Phys. Rev.* **134**, A981.
5   P. Weiss and R. Forrer, 1926, *Ann. Phys.* (France) **5**, 153.
6   K. A. Müller and W. Berlinger, 1971, *Phys. Rev. Lett.* **26**, 13.
7   Ref. 8.1, p. 119.
8   P. H. Hammond, *Feedback Theory and its Applications*, p. 67 (London: English University Press, 1958).
9   Minorsky, Ref. 1, chapter 16.
10   R. M. May, 1976, *Nature* **261**, 459.
11   M. Ya. Azbel' and E. A. Kaner, 1956, *J. Exp. Theor. Phys.* **30**, 811.
     W. M. Walsh, Resonances both Temporal and Spatial, in *Solid State Physics* (vol. 1 Electrons in Metals) ed. J. F. Cochran and R. R. Haering (New York: Gordon and Breach, 1968).
12   W. M. Walsh, ref. 11, p. 158.
13   A. B. Pippard, 1947, *Proc. Roy. Soc.* A **191**, 385.

Chapter 10: Parametric excitation

1   J. W. Strutt, Baron Rayleigh, 1887, *Phil. Mag.* **24**, 145.
2   E. T. Whittaker and G. N. Watson, *Modern Analysis*, 4th edition, p. 412 (Cambridge University Press, 1927).
3   Ref. 2, chapter 19.
4   B. J. Robinson, 1961, *Institute of Electrical Engineers Monograph* 480 E.
5   T. H. Beeforth and H. J. Goldsmid, *Physics of Solid State Devices*, p. 48 (London: Pion, 1970).

Chapter 11: Maintained oscillators

1   D. C. Gall, 1942, *J. Inst. Elec. Eng.* **89**, 434.
2   H. Nyquist, 1932, *Bell Syst. Tech. J.* **11**, 126. *See also* ref. 9.8, p. 71.
3   A. Hund, *Phenomena in High-frequency Systems*, chapter 2 (New York: McGraw-Hill, 1936).
4   G. B. Clayton, *Operational Amplifiers* (London: Butterworths, 1971).
5   Ref. 2.2.
6   W. J. Gazeley, *Clock and Watch Escapements*, p. 23 (London: Heywood, 1956).

7   J. H. Simpson and R. S. Richards, *Physical Principles and Applications of Junction Transistors*, p. 176 (Oxford: Clarendon, 1962).
    Ref. 10.5, p. 45.
8   A. Liénard, 1928, *Rev. Gén. de l'Electricité*, **23**, 901.
9   J. R. Bristow, 1946, *Proc. Roy. Soc.* A **189**, 88.
10 Ref. 7.3, p. 232.
11 M. Sargent, M. O. Scully and W. E. Lamb, *Laser Physics*, chapter 5 (Reading, Mass.: Addison-Wesley, 1974).
12 J. P. Gordon, H. Z. Zeiger and C. H. Townes, 1954, *Phys. Rev.* **95**, 282.
13 H. Abraham and E. Block, 1919, *Ann. Phys.* (Germany) **12**, 237.
14 *Encyclopaedia Britannica*, 15th edition – Macropaedia **8**, 132; Geysers and Fumaroles (Chicago, 1974).
15 H. Kalmus, 1941, *Nature* **148**, 626.

Chapter 12: Coupled vibrators

1   Unpublished observation by Dr N. Elson.
2   P. J. Wheatley, *Molecular Structure*, 2nd edition, p. 66 (Oxford: Clarendon Press, 1968).
3   G. Herzberg, *Molecular Spectra and Molecular Structure*, vol. 2, p. 215 (New York: Van Nostrand, 1945).
4   Ref. 8.1, p. 37.
5   Minorsky, ref. 9.1, chapter 18.
6   Ref. 11.11, chapter 11.
7   J. R. Waldram, 1976, *Rep. Prog. Phys.* **39**, 751.
8   B. D. Josephson, 1962, *Phys. Lett.* **1**, 251.

Index